Herbal Products

FORENSIC
SCIENCE AND MEDICINE

Steven B. Karch, MD, SERIES EDITOR

HERBAL PRODUCTS: *TOXICOLOGY AND CLINICAL PHARMACOLOGY, SECOND EDITION,*
edited by **Timothy S. Tracy and Richard L. Kingston,** 2007

CRIMINAL POISONING: *INVESTIGATIONAL GUIDE FOR LAW ENFORCEMENT, TOXICOLOGISTS,*
FORENSIC SCIENTISTS, AND ATTORNEYS, SECOND EDITION, by **John H. Trestrail, III,**
2007

FORENSIC PATHOLOGY OF TRAUMA: *COMMON PROBLEMS FOR THE PATHOLOGIST,*
by **Michael J. Shkrum and David A. Ramsay,** 2007

MARIJUANA AND THE CANNABINOIDS, edited by **Mahmoud A. ElSohly,** 2006

SUDDEN DEATHS IN CUSTODY, edited by **Darrell L. Ross and Theodore C. Chan,**
2006

THE FORENSIC LABORATORY HANDBOOK: *PROCEDURES AND PRACTICE,* edited by **Ashraf**
Mozayani and Carla Noziglia, 2006

DRUGS OF ABUSE: *BODY FLUID TESTING,* edited by **Raphael C. Wong and Harley Y.**
Tse, 2005

A PHYSICIAN'S GUIDE TO CLINICAL FORENSIC MEDICINE: *SECOND EDITION,* edited by
Margaret M. Stark, 2005

FORENSIC MEDICINE OF THE LOWER EXTREMITY: *HUMAN IDENTIFICATION AND TRAUMA*
ANALYSIS OF THE THIGH, LEG, AND FOOT, by **Jeremy Rich, Dorothy E. Dean,**
and Robert H. Powers, 2005

FORENSIC AND CLINICAL APPLICATIONS OF SOLID PHASE EXTRACTION, by **Michael J.**
Telepchak, Thomas F. August, and Glynn Chaney, 2004

HANDBOOK OF DRUG INTERACTIONS: *A CLINICAL AND FORENSIC GUIDE,* edited by
Ashraf Mozayani and Lionel P. Raymon, 2004

DIETARY SUPPLEMENTS: *TOXICOLOGY AND CLINICAL PHARMACOLOGY,* edited by **Melanie**
Johns Cupp and Timothy S. Tracy, 2003

BUPRENOPHINE THERAPY OF OPIATE ADDICTION, edited by **Pascal Kintz and Pierre**
Marquet, 2002

BENZODIAZEPINES AND GHB: *DETECTION AND PHARMACOLOGY,* edited by **Salvatore J.**
Salamone, 2002

ON-SITE DRUG TESTING, edited by **Amanda J. Jenkins and Bruce A. Goldberger,**
2001

BRAIN IMAGING IN SUBSTANCE ABUSE: *RESEARCH, CLINICAL, AND FORENSIC APPLICATIONS,*
edited by **Marc J. Kaufman,** 2001

TOXICOLOGY AND CLINICAL PHARMACOLOGY OF HERBAL PRODUCTS, edited by **Melanie**
Johns Cupp, 2000

CRIMINAL POISONING: *INVESTIGATIONAL GUIDE FOR LAW ENFORCEMENT, TOXICOLOGISTS,*
FORENSIC SCIENTISTS, AND ATTORNEYS, by **John H. Trestrail, III,** 2000

HERBAL PRODUCTS

Toxicology and Clinical Pharmacology

SECOND EDITION

Edited by

Timothy S. Tracy, PhD

Department of Experimental and Clinical Pharmacology
University of Minnesota
Minneapolis, MN

Richard L. Kingston, PharmD

SafetyCall International, PLLC
Clinical Services
Bloomington, MN

HUMANA PRESS ✴ TOTOWA, NEW JERSEY

© 2007 Humana Press Inc.
999 Riverview Drive, Suite 208
Totowa, New Jersey 07512
www.humanapress.com

The content and opinions expressed in this book are the sole work of the authors and editors, who have warranted due diligence in the creation and issuance of their work. The publisher, editors, and authors are not responsible for errors or omissions or for any consequences arising from the information or opinions presented in this book and make no warranty, express or implied, with respect to its contents.

This publication is printed on acid-free paper. ∞

ANSI Z39.48-1984 (American Standards Institute) Permanence of Paper for Printed Library Materials.

Production Editor: Amy Thau

Cover design by Nancy K. Fallatt

For additional copies, pricing for bulk purchases, and/or information about other Humana titles, contact Humana at the above address or at any of the following numbers: Tel: 973-256-1699; Fax: 973-256-8341; E-mail: orders@humanapr.com, or visit our Website at www.humanapress.com

Photocopy Authorization Policy:

Printed in the United States of America. 10 9 8 7 6 5 4 3 2 1

eISBN 10-digit: 1-59745-383-8
eISBN 13-digit: 978-1-59745-383-7

Library of Congress Cataloging in Publication Data
Herbal products : toxicology and clinical pharmacology / edited by
Timothy S. Tracy, Richard L. Kingston. -- 2nd ed.
    p. ; cm. -- (Forensic science and medicine)
  Rev. ed. of: Toxicology and clinical pharmacology of herbal
products / edited by Melanie Johns Cupp. c2000.
  Companion to: Dietary supplements / edited by Melanie Johns Cupp
and Timothy S. Tracy. c2003.
  Includes bibliographical references and index.
  ISBN 10-digit: 1-58829-313-0 (alk. paper)
  ISBN 13-digit: 978-1-58829-313-8 (alk. paper)
 1. Herbs--Toxicology. 2. Materia medica, Vegetable--Toxicology.
I. Tracy, Timothy S. II. Kingston, Richard L. III. Toxicology and
clinical pharmacology of herbal products. IV. Dietary supplements.
V. Series.
  [DNLM: 1. Plants, Medicinal. 2. Phytotherapy--adverse effects.
3. Plant Extracts--pharmacology. 4. Plants, Medicinal--adverse
effects. QV 766 H5347 2007]
  RA1250.T68 2007
  615.9'52--dc22
                        2006017680

Preface

Herbalists and laypersons have used herbs for centuries. Interest in and use of herbal products was revitalized in the late 1990s with the passage of the Dietary Supplement Health and Education Act, which allowed dietary supplements to be marketed without enduring the FDA-approval process required of drugs. Thus, despite the widespread use of herbal products, information about their safety and efficacy is generally sparse compared with the information available about prescription drugs. *Toxicology and Clinical Pharmacology of Herbal Products* was published in 2000 to help fill this information void. Since its publication, additional scientific information has come to light, and the public's interest in particular herbs has changed. *Herbal Products: Toxicology and Clinical Pharmacology, Second Edition* updates the information presented in *Toxicology and Clinical Pharmacology of Herbal Products*. Herbs were chosen for inclusion in the current volume based on their popularity, toxicity, and quantity and quality of information available. A companion volume, *Dietary Supplements: Toxicology and Clinical Pharmacology*, covers nonherbal dietary supplements.

The aim of this book is to present, in both comprehensive and summative formats, objective information on herbal supplements from the most reliable sources, with an emphasis on information not readily available elsewhere (i.e., detailed descriptions of case reports of adverse effects, pharmacokinetics, interactions, etc.). It is not designed to be a "prescribers handbook;" the intended audience is both forensic and health care professionals, particularly researchers and clinicians interested in more detailed or context-oriented clinical information than is available in most "herbal" or "natural product" references.

Although information about dietary supplements is widely available on the Internet, it is usually provided by product distributors, and is designed to sell products rather than to provide objective information about product efficacy and toxicity. Even reviews of dietary supplements in journals, newsletters, books, and electronic databases can be biased or incorrect. In compiling information to be included in *Herbal Products: Toxicology and Clinical Pharmacology, Second Edition*, emphasis was placed on the use of original studies published in reputable, peer-reviewed journals. Older studies, as well as more current literature, were utilized for completeness, with an emphasis on newer literature and double-blind, controlled trials. Where appropriate, information was obtained from meta-analyses, systematic reviews, or other high-quality reviews, such as those authored by recognized experts. Case reports of adverse effects and interactions, although anecdotal in nature, were used to identify and describe uncommon but potentially serious adverse events that may not have been noted in controlled studies because of small sample size or short duration.

Each of the chapters in this volume includes an Introduction, which contains a review of the product's history and a description of the plant. This is followed by sections on Commonly Promoted Uses, Sources and Chemical Composition, and descriptions of Products Available, which is kept general because of the myriad and ever-changing products on the market. Product quality is also discussed in this section. The Pharmacological/Toxicological Effects section focuses on in vitro data and animal studies chosen to provide an explanation for the herb's mechanism of action, clinical effects in humans, and rationale for clinical studies. It should be noted that because of the nature of herbal supplement claims (*see* Regulatory Status section), some promoted product uses might not have been studied in humans; conversely, known pharmacological and therapeutic effects might not be promoted commercially as a result of limitations in the ability of manufacturers to make "health claims" related to known pharmacological effects of various herbs. As a result, there is generally a mismatch among the nature of the information presented in the Commonly Promoted Uses and Pharmacological/Toxicological Effects sections. However, emphasis is placed on inclusion of basic science data and clinical studies that relate to the promoted uses.

The Pharmacokinetics section of each chapter covers absorption, tissue distribution, elimination, and body fluid concentrations. Such pharmacokinetic information is usually not included in other sources and may be useful in forensic investigations or in the clinical setting regarding use of the product in patients with renal or hepatic insufficiency. A section on Adverse Effects and Toxicity follows and includes detailed information on case reports of adverse reactions to the herb. The Interactions section includes discussions of interactions between the supplement and drugs or foods. The Reproduction section follows and is generally limited because of lack of information. Each chapter ends with a discussion of Regulatory Status of the product. The amount of information included in each of these sections varies according to availability.

Adverse reactions to herbals appear uncommon compared with those attributed to prescription drugs. This may be a function of health care and forensic professionals' unfamiliarity with the products' pharmacology and toxicology or assumption that the products are "natural" and therefore safe. Thus, an adverse reaction may go unrecognized or be attributed to a prescription medication.

It is hoped that the information in *Herbal Products: Toxicology and Clinical Pharmacology, Second Edition* will be used to solve clinical or forensic problems involving dietary supplements, promote dialogue between health care professionals and patients, and stimulate intellectual curiosity about these products, fostering further research into their therapeutic and adverse effects.

Timothy S. Tracy, PhD
Richard L. Kingston, PharmD

Contents

Contents

Contributors

MARGARET ARTZ • Senior Pharmacy Analyst, Department of Research and Development, Ingenix, Eden Prairie, MN

DANIEL BERKNER • Clinical Toxicologist, SafetyCall International, PLLC, Bloomington, MN

DEAN FILANDRINOS • Vice President of Operations and Senior Clinical Toxicologist, SafetyCall International, PLLC, Bloomington, MN; and Department of Experimental and Clinical Pharmacology, University of Minnesota, Minneapolis, MN

DOUGLAS D. GLOVER • Professor Emeritus, Department of Obstetrics and Gynecology, West Virginia University School of Medicine, Morgantown, WV

ILA M. HARRIS • Department of Pharmaceutical Care and Health Systems, University of Minnesota, Minneapolis, MN

LESLIE HELOU • Department of Pharmaceutical Care and Health Systems, University of Minnesota, Minneapolis, MN

BRIAN J. ISETTS • Department of Pharmaceutical Care and Health Systems, University of Minnesota, Minneapolis, MN

STEVEN B. KARCH • Consultant Pathologist/Toxicologist, Berkeley, CA

RICHARD L. KINGSTON • SafetyCall International, PLLC, Principal and Senior Clinical Toxicologist, Bloomington, MN

KATIE L. MEYERS • Pharmacy Department, Children's Hospitals and Clinics of Minnesota, Minneapolis, MN

LEO SIORIS • President, Clinical Services and Senior Clinical Toxicologist, SafetyCall International, PLLC, President, Clinical Services, Bloomington, MN; and Department of Experimental and Clinical Pharmacology, University of Minnesota, Minneapolis, MN

TIMOTHY S. TRACY • Department of Experimental and Clinical Pharmacology, University of Minnesota, Minneapolis, MN

ANDERS WESTANMO • Department of Pharmacy, Veterans Administration Hospital, Minneapolis, MN

THOMAS R. YENTSCH • Fairview-University Hospital, Minneapolis, MN

Chapter 1

Ma Huang and the Ephedra Alkaloids

Steven B. Karch

Summary

Ephedra has been used as a natural medicine for thousands of years by numerous cultures with very little concern about toxicity. Its most recent popularity is related to its purported "weight loss" or "performance enhancing" attributes. In spite of that in 2004, concerns over safety resulted in the banning of all over-the-counter (OTC) sales of ephedra-containing dietary supplements by the Food and Drug Administration.

All ephedra plants contain phenylalanine-derived alkaloids, including ephedrine, pseudoephedrine, methylephedrine, and trace amounts of phenylpropanolamine. Previously marketed herbal supplements typically stated total ephedra alkaloid content, although actual levels of individual alkaloid varied depending on raw material and production runs.

A double-blind, placebo-controlled trial by Boozer et al. examined issues of long-term safety and efficacy of ephedra, demonstrating its ability to reduce body weight and body fat while improving blood lipids without significant adverse events. Although other studies have documented a favorable adverse effect profile for appropriately administered doses of ephedra-containing supplements, there have been numerous anecdotal reports of adverse effects. Abuse and misuse of ephedra-containing products likely contributed to spontaneously reported adverse effects and increased concerns over safety.

As with other sympathomimetic agents, theoretical drug interactions with ephedra alkaloids are possible. Despite this potential, only a handful of adverse drug interactions have been reported. This is especially pertinent when considering the extensive use of both ephedra-containing supplements and ephedrine- or pseudoephedrine-containing OTC products. The most notable interaction exists between nonselective monoamine oxidase inhibitors and ephedra- or ephedrine-containing products.

From Forensic Science and Medicine:
Herbal Products: Toxicology and Clinical Pharmacology, Second Edition
Edited by: T. S. Tracy and R. L. Kingston © Humana Press Inc., Totowa, NJ

With the ban of ephedra-containing dietary supplements and severe restrictions in access to ephedrine-containing OTC products, the landscape of clinical use associated with agents of this nature has been dramatically changed forever. Interest in further clinical study will likely be severely limited.

Key Words: Herbal stimulants; weight loss; phenylalanine; alkaloids; bronchodilator; athletic performance.

1. HISTORY

Ephedra, and other medicinal plants have been identified at European neanderthal burial sites dating from 60,000 BCE *(1)*. Thousands of years later, Pliny accurately described the medicinal uses of ephedra. But thousands of years before Pliny, traditional Chinese healers used ephedra extracts. Chinese texts from the 15th century recommended ephedra as an antipyretic and antitussive. In Russia, around the same time, extracts of ephedra were used to treat joint pain; and though recent laboratory studies confirm that ephedra might be useful for that purpose *(2)*, additional trials and studies have not been forthcoming. In the 1600s, Indians and Spaniards in the American Southwest used ephedra as a treatment for venereal disease *(3)*. That idea might also have had some merit, as some studies show that ephedra contains compounds with antibiotic activity called transtorines *(4)*. Whether the transtorines will prove to be clinically useful has not been determined.

In 1885, Nagayoshi Nagi, a German-trained, Japanese-born chemist, isolated and synthesized ephedrine. Nagi's original observations were confirmed by Merck chemists 40 years later *(5)*. Merck's attempts at commercializing ephedrine were unsuccessful, at least until 1930, when Chen and Schmidt published a monograph recommending ephedrine as the treatment of choice for asthma *(3)*. During the 1920s and 1930s, epinephrine was the only effective oral agent for treating asthma. Epinephrine, which had been available since the early 1900s was (and still is) an effective bronchodilator, but it has to be given by injection, or administered with special nebulizers. Ephedrine was nearly as effective as epinephrine, and could be taken orally. As a result, ephedrine became the first-line drug against asthma. It was displaced from that front during the late 1970s and early 1980s, when aerosolized synthetic β-agonists were introduced.

Unlike most of the other alkaloids contained in ephedra (methylephedrine and cathinone are both psychoactive, but the amount contained in unadulterated ephedra is too low to be of clinical significance), ephedrine is also a potent central nervous system (CNS) stimulant *(6)*. Injections of ephedrine, called philopon (which means "love of work") were given to Japanese kamikaze pilots during World War II. A major epidemic of ephedrine abuse occurred in postwar Japan, when stockpiles of ephedrine accumulated for use by the Army were dumped on the black market. Abusers in Tokyo, and other large Japanese cities, injected themselves with ephedrine (then referred to as hirapon), in much the same way that methamphetamine is injected today *(7)*. In the Philippines, a mixture of ephedrine and caffeine called shabu was traditionally smoked for its stimulating effect. In the late 1980s, shabu smoking gave way to the practice of smoking methamphetamine ("ice"). In what is perhaps a tribute to the past, some "ice" is sold under the philopon name.

The chemistry and nomenclature of these compounds are somewhat confusing, and are best understood by reference to the synthetic route used by plants to make ephedrine. All ephedra plants contain phenylalanine-derived alkaloids. Plants use phenylalanine as a precursor, but incorporate only seven of its carbon atoms. Phenylalanine is metabolized to benzoic acid, which is then acetylated and decarboxylated to form pyruvic acid. Transamination, results in the formation of forms (–)-cathinone.

Reduction of one carbonlyl group leads to the formation of either (–)-norephedrine (phenylpropanolamine is the name used to refer to the synthetic mixture of ± norephedrine), or norpseudoephedrine (called cathine). N-methylation of (–)-norephedrine results in the formation of (l) ephedrine. N-methylation of cathine leads to the formation of (+)-pseudoephedrine *(8)*.

2. CURRENT PROMOTED USES

Physicians routinely used intravenous ephedrine for the prophylaxis and treatment of hypotension caused by spinal anesthesia particularly during caesarean section *(9)*. In the past, ephedrine was used to treat Stokes–Adams attacks (complete heart block), and was also recommended as a treatment for narcolepsy. Over the years, ephedrine has been replaced by other, more effective agents *(10)*, and the advent of highly selective β-agonists has mostly eliminated the need to use ephedrine in treating asthma.

European medical researchers have, for several years, used ephedrine to help promote weight loss, at least in the morbidly obese *(11,12)*, and nutritional supplements containing naturally occurring ephedra alkaloids are sold in the United States for the same purpose. Clinical trials confirm that, taken as directed, use of these supplements does result in weight loss, though whether such losses are sustained has not been determined *(13,14,15)*.

Prior to its banning by the Food and Drug Administration (FDA) in 2004, ephedra was found in many "food supplements," used by bodybuilders. Generally, it was compounded with other ingredients such as vitamins, minerals, and amino acids in products, which are said to increase muscle mass and enhance endurance *(16)*. Performance improvement secondary to ephedrine ingestion has been established in a controlled clinical trials *(17,18,19)*, and use of ephedrine has been prohibited by the International Olympic Committee.

Ephedra was also sold in combination with many other herbs in obscure combinations. Labels frequently listed 10 or 15 different herbs, but, analysis usually disclosed only the ephedra alkaloids and caffeine as present in sufficient quantities to be physiologically active. After several well-publicized accidental deaths, products clearly intended for abuse, such as "herbal ecstasy," and other "look-alike drugs" (products usually containing ephedrine or phenylpropanolamine designed to look like illicit methamphetamine, but in concentrations higher than recommended by industry or the FDA) were withdrawn from the market. Labels on these products were frequently misleading. For example, one might suppose that a product called "Ephedrine 60™ " contained 60 mg of ephedrine when, in fact, the actual ephedrine content was 25 mg.

3. SOURCES AND CHEMICAL COMPOSITION

Ephedra (*ephérdre du valais* in French and *Walliser meerträubchen* in German) is a small perennial shrub with thin stems. It rarely grows to more than a foot in height, and at first glance, the plant looks very much like a small broom. Different, closely related species are found in Western Europe, southeastern Europe, Asia, and even the Americas. Some of the better known species include *Ephedra sinica* and *E. equisentina* from China (collectively known as ma haung), as well as *E. geriardiana, E. intermedia,* and *E. major,* which grow in India and Pakistan, and countless other members of the family Ephedraceae that grow in Europe and the United States (*E. distachya, E. vulgaris*) *(20)*.

Ephedra species vary widely in their ephedrine content. One of the most common Chinese cultivars, known as "China 3," contains 1.39% ephedrine, 0.361% pseudoephedrine, and 0.069% methylephedrine *(21)*. This mix is fairly typical for commercially grown ephedra plants. Noncommercial varieties of ephedra may contain no ephedrine at all *(22)*, while others may contain more pseudoephedrine than ephedrine. Depending on the variety, trace amounts of phenylpropanolamine, (–) norephedrine, and methylephedrine may also be present, however (+) norephedrine does not occur naturally, and its presence is proof of adulteration.

Labels on herbal supplements listed total ephedra alkaloid content, usually 10 or 11 mg per serving. Depending on the raw materials used, different production runs of the same product contained ephedrine and pseudoephedrine in varying proportions. Occasionally, supplement makers were accused of adulterating their product by adding synthetic ephedrine or pseudoephedrine. Unlike with (+)-norephedrine, these compounds occur naturally and product adulteration should not have been alleged just because alkaloids other than ephedrine were detected in trace amounts, or because the ratio of ephedrine to pseudoephedrine was close to, or even greater than, 1:1. Of course, if one of the minor alkaloids, such as methylephedrine, were found to be present in concentrations approaching those of ephedrine, the ratio could only be explained by adulteration.

4. PRODUCTS AVAILABLE

Prior to its ban in 2004, no one government agency was tasked with tracking production of ephedrine-containing products. Nor were these products indexed by any industry or trade organization. Ephedrine-containing supplement products were mostly purchased at health food stores or over the Internet. Claims made by some of the Internet vendors were quite outrageous and totally unsupported by any scientific research. The large supplement makers, of course, had web pages, many of which contained, or had links to, the most recent peer review studies. But in addition to the established names, hundreds of other, smaller manufacturers also advertised and sold over the Internet. These companies came into and went out of existence so rapidly that a detailed listing of their web sites would likely be outdated before the links were published. Even today, a simple search using the word "ephedrine," will disclosed numerous off-shore vendors, along with numbers of attorneys soliciting for ephedra-related class action legal cases.

In addition to selling their own proprietary mixture, many of these same web sites sold the same popular products as the herbal and general retail outlets, such as a previous Twin Labs best seller "Ripped Fuel™ ," which contained ephedrine in the form of ma huang, combined with guarana, L-carnitine, and chromium picolinate. Metabolife 356™ contained guarana (40 mg caffeine), 12 mg ephedrine as ma huang, chromium picolinate 75 mg, and several other ingredients. Ever since ephedrine became the precursor of choice for making methamphetamine, federal regulators have severely restricted bulk sales of ephedrine, but these restrictions have been bypassed in some cases by illegally ordering from a foreign web site *(23)*.

In most products, ephedrine content ranged anywhere from 12 to 80 mg per serving, with the majority of products falling into the lower range. Industry standards called for a total dose of ephedrine of less than 100 mg/day. The FDA, however, allowed a maximum daily dose of 150 mg/day of synthetic ephedrine. Unless fortified, the expected ephedrine content of ma huang capsules was generally less than 10%. Thus, a capsule said to contain 1000 g of ephedra would probably have contained no more than 80 mg of ephedrine.

In the United States, (+)-norpseudoephedrine, in its pure form, is considered a Schedule IV controlled substance. However, because of the small amounts of this alkaloid in ephedra plants or extracts, the Drug Enforcement Administration (DEA) had never stated or proposed that ephedra products were subject to the scheduling requirements of the Controlled Substances Act. Quite the contrary, DEA published a proposed rule in 1998 that stated DEA's intent to exempt legitimate ephedra products in finished form from regulation even as "chemical mixtures." Other regulatory sanctions and actions on ephedra rendered action on this regulation moot.

5. PHARMACOLOGICAL EFFECTS

Studies have shown that resultant effects are similar, regardless of whether pure synthetic ephedrine or naturally occurring ephedra is ingested *(24,25)*. There are, however, significant enantioselective differences between the enantomers in both pharmacokinetic and pharmacodynamic effects. All of the ephedra alkaloids have important effects on the cardiovascular and respiratory systems, but not to the same degree.

Ephedrine, the predominant alkaloid in ephedra, is both an α and β stimulant. It directly stimulates α_2 and β_1; receptors and, because it also causes the release of norepinephrine from nerve endings, it also acts as a β_2 stimulant. The resultant physiological changes are variable, depending on receptor distribution and receptor regulation *(26)*. Tolerance to ephedrine's β agonist actions emerges rapidly, which is why ephedrine is no longer the preferred agent for treating asthma; receptor downregulation quickly occurs and the bronchodilator effects are lost *(27,28)*.

Receptor distribution probably explains why ephedrine has no effect on diastolic pressure, and only minimal effect on systolic. β_2 Stimulation of vessels in peripheral muscles results in peripheral vasodilation and "diastolic runoff," which more than cancels ephedrine's other inotropic effects *(29)*. The absence of any significant effect on blood pressure was firmly established during the late 1970s and early 1980s in dozens of double-blind, placebo-controlled studies performed to compare the effectiveness of ephedrine with that of newly synthesized adrenergic agents *(30–60)*.

The pharmacokinetic and toxicokinetic behavior of any isomer cannot be used to predict that of any other ephedrine isomer. The (+) isomer of methamphetamine, for example, is a potent CNS stimulant, but the (–)-isomer is merely a decongestant. There is a tendency in the literature to lump together all "ephedrine alkaloids" and use the term "class effect" to assume that all the different drugs in that class exert the same effects on the same biological targets. In fact, some of the drugs in the class will be similar in some regards and different in others.

The affinity of the various ephedrine isomers for human β-receptors has been measured and compared (as indicated by the amount of cyclic adenosine monophosphate produced compared to that of isoproterenol) in tissue culture. Activity of the different isomers is highly stereoselective, i.e., the different isomers had very different receptor-binding characteristics. For β_{1}-receptors, maximal response (relative to isoproterenol = 100%) was greatest for ephedrine (68% for 1R,2S-ephedrine and 66% for the 1S,2R-ephedrine isomer). Both of the pseudoephedrine isomers had much lower affinities (53%). When binding to β_2-receptors was measured, the rank order of potency for 1R,2S-ephedrine was 78%, followed by 1R,2R-pseudoephedrine (50%), followed by 1S,2S-pseudoephedrine (47%). The 1S,2R-ephedrine isomer had only 22% of the activity exerted by isoproterenol, but was the only isomer that showed any significant agonist activity on human β_3-receptors (31%) *(61)*. Stimulation of β_3-receptors, which are thought to be located only in fat cells, may account for ephedrine's ability to cause weight loss *(62–64)*.

Ephedrine is also an α agonist and, as such, is capable of stimulating bladder smooth muscle. At one time, it was used to promote urinary continence *(65,66)*. In animal models, when compared to norepinephrine, ephedrine is a relatively weak α-adrenergic agonist, possessing less than one-third the activity of norepinephrine *(67)*. Ephedrine's usefulness as a bronchodilator is limited by the number of β-receptors on the bronchi. The number of β-

receptors located on human lymphocytes (which correlates with the number found in the lungs) decreases rapidly after the administration of ephedrine; the density of binding sites drops to 50% after 8 days of treatment and returns to normal 5 to 7 days after the drug has been withdrawn *(27)*.

6. CLINICAL STUDIES

6.1. Bronchodilation

Banner et al. summarized studies where the effects of ephedrine and ephedra were compared to placebo in controlled studies in humans. None of the controlled trials disclosed any evidence of cardiovascular toxicity when ephedrine was given in doses as high as 1 mg/kg, even when it was administered to severe asthmatics with known cardiac arrhythmias *(57)*. The trial reported by Banner et al. studied the respiratory and circulatory effects of orally administered ephedrine sulfate, 25 mg, aminophylline, 400 mg, terbutaline sulfate, 5 mg, and placebo in 20 patients with ventricular arrhythmia by a double-blind crossover method. The study was comprised of 20 patients, with an average age of 60 years and a preexisting history of both asthma and heart disease (as evidence by the presence of frequent premature ventricular contractions). The bronchodilator effect of terbutaline was similar to that of aminophylline over 4 hours but superior to ephedrine at hour 4. Both terbutaline and ephedrine exhibited chronotropic effects, with the effect of terbutaline greater than that of ephedrine at hour 4. The effect of aminophylline on heart rate (HR) did not differ from placebo. Only terbutaline was associated with an increase in ventricular ectopic beats. Ventricular tachycardia occurred in three patients treated with terbutaline and in one patient with ephedrine (which occurred before he was given ephedrine). There were no significant changes in blood pressure. Orally administered terbutaline should not be regarded as safer than orally administered ephedrine or aminophylline in patients with arrhythmias.

In 1992, Astrup studied the effects of ephedrine and caffeine in a group of obese patients *(68)*. In a randomized, placebo-controlled, double-blind study, 180 obese patients were treated by diet (4.2 mJ/day) and either an ephedrine/caffeine combination (20 mg/200 mg), ephedrine (20 mg), caffeine (200 mg), or placebo three times a day for 24 weeks. Withdrawals were distributed equally in the four groups, and 141 patients completed the trial. Mean weight losses was significantly greater with the combination than with placebo from week 8 to week 24 (ephedrine/caffeine, 16.6 ± 6.8 kg vs placebo, 13.2 ± 6.6

kg [mean ± standard deviation {SD}], $P = 0.0015$). Weight loss in both the ephedrine and the caffeine groups was similar to that of the placebo group. Side effects (tremor, insomnia, and dizziness) were transient and after 8 weeks of treatment they had reached placebo levels. Systolic and diastolic blood pressure fell similarly in all four groups.

6.2. Weight Loss

The most recent of the studies examining weight control were designed to address concerns about long-term safety and efficacy for weight loss using a mixture containing 90 mg of ephedrine (from ephedra) and 192 mg of caffeine, derived from cola nuts *(15)*. A 6-month randomized, double-blind, placebo-controlled trial was performed, in which a total of 167 subjects (body mass index $31.8 ± 4.1 \text{ kg/m}^2$) were randomized to receive either placebo ($n = 84$) or herbal treatment ($n = 83$). The primary outcome measurements were changes in blood pressure, heart function, and body weight. Secondary variables included body composition and metabolic changes. It was found that herbal vs placebo treatment decreased body weight ($-5.3 ± 5.0$ vs $-2.6 ± 3.2$ kg, $P < 0.001$), body fat ($-4.3 ± 3.3$ vs $-2.7 ± 2.8$ kg, $P = 0.020$), and low-density lipoprotein cholesterol ($-8 ± 20$ vs $0 ± 17$ mg/dL, $P = 0.013$), and increased high-density lipoprotein cholesterol ($+2.7 ± 5.7$ vs $-0.3 ± 6.7$ mg/dL, $P = 0.004$). Herbal treatment produced small changes in blood pressure variables ($+3$ to -5 mmHg, $P≤0.05$), and increased HR ($4 ± 9$ vs $-3 ± 9$ beats per minute, $P < 0.001$), but cardiac arrhythmias were not increased ($P > 0.05$). By self-report, dry mouth ($P < 0.01$), heartburn ($P < 0.05$), and insomnia ($P < 0.01$) were increased and diarrhea decreased ($P < 0.05$). Irritability, nausea, chest pain, and palpitations did not differ, nor did numbers of subjects who withdrew. CONCLUSIONS: In this 6-month placebo-controlled trial, herbal ephedra/caffeine (90/192 mg/day) promoted body-weight and body-fat reduction and improved blood lipids without significant adverse events.

6.3. Athletic Performance

In a series of studies, Bell et al. assessed the effects of ephedrine mixtures on performance, and found measurable improvement. One and one-half hours after ingesting a placebo (P), caffeine (C) (4 mg/kg), ephedrine (E) (0.8 mg/kg), or caffeine and ephedrine, 12 subjects performed a 10-km run while wearing a helmet and backpack weighing 11 kg. The trials were performed in a climatic suite at 12–13°C, on a treadmill where the speed was regulated by the subject. VO_2, VCO_2, $V(E)$, HR, and rating of perceived exertion were measured during the run at 15 and 30 minutes, and again when the individual reached 9 km. Blood was sampled at 15 and 30 minutes and again at the end of the run and assayed for lactate, glucose, and catecholamines. Run times (mean ± SD), in minutes, were for C (46.0 ± 2.8), E (45.5 ± 2.9), C + E (45.7 ± 3.3), and P (46.8 ± 3.2). The run times for the E trials (E and C + E) were significantly reduced compared with the non-E trials (C and P). Pace was increased for the E trials compared with the non-E trials over the last 5 km of the run. VO_2 was not affected by drug ingestion. HR was elevated for the ephedrine trials (E and C + E), but the respiratory exchange ratio (a measure of maximal exertion) remained similar for all trails. Caffeine increased the epinephrine and norepinephrine response associated with exercise and also increased blood lactate, glucose, and glycerol levels. Ephedrine reduced the epinephrine response but increased dopamine and free fatty acid levels. Bell concluded previously that the effects of caffeine, when taken with ephedrine, were not additive, and that all of the observed improvement could be accounted for by the presence of ephedrine *(19)*.

7. PHARMACOKINETICS

Phenylpropanolamine is readily and completely absorbed, but pseudoephedrine, with a bioavailability of only approx 38%, is subject to gut wall metabolism, and absorption may be erratic *(69)*. Pure ephedrine is well absorbed from the stomach, but absorption is much slower when it is given as a component of ma huang, rather than in its pure form *(70)*. Ephedrine ingested in the form of ma huang has a t_{max} of nearly 4 hours, compared to only 2 hours when pure ephedrine is given. Like its enantiomers, ephedrine is eliminated in the urine largely as unchanged drug, with a half-life of approx 3–6 hours.

The rate at which any of the enantiomers is eliminated depends upon the urinary pH. At high pHs, excretion time is prolonged. At low pH ranges, excretion is accelerated. In controlled laboratory studies, where volunteer subjects were given either bicarbonate or ammonium chloride, the higher the urine pH, the more slowly the ephedrine and pseudoephedrine were excreted. Conversely, when the urine pH is low, excretion is accelerated *(71)*. The importance of these observations is hard to assess, because without the addition of bicarbonate, urine pH values in the general population rarely approach 8.0. A study of pseudoephedrine pharmacokinetics in 33 volunteers who were not treated with drugs to alter urine pH found that these parameters could not be correlated to urine pH, mainly because there was little difference in pH between the different participants *(72)*. Excretion patterns may be much more rapid in children, and a greater dosage may be required to achieve therapeutic effects. Patients with renal impairment are at special risk for toxicity.

Peak concentrations for the other enantiomers, specifically phenylpropanolamine and pseudoephedrine, occur earlier (0.5 and 2 hours, respectively) than for ephedrine, but all three drugs are extensively distributed into extravascular sites (apparent volume of distribution between 2.6 and 5.0 L/kg). No protein-binding data in humans are available. Peak ephedrine levels after ingestion of 400 mg of ma huang, containing 20 mg of ephedrine, resulted in blood concentrations of 81 ng/mL—essentially no different than the peak ephedrine levels observed after giving an equivalent amount of pure ephedrine *(70,25)*. In another study, 50 mg of ephedrine given orally to six healthy, 21-year-old women produced mean peak plasma concentrations of 168 ng/mL, 127 min after ingestion, with a half-life of slightly more than 9 hours *(73)*. The results are comparable to those obtained in studies done nearly 30 years earlier *(74)*.

Very high levels of methylephedrine have been observed in Japanese polydrug abusers taking a cough medication called BRON. Concentrations of methylephedrine less than 0.3 mg/L, the range generally observed in individuals taking BRON for therapeutic rather than recreational purposes *(75)*, appear to be nontoxic and devoid of measurable effects. Methylephedrine is a minor component of most ephedra plants, but in Japan (where, unlike in the United States, methylephedrine is legally sold) it is produced synthetically, and is used in cough and cold remedies, especially BRON *(76–78)*. In terms of catecholamine stimulation, methylephedrine appears comparable to ephedrine; however, it does not react with most standard urine screening tests for ephedrine *(75)*. This can be a cause of some forensic confusion, because 10–15% of a given dose of methylephedrine is converted to ephedrine *(75)*.

Although the issue has been raised in litigation, the amounts of methylephedrine and norephedrine contained in naturally occurring ephedra are so low as to be of no clinical consequence. For example, the study by Gurley et al. found that most of the commercial products tested had no methylephedrine whatsoever, but when it was present, it was usually in quantities of less than 1 mg per serving (range 0.2 to 2.2 mg). If the volume of distriution (Vd) of methylephedrine is assumed to be 3.5, approximately the same as ephedrine, then a 70-kg man ingesting a 2-mg serving of methylephedrine would produce a blood concentration of (dose = kg weight ×

blood concentration × Vd) 0–0.06 mg, undoubtedly below most laboratories' minimum level of detection, and a clinically insignificant finding. Similar considerations apply to the small amounts of norephedrine found in these products.

8. ADVERSE EFFECTS AND TOXICITY

8.1. General Overview

Two journal articles analyzing adverse event reporters (AERs) have been published in the peer-reviewed literature, and both reports have received wide publicity *(79,80)*. The reports are, however, of limited use in assessing toxicity, because they are comprised of passively collected anecdotal data, which is often incomplete and unreliably reported. For example, one of the FDA ephedrine AERs "analyzed" in an article published in the *New England Journal of Medicine* described the sudden death of a teenage girl who had been born with a lethal cardiac malformation who died while playing volleyball *(79)*. Postmortem blood and tissue tested negative for ephedrine, and the article failed to mention the existence of the cardiac malformation. In other AERs, massive doses of ephedrine were consumed (as with products intended for abuse, such as "herbal ecstasy," now withdrawn from the market). Toxicology testing was rarely performed in any of these cases, and it is not known with any certainty whether ephedrine was even taken. Even the authors of the two papers concede that anecdotal reports cannot be used to prove causality, stating that "Our report does not prove causation, nor does it provide quantitative information with regard to risk" *(79)*. There is little point in reviewing material that cannot be used to prove causality, and it is not included in the summaries that follow, which are comprised only of published, peer-review case reports, epidemiological surveys, and controlled clinical trials. An additional review of the utility of spontaneously reported adverse events involving supplements and, more specifically, ephedra was published by Kingston et al. *(81)*. The review discussed the limitations of spontaneously reported data in assessing supplement safety and determining causality between exposure and adverse effects.

Despite conflicting data regarding the safety of ephedra from clinical studies and conclusions drawn from spontaneously reported adverse events, FDA banned the sale of ephedra-containing supplements in 2004 (*see* Regulatory Status).

8.2. Neurological Disorders

Many strokes attributed to ephedrine have actually been caused by the ingestion of ephedrine enantiomers, pseudoephedrine *(82–85)*, phenylpropanolamine *(86–93)*, and even methylephedrine *(77)*. Two cases of ischemic stroke have been reported *(94,95)*, but in neither case was their any toxicological testing to confirm the use of ephedrine. A decade-old report described the autopsy findings in three individuals with intracerebral hemorrhage and positive toxicology testing for ephedrine; however, one had hypertensive cerbrovasular disease and the other had a demonstrable ruptured aneurysm *(96)*. Intracerebral hemorrhage has also been described in suicide and attempted suicide victims who took overdoses of pseudoephedrine *(97,98)*. There is also a report describing a patient who developed described arteritis following the intravenous administration of ephedrine during a surgical procedure *(99)*. On the other hand, a large study to assess risk factors for stroke in young people (age 20–49) over a 1-year period was carried out in Poland, a country where ephedra-based products are widely used. Nearly one-half the cases of stroke were associated with preexisting hypertension, another 15% had hyperlipidemia, and 6% were diabetic *(100)*. None of the individuals were ephedrine users.

Sometimes, especially in Japan and the Philippines, ephedrine is taken specifically as a psychostimulant. In Japan, BRON, the OTC cough medication containing methylephedrine, dihydrocodeine, caffeine, and chlorpheniramine, is very widely abused, and transient psychosis commonly results *(76–78)*. Reports of ephedrine-related psychosis following prolonged, heavy use are fairly common *(101–105)*. In general, psychosis is only seen in ephedrine users ingesting more than 1000 mg/day, and it resolves rapidly once the drug is withdrawn *(106)*. Ephedrine psychosis closely resembles psychosis induced by amphetamines: paranoia with delusions of persecution and auditory and visual hallucinations, even though consciousness remains unclouded. Typically, patients with ephedrine psychosis will have ingested more than 1000 mg/day. Recovery is rapid after the drug is withdrawn *(103)*. The ephedrine content per serving of most food supplements is on the order of 10–20 mg, making it extremely unlikely that, in recommended doses, use of any of the products would lead to neurological symptoms.

8.3. Renal Disorders

Reports, particularly in the European literature, have described the occurrence of renal calculi in chronic ephedrine users *(107–111)*. A review of cases from a large commercial laboratory specializing in the analysis of kidney stones found that 200 out of 166,466, or 0.064%, of stones analyzed by that laboratory, contained either ephedrine or pseudoephedrine. Unfortunately, the analytic technique used could not distinguish ephedrine from pseudoephedrine, and because pseudoephedrine is used so much more widely than ephedrine, it seems that the risk of renal calculus associated with ephedrine use must be quite small *(110)*. There have been no new reports of ephedrine-related nephrolithiasis since 1999. Direct toxicity, with altered renal function and demonstrable kidney lesions related to ephedrine use, has never been demonstrated. Urinary retention, occurring as a consequence of drug overdose, was occasionally reported *(112,113)*, but additional cases have not been described in more than a decade. The FDA and Commission E both warn against the possibility of urinary retention in patients with prostatic enlargement, but the theoretical basis for this concern is unclear, and, in any case, retention in patients with prostate disease has not been reported.

Small amounts of ephedrine are oxidized in to norephedrine and norpseudoephedrine in the liver *(24,73)*. In patients with diminished renal function, these drugs may accumulate and have the potential to cause serious toxicity. None of the ephedrine enantiomers are easily removed by dialysis, and treatment of overdose remains supportive, using pharmacological antagonists to counter the α- and β-adrenergic effects of these drugs *(114)*. Because excretion is pH-dependent, patients with renal tubular acidosis are also at risk *(115)*. The FDA reports having received a number of accounts of hematuria after use of ephedra-based products, but no such cases have ever appeared in the peer-reviewed literature, and review of the reports published by the FDA shows that all of the affected individuals were taking multiple remedies, some capable of causing interstitial nephritis.

8.4. Cardiovascular Diseases

Ephedrine and pseudoephedrine share properties with cocaine and with the amphetamines because they: (1) stimulate β-receptors directly, and (2) also cause the increased release of norepinephrine. Chronic exposure to abnormally high levels of circulating catecholamines can damage the heart. This is certainly the case with cocaine and methamphetamine *(116,117)*, but ephedrine-related cardiomyopathy is an extremely rare occurrence, occurring only in individuals who take massive amounts of drug for prolonged periods of time. Only two papers have ever been published on the subject *(118,119)*.

The two existing reports are uninterpretable, because histological findings were not described in either report, and angiography was not performed, thereby making it impossible to actually establish the diagnosis of cardiomyopathy.

Similar considerations apply to the relationship (if any) between myocardial infarction and ephedrine use. The report by Cockings and Brown described a 25-year-old drug abuser who injected himself with an unknown amount of cocaine intravenously *(120)*. The only other published reports involved a woman in labor who was receiving other vasoactive drugs *(121)*; and two pseudoephedrine users, one of whom was also taking bupropion, who developed coronary artery spasm *(122,123)*. Three cases of ephedra-related coronary spasm in anesthetized patients have also been reported, but multiple agents were administred in all three cases, and the normal innervation of the coronary arteries was disrupted in two of the cases where a high spinal anesthetic had been administered *(121,124)*. One case of alleged ephedrine-related hypersensitivity myocarditis has been reported *(125)*, but the patient was taking many other herbal supplements, and the responsible agent is not known with certainty. Although there are no reasons why ephedra alkaloids should not cause allergic reactions, the incidence appears to be extremely low.

Although clinical trials or epidemiological studies are lacking, it has been suggested that maternal use of OTC cold medication may result in fetal arrhythmias *(126,127)*, but linkage between ephedrine and isomers and arrhythmia has never been demonstrated. The literature contains one case report *(128)* describing arrhythmias occurring in a 14-year-old who overdosed on cold medications. The child had taken a total of 3300 mg of caffeine, 825 mg of phenylpropanolamine, and 412 mg of ephedrine. Clearly, large doses of ephedrine, and its enantiomers, are capable of exerting toxicity.

The paucity of peer-reviewed studies describing cardiovascular complication with ephedra alkaloids suggests that few such cases are occurring. This notion is support by the studies of Porta et al., who performed a follow-up study of more than 100,000 persons below age 65 years who filled a total of 243,286 prescriptions for pseudoephedrine. No hospitalizations could be attributed to the drug. There were no admissions within 15 days of filling a prescription for pseudoephedrine for cerebral hemorrhage, thrombotic stroke, or hypertensive crisis. There were a small number of hospitalizations for myocardial infarction, seizures, and neuropsychiatric disorders, but the rate of such admissions among the pseudoephedrine users was close to the expected rate in the population at large.

8.5. Workplace Drug Testing

Ephedra alkaloids, even when used in the recommended amounts, can cause positive urine screening tests for methamphetamine *(129,130)*, sometimes yielding surprisingly high concentrations.

8.6. Postmortem Toxicology

Very few fatalities have ever been reported (or studied), but it appears that the therapeutic index for ephedrine is very great. A 1997 case report described a 28-year-old woman with two prior suicide attempts, who died after ingesting amitriptyline and ephedrine. The blood ephedrine concentration was 11,000 ng/mL, and the liver concentration was twice that value (kidney, 14 mg/kg; brain, 8.9 mg/kg). The amitriptyline concentration was 0.33 mg/kg in blood and 7.8 mg/kg in liver *(131)*. Values in a second case report (where methylephedrine concentrations were nearly 6000 ng/mL) may or may not be relevant to the problem of ephedrine toxicity, as the individual in question took massive quantities of a calcium channel blocker, and it is not known whether methylephedrine exerts all the same effects as ephedrine *(132)*. Baselt and Cravey mention the case of a young woman who died several hours after ingesting 2.1 g of ephedrine combined with 7.0 g of caffeine, but tissue findings were not described. Her blood ephedrine level was 5 mg/L, whereas the concentration in the liver was 15 mg/kg *(133)*.

A report from the European literature describes the findings in a 19-year-old woman who committed suicide by taking 40 Letigen® tablets (200 mg of caffeine and 20 mg of ephedrine) amounting to 10 g of caffeine and 1 g of ephedrine. She developed severe toxic manifestations from the heart, CNS, muscles, liver, and kidneys leading to several cardiac arrests, and died subsequently of cerebral edema and incarceration on the fourth day of hospitalization. Postmortem blood concentrations were not given *(134)*.

Pseudoephedrine concentrations, but not measurements for ephedrine or any of the other enantiomers, have been published by the National Association of Medical Examiners in their Annual Registry report. In 15 children diagnosed with sudden infant death syndrome, the mean blood pseudoephedrine concentration was 3.55 mg/L, the median 2.3 mg/L, with a range of 0.07–13.0 mg/L (SD = 3.36 mg/L). The authors of the study take pains to point out that "The data do not allow definitive statements about the toxicity of pseudoephedrine at a given concentration" *(129)*.

In the only autopsy study yet published *(135)*, all autopsies in the San Francisco Medical Examiner's jurisdiction from 1994 to 2001 where ephedrine or any its isomers (E+) were detected were reviewed. Cases where ephedrine or its isomers were detected were compared with those in a control group of drug-free trauma victims. Of 127 ephedrine-positive cases identified, 33 were the result of trauma. Decedents were mostly male (80.3%) and mostly Caucasian (59%). Blood ephedrine concentrations were less than 0.49 mg/L in 50% of the cases, with a range of 0.07–11.73 mg/L in trauma victims, and 0.02–12.35 mg/L in nontrauma cases. Norephedrine was present in the blood of only 22.8% (mean concentration of 1.81 mg/L, SD=3.14 mg/L) and in the urine of 36.2% of the urine specimens, with a mean concentration of 15.6 mg/L, SD=21.50 mg/L). Pseudoephedrine (PE) was detected in the blood of 6.3% (8/127). More than 88% (113/127) of the decedents who tested positive for ephedrine or one of its isomers also tested positive for other drugs, the most common being cocaine (or its metabolites) and morphine. The most frequent pathological diagnoses were hepatic steatosis (27/127) and nephrosclerosis (22/127). Left ventricular hypertrophy was common, and coronary artery disease was detected in nearly one-third of the cases. The most common findings in the ephedrine-positive deaths reviwed were those generally associated with chronic stimulant abuse. There were no cases of heat stroke and no cases of rhabdomyolysis.

8.7. Methamphetamine Manufacture

Either (–)-ephedrine or (+)-pseudoephedrine can be used to make methamphetamine by reductive dehalogenation using red phosphorus as a catalyst. If (–)-ephedrine is used as the starting material, the process will generate (+)-methamphetamine. If psuedoephedrine is used, the result will be dextromethamphetamine *(136)*. As this synthetic route has become nearly universal, both state and federal governments have enacted laws limiting the amount of pure ephedrine or pseudoephedrine that can be purchased.

9. DRUG INTERACTIONS

The ephedra alkaloids are all sympathomimetic amines, which means that a host of drug interactions are theoretically possible. In fact, only a handful of adverse drug interactions have been reported in the peer-reviewed literature. The most important of these involve the monoamine oxidase inhibitors (MAOI). Irreversible, nonselective MAOIs have been reported to adversely interact with indirectly acting sympathomimetic amines present in many cough and cold medicine. In controlled trials with individuals taking moclobemide, ephedrine's effects on pulse and blood pressure were potentiated, but only at higher doses than those currently provided in health supplements *(137)*. Ephedrine-MAOI interaction may, on occasion, be severe enough to mimic pheochromocytoma *(138)*. In addition, there is decreased metabolic clearance of pseudoephedrine when MAOIs are administered concurrently *(139)*. At least one case report suggests that selective serotonin reuptake inhibitor antidepressants can react with pseudoephedrine, leading to the occurrence of "serotonin syndrome" *(140)*. Bromocriptine, the ergot-derived dopamine agonist can interact with pseudoephedrine, and would presumably interact with ephedrine as well *(141)*. Surgical patients being treated with clonidine have an enhanced pressor response to ephedrine, apparently a result of clonidine-induced potentiation of α_1-adrenoceptor-mediated vasoconstriction *(142,143)*. In some clinical trials, the coadministration of ephedrine with morphine has been shown to increase analgesia *(144)*, but this approach to pain relief remains somewhat controversial.

10. REPRODUCTION

Use of ephedra-containing products is likely unsafe during pregnancy because of reports of psychoses and cardiovascular effects *(79,104,145)*.

11. REGULATORY STATUS

In 2004, the FDA issued a final rule prohibiting the sale of dietary supplements containing ephedrine alkaloids (ephedra), citing concerns over safety and potential risk of illness or injury.

The FDA reviewed evidence about ephedra's pharmacology: peer-reviewed scientific literature on ephedra's safety and effectiveness, adverse event reports, and a seminal report by the RAND Corporation, an independent scientific institute. Spontaneously reported adverse effects with high-profile sports figures and others raised public awareness and fueled the debate over safety. Subsequent to the ban, various trade groups and supplement companies have criticized the ban, and an appeal of the decision with temporary suspension of sanctions in some jurisdictions, pending further review, has occurred. Regardless of the regulatory outcome, reintroduction of OTC ephedra-containing supplements is not likely to occur.

Although banned in the United States, use of ephedra in other countries is likely to continue.

REFERENCES

1. Lietava J. Medicinal plants in a Middle Paleolithic grave Shanidar IV? J Ethnopharmacol 1992;35(3):263–266.
2. Kasahara Y, Hikino H, Tsurufuji S, Watanabe M, Ohuchi K. Antiinflammatory actions of ephedrines in acute inflammations. Planta Med 1985;(4):325–331.

3. Grinspoon L, Hedblom P, eds. *The speed culture: amphetamine use and abuse in America*. Cambridge: Harvard University Press, 1975.
4. al-Khalil S, Alkofahi A, el-Eisawi D, al-Shibib A. Transtorine, a new quinoline alkaloid from Ephedra transitoria. J Nat Prod 1998;61(2):262–263.
5. Holmstedt B. Historical perspective and future of ethnopharmacology. J Ethnopharmacol 1991;32(1–3):7–24.
6. Martin WR, Sloan JW, Sapira JD, Jasinski DR. Physiologic, subjective and behavioral effects of amphetamine, methamphetamine, ephedrine, phenmetrazine and methylphenidate in man. Clin Pharmacol Ther 1971;12:245–248.
7. Deverall RLG. Red China's dirty drug war; the story of the opium, heroin, morphine and philopon traffic. Tokyo, 1954.
8. Dewick PM, ed. *Medicinal natural products: a biosynthetic approach*. New York: Wiley, 1997.
9. Yap JC, Critchley LA, Yu SC, Calcroft RM, Derrick JL. A comparison of three fluid-vasopressor regimens used to prevent hypotension during subarachnoid anaesthesia in the elderly. Anaesth Intensive Care 1998;26(5):497–502.
10. Pomerantz B, O'Rourke R. The Stokes-Adams syndrome. Am J Med 1969;46(6):941–960.
11. Astrup A, Lundsgaard C. What do pharmacological approaches to obesity management offer? Linking pharmacological mechanisms of obesity management agents to clinical practice. Exp Clin Endocrinol Diabetes 1998;106(Suppl 2):29–34.
12. Ramsey JJ, Colman RJ, Swick AG, Kemnitz JW. Energy expenditure, body composition, and glucose metabolism in lean and obese rhesus monkeys treated with ephedrine and caffeine. Am J Clin Nutr 1998;68(1):42–51.
13. Greenway F, Herber D, Raum W, Herber D, Morales S. Double-blind, randomized, placebo controlled clinical trials with non-prescription medications for the treatment of obesity. Obes Res 1999;7(4):370–378.
14. Boozer CN, Nasser JA, Heymsfield SB, Wang V, Chen G, Solomon JL. An herbal supplement containing Ma Huang-Guarana for weight loss: a randomized, double-blind trial. Int J Obes Relat Metab Disord 2001;25(3):316–324.
15. Boozer CN, Daly PA, Homel P, et al. Herbal ephedra/caffeine for weight loss: a 6-month randomized safety and efficacy trial. Int J Obes Relat Metab Disord 2002;26(5):593–604.
16. Clarkson PM, Thompson HS. Drugs and sport. Research findings and limitations. Sports Med 1997;24(6):366–384.
17. Bell DG, Jacobs I, McLellan TM, Zamecnik J. Reducing the dose of combined caffeine and ephedrine preserves the ergogenic effect. Aviat Space Environ Med 2000;71(4):415–419.
18. Bell DG, Jacobs I, Ellerington K. Effect of caffeine and ephedrine ingestion on anaerobic exercise performance. Med Sci Sports Exerc 2001;33(8):1399–1403.
19. Bell DG, McLellan TM, Sabiston CM. Effect of ingesting caffeine and ephedrine on 10-km run performance. Med Sci Sports Exerc 2002;34(2):344–349.
20. Namba T, Kubo M, Kanai Y, Namba K, Nishimura H, Qazilbash NA. Pharmacognostical studies of Ephedra plants Part I—The comparative histological studies on

Ephedra rhizomes from Pakistan and Afghanistan and Chinese crude drugs "Ma-Hung-Gen". Planta Med 1976;29(3):216–225.

21. Sagara K, Oshima T, Misaki T. A simultaneous determination of norephedrine, pseudoephedrine, ephedrine and methylephedrine in Ephedrae Herba and oriental pharmaceutical preparations by ion-pair high-performance liquid chromatography. Chem Pharm Bull (Tokyo) 1983;31(7):2359–2365.

22. Zhang JS, Tian Z, Lou ZC. [Quality evaluation of twelve species of Chinese Ephedra (ma huang)]. Yao Hsueh Hsueh Pao 1989;24(11):865–871.

23. Anonymous. Implementation of the Comprehensive Methamphetamine Control Act of 1996; regulation of pseudoephedrine, phenylpropanolamine, and combination ephedrine drug products and reports of certain transactions to nonregulated persons. Final rule. Fed Regist 2002;67(60):14,853–14,862.

24. Gurley BJ, Gardner SF, White LM, Wang PL. Ephedrine pharmacokinetics after the ingestion of nutritional supplements containing Ephedra sinica (ma huang). Ther Drug Monit 1998;20(4):439–445.

25. Haller CA, Jacob P III, Benowitz NL. Pharmacology of ephedra alkaloids and caffeine after single-dose dietary supplement use. Clin Pharmacol Ther 2002;71(6):421–432.

26. Webb AA, Shipton EA. Re-evaluation of i.m. ephedrine as prophylaxis against hypotension associated with spinal anaesthesia for Caesarean section. Can J Anaesth 1998;45(4):367–369.

27. Neve KA, Molinoff PB. Effects of chronic administration of agonists and antagonists on the density of beta-adrenergic receptors. Am J Cardiol 1986;57(12):17F–22F.

28. Williams BR, Barber R, Clark RB. Kinetic analysis of agonist-induced down regulation of the beta(2)-adrenergic receptor in BEAS-2B cells reveals high- and low-affinity components. Mol Pharmacol 2000;58(2):421–30.

29. Saupe KW, Smith CA, Henderson KS, Dempsey JA. Diastolic time: an important determinant of regional arterial blood flow. Am J Physiol 1995;269(3 Pt 2):H973–979.

30. Muittari A, Mattila MJ, Tiitinen H. The combined action of broncholytic agents in asthmatic patients: orciprenaline, isoprenaline, and ephedrine. Ann Med Intern Fenn 1968;57(1):31–35.

31. Muittari A, Mattila MJ, Patiala J. Bronchodilator action of drug combinations in asthmatic patients: decloxizine, orciprenaline, and ephedrine. Ann Allergy 1969;27(6):274–279.

32. Muittari A, Mattila MJ. Bronchodilator action of drug combinations in asthmatic patients. Ephedrine, theophylline and tranquilizing drugs. Curr Ther Res Clin Exp 1971;13(6):374–385.

33. Dulfano MJ, Glass P. Evaluation of a new B 2 adrenergic receptor stimulant, terbutaline, in bronchial asthma. II. Oral comparison with ephedrine. Curr Ther Res Clin Exp 1973;15(4):150–157.

34. Kerr A, Gebbie T. Comparison of orciprenaline, ephedrine and methoxyphenamine as oral bronchodilators. N Z Med J 1973;77(492):320–322.

35. Weinberger MM, Bronsky EA. Evaluation of oral bronchodilator therapy in asthmatic children. Bronchodilators in asthmatic children. J Pediatr 1974;84(3):421–427.

36. Bierman CW, Pierson WE, Shapiro GG. Exercise-induced asthma. Pharmacological assessment of single drugs and drug combinations. JAMA 1975;234(3):295–298.
37. Bierman CW, Pierson WE, Shapiro GG. The pharmacological assessment of single drugs and drug combinations in exercise-induced asthma. Pediatrics 1975;56(5 Pt-2 Suppl):919–922.
38. Bye C, Hill HM, Hughes DT, Peck AW. A comparison of plasma levels of L(+) pseudoephedrine following different formulations, and their relation to cardiovascular and subjective effects in man. Eur J Clin Pharmacol 1975;8(1):47–53.
39. Chervinsky P, Chervinsky G. Metaproterenol tablets: their duration of effect by comparison with ephedrine. Curr Ther Res Clin Exp 1975;17(6):507–518.
40. Geumei A, Miller WF, Paez PN, Gast LR. Evaluation of a new oral beta2 adrenoceptor stimulant bronchodilator, terbutaline. Pharmacology 1975;13(3):201–211.
41. Nelson HS, Black JW, Branch LB, et al. Subsensitivity to epinephrine following the administration of epinephrine and ephedrine to normal individuals. J Allergy Clin Immunol 1975;55(5):299–309.
42. Tashkin DP, Meth R, Simmons DH, Lee YE. Double-blind comparison of acute bronchial and cardiovascular effects of oral terbutaline and ephedrine. Chest 1975;68(2):155–161.
43. Weinberger M, Bronsky E, Bensch GW, Bock GN, Yecies JJ. Interaction of ephedrine and theophylline. Clin Pharmacol Ther 1975;17(5):585–592.
44. Blumberg MZ, Tinkelman DG, Ginchansky EJ, Blumberg BS, Taylor JC, Avner SE. Terbutaline and ephedrine in asthmatic children. Pediatrics 1977;60(1):14–19.
45. Simi WW, Miller WC. Clinical investigation of fenoterol, a new bronchodilator, in asthma. J Allergy Clin Immunol 1977;59(2):178–181.
46. Tinkelman DG, Avner SE. Ephedrine therapy in asthmatic children. Clinical tolerance and absence of side effects. JAMA 1977;237(6):553–557.
47. Tinkelman DG, Avner SE. Assessing bronchodilator responsiveness. J Allergy Clin Immunol 1977;59(2):109–114.
48. Bush RK, Smith AM, Welling PG, Alcala JC, Lee TP. A comparison of a theophylline-ephedrine combination with terbutaline. Ann Allergy 1978;41(1):13–17.
49. Chervinsky P. The development of drug tolerance during long-term beta2 agonist bronchodilator therapy. Chest 1978;73(6 Suppl):1001–1002.
50. Cohen BM. The cardiorespiratory effects of oral terbutaline and an ephedrine-theophylline-phenobarbital combination: comparison in patients with chronic obstructive ventilatory disorders. Ann Allergy 1978;40(4):233–239.
51. Eggleston PA, and McMahan S. A. The effects of fenoterol, ephedrine and placebo on exercise-induced asthma. Chest 1978;73(6 Suppl):1006–1008.
52. Falliers CJ. Controlled assessment of beta2 adrenergic therapy for childhood asthma. Chest 1978;73(6 Suppl):1008–1010.
53. Falliers CJ, Cato AE, Harris JR. Controlled assessment of oral bronchodilators for asthmatic children. J Int Med Res 1978;6(4):326–336.
54. Miller WC. Long-term beta2 bronchodilator therapy and the question of tolerance. Chest 1978;73(6 Suppl):1000–1001.
55. Stenius B, Haahtela T, Poppius H. Controlled comparison of Nuelin, Theodrox and Marax in asthmatic out-patients. Eur J Clin Pharmacol 1978;13(3):179–183.

56. VanArsdel PP Jr, Schaffrin RM, Rosenblatt J, Sprenkle AC, Altman LC. Evaluation of oral fenoterol in chronic asthmatic patients. Chest 1978;73(6 Suppl):997–998.
57. Banner AS, Sunderrajan EV, Agarwal MK, Addington WW. Arrhythmogenic effects of orally administered bronchodilators. Arch Intern Med 1979;139(4):434–437.
58. James TD, Lyons HA. A comparative study of bronchodilator effects of carbuterol and ephedrine. JAMA 1979;241(7):704–705.
59. Muittari A, Mattila MJ. Objective and subjective assessment of ephedrine combinations in asthmatic outpatients. Ann Clin Res 1979;11(3):87–89.
60. Pierson DJ, Hudson LD, Stark K, Hedgecock M. Cardiopulmonary effects of terbutaline and a bronchodilator combination in chronic obstructive pulmonary disease. Chest 1980;77(2):176–182.
61. Vansal SS, Feller DR. Direct effects of ephedrine isomers on human beta-adrenergic receptor subtypes. Biochem Pharmacol 1999;58(5):807–810.
62. Astrup A. Thermogenic drugs as a strategy for treatment of obesity. Endocrine 2000;13(2):207–212.
63. Bray GA. Reciprocal relation of food intake and sympathetic activity: experimental observations and clinical implications. Int J Obes Relat Metab Disord 2000;24(Suppl 2):S8–S17.
64. Cheng JT, Liu IM, Yen ST, Juang SW, Liu TP, Chan P. Stimulatory effect of D-ephedrine on beta3 adrenoceptors in adipose tissue of rats. Auton Neurosci 2001;88(1–2):1–5.
65. Castleden CM, Duffin HM, Briggs RS, Ogden BM. Clinical and urodynamic effects of ephedrine in elderly incontinent patients. J Urol 1982;128(6):1250–1252.
66. Kadar N, Nelson JH Jr. Treatment of urinary incontinence after radical hysterectomy. Obstet Gynecol 1984;64(3):400–405.
67. Alberts P, Bergstrom PA, Fredrickson MG. Characterisation of the functional alpha-adrenoceptor subtype in the isolated female pig urethra. Eur J Pharmacol 1999;371(1):31–38.
68. Astrup A, Breum L, Toubro S, Hein P, Quaade F. The effect and safety of an ephedrine/caffeine compound compared to ephedrine, caffeine and placebo in obese subjects on an energy restricted diet: a double blind trial. Int J Obes Relat Metab Disord 1992;16(4):269–277.
69. Kanfer IR, Dowse R, Vuma V. Pharmacokinetics of oral decongestants. Pharmacotherapy 1993;13(6 Pt 2):116S–128S; discussion 143S–146S.
70. White LM, Gardner SF, Gurley BJ, Marx MA, Wang PL, Estes M. Pharmacokinetics and cardiovascular effects of ma-huang (Ephedra sinica) in normotensive adults. J Clin Pharmacol 1997;37(2):116–122.
71. Kuntzman RG, Tsai I, Brand L, Mark LC. The influence of urinary pH on the plasma half-life of pseudoephedrine in man and dog and a sensitive assay for its determination in human plasma. Clin Pharmacol Ther 1971;12(1):62–67.
72. Dickerson J, Perrier D, Mayersohn M, Bressler R. Dose tolerance and pharmacokinetic studies of L(+) pseudoephedrine capsules in man. Eur J Clin Pharmacol 1978;14:239–259.

73. Vanakoski J, Seppala T. Heat exposure and drugs. A review of the effects of hyperthermia on pharmacokinetics. Clin Pharmacokinet 1998;34(4):311–322.
74. Wilkinson G, Beckett A. Absorption, metabolism and excretion of the ephedrine in man. II. J Pharm Sci 1968;57:1933–1938.
75. Kunsman GW, Jones R, Levine B, Smith ML. Methylephedrine concentrations in blood and urine specimens. J Anal Toxicol 1998;22(4):310–313.
76. Tokunaga I, Takeichi S, Kujime T, Maeiwa M. Electroencephalographical analysis of acute drug intoxication — SS Bron solution-W. Arukoru Kenkyuto Yakubutsu Ison 1989;24(6):471–479.
77. Ishigooka J, Yoshida Y, Murasaki M. Abuse of "BRON": a Japanese OTC cough suppressant solution containing methylephedrine, codeine, caffeine and chlorpheniramine. Prog Neuropsychopharmacol Biol Psychiatry 1991;15(4):513–521.
78. Levine B, Caplan YH, Kauffman G. Fatality resulting from methylphenidate overdose. J Anal Toxicol 1986;10(5):209–210.
79. Haller CA, Benowitz NL. Adverse cardiovascular and central nervous system events associated with dietary supplements containing ephedra alkaloids. N Engl J Med 2000;343(25):1833–1838.
79a. Haller CA and Benowitz NL. Dietary supplements containing ephedra alkaloids, correspondence. N Engl J Med 2000;344(April 5):1095–1097.
80. Samenuk D, Link MS, Homoud MK, et al. Adverse cardiovascular events temporally associated with ma huang, an herbal source of ephedrine. Mayo Clin Proc 2002;77(1):12–16.
81. Kingston R, Blumenthal M. A rational perspective on adverse events reports on herbs: misinterpretation of adverse reactions tabulated in the TESS Annual Report of the American Association of Poison Control Centers as they relate to ephedra dietary supplements. HerbalGram 2003;60:48–53.
82. Pentel P. Toxicity of over-the-counter stimulants. JAMA 1984;252(14):1898–1903.
83. Stoessl AJ, Young GB, Feasby TE. Intracerebral haemorrhage and angiographic beading following ingestion of catecholaminergics. Stroke 1985;16(4):734–736.
84. Roberge RJ, Hirani KH, Rowland PL III, Berkeley R, Krenzelok EP. Dextromethorphan- and pseudoephedrine-induced agitated psychosis and ataxia: case report. J Emerg Med 1999;17(2):285–288.
85. Wilson H, Woods D. Pseudoephedrine causing mania-like symptoms. N Z Med J 2002;115(1148):86.
86. Johnson DA, Etter HS, Reeves DM. Stroke and phenylpropanolamine use. Lancet 1983;2(8356):970.
87. Glick R, Hoying J, Cerullo L, Perlman S. Phenylpropanolamine: an over-the-counter drug causing central nervous system vasculitis and intracerebral hemorrhage. Case report and review. Neurosurgery 1987;20(6):969–974.
88. Lake CR, Zaloga G, Bray J, Rosenberg D, Chernow B. Transient hypertension after two phenylpropanolamine diet aids and the effects of caffeine: a placebo-controlled follow-up study. Am J Med 1989;86(4):427–432.

89. Thomas SH, Clark KL, Allen R, Smith SE. A comparison of the cardiovascular effects of phenylpropanolamine and phenylephrine containing proprietary cold remedies. Br J Clin Pharmacol 1991;32(6):705–711.

90. Ryu SJ, Lin SK. Cerebral arteritis associated with oral use of phenylpropanolamine: report of a case. J Formos Med Assoc 1995;94(1-2):53–55.

91. Tapia J. [Cerebral hemorrhage associated with the use of phenylpropanolamine. Clinical cases]. Rev Med Chil 1996;124(12):1499–1503.

92. Chung YT, Hung DZ, Hsu CP, Yang DY, Wu TC. Intracerebral hemorrhage in a young woman with arteriovenous malformation after taking diet control pills containing phenylpropanolamine: a case report. Zhonghua Yi Xue Za Zhi (Taipei) 1998;61(7):432–435.

93. Kernan WN, Viscoli CM, Brass LM, et al. Phenylpropanolamine and the risk of hemorrhagic stroke. N Engl J Med 2000;343(25):1826–1832.

94. Vahedi K, Domigo V, Amarenco P, Bousser MG. Ischaemic stroke in a sportsman who consumed MaHuang extract and creatine monohydrate for body building. J Neurol Neurosurg Psychiatry 2000;68(1):112–113.

95. du Boisgueheneuc F, Lannuzel A, Caparros-Lefebvre D, De Broucker T. [Cerebral infarction in a patient consuming MaHuang extract and guarana]. Presse Med 2001;30(4):166–167.

96. Bruno A, Nolte KB, Chapin J. Stroke associated with ephedrine use. Neurology 1993;43(7):1313–1316.

97. Loizou LA, Hamilton JG, Tsementzis SA. Intracranial haemorrhage in association with pseudoephedrine overdose. J Neurol Neurosurg Psychiatry 1982;45(5):471–472.

98. LeBayon A, Castelnovo G, Briere C, Labauge P. [Cerebromeningeal hemorrhage after voluntary overdose of pseudoephedrine]. Presse Med 2000;29(31):1702.

99. Mourand I, Ducrocq X, Lacour JC, Taillandier L, Anxionnat R, Weber M. Acute reversible cerebral arteritis associated with parenteral ephedrine use. Cerebrovasc Dis 1999;9(6):355–357.

100. Jovanovic Z. [Risk factors for stroke in young people]. Srp Arh Celok Lek 1996;124(9–10):232–235.

101. Herridge CF, a'Brook MF. Ephedrine psychosis. Br Med J 1968;2(598):160.

102. Roxanas MG, Spalding J. Ephedrine abuse psychosis. Med J Aust 1977;2(19):639–640.

103. Shufman NE, Witztum E, Vass A. [Ephedrine psychosis]. Harefuah 1994;127(5–6):166–168, 215.

104. Doyle H, Kargin M. Herbal stimulant containing ephedrine has also caused psychosis. BMJ 1996;313(7059):756.

105. Tormey WP, Bruzzi A. Acute psychosis due to the interaction of legal compounds—ephedra alkaloids in "vigueur fit" tablets, caffeine in "red bull" and alcohol. Med Sci Law 2001;41(4):331–336.

106. Kalix P. The pharmacology of psychoactive alkaloids from ephedra and catha. J Ethnopharmacol 1991;32(1–3):201–208.

107. Bories H. [Kidney stones]. Infirm Fr 1976;(179):13–18.

108. Schweisheimer W. [Kidney stones]. Krankenpflege (Frankf) 1976;30(6):194–195.

109. Blau JJ. Ephedrine nephrolithiasis associated with chronic ephedrine abuse. J Urol 1998;160(3 Pt 1):825.
110. Powell T, Hsu FF, Turk J, Hruska K. Ma-huang strikes again: ephedrine nephrolithiasis. Am J Kidney Dis 1998;32(1):153–159.
111. Assimos DG, Langenstroer P, Leinbach RF, Mandel NS, Stern JM, Holmes RP. Guaifenesin- and ephedrine-induced stones. J Endourol 1999;13(9):665–667.
112. Glidden RS, DiBona FJ. Urinary retention associated with ephedrine. J Pediatr 1977;90(6):1013–1014.
113. Lindberg AW. [Urinary retention caused by Elsinore pills]. Ugeskr Laeger 1988;150(35):2086–2087.
114. Lyon CC, Turney JH. Pseudoephedrine toxicity in renal failure. Br J Clin Pract 1996;50(7):396–397.
115. Brater DC, Kaojarern S, Benet LZ, et al. Renal excretion of pseudoephedrine. Clin Pharmacol Ther 1980;28(5):690–694.
116. Karch SB, Stephens B, Ho CH. Relating cocaine blood concentrations to toxicity—an autopsy study of 99 cases. J Forensic Sci 1998;43(1):41–45.
117. Karch SB, Stephens BG, Ho CH. Methamphetamine-related deaths in San Francisco: demographic, pathologic, and toxicologic profiles. J Forensic Sci 1999;44(2):359–368.
118. To LB, Sangster JF, Rampling D, Cammens I. Ephedrine-induced cardiomyopathy. Med J Aust 1980;2(1):35–36.
119. Gaultieri J. Cardiomyopathy in a heavy ephedrine abuser. J Tox Clin Tox 1996;34:581–582.
120. Cockings JG, Brown M. Ephedrine abuse causing acute myocardial infarction. Med J Aust 1997;167(4):199–200.
121. Menegakis NE, Amstey MS. Case report of myocardial infarction in labor. Am J Obstet Gynecol 1991;165(5 Pt 1):1383–1384.
122. Wiener I, Tilkian AG, Palazzolo M. Coronary artery spasm and myocardial infarction in a patient with normal coronary arteries: temporal relationship to pseudoephedrine ingestion. Cathet Cardiovasc Diagn 1990;20(1):51–53.
123. Derreza H, Fine MD, Sadaniantz A. Acute myocardial infarction after use of pseudoephedrine for sinus congestion. J Am Board Fam Pract 1997;10(6):436–438.
124. Hirabayashi Y, Saitoh K, Fukuda II, Mitsuhata H, Shimizu R. Coronary artery spasm after ephedrine in a patient with high spinal anesthesia. Anesthesiology 1996;84(1):221–224.
125. Zaacks SM, Klein L, Tan CD, Rodriguez ER, Leikin JB. Hypersensitivity myocarditis associated with ephedra use. J Toxicol Clin Toxicol 1999;37(4):485–489.
126. Anastasio GD, Harston PR. Fetal tachycardia associated with maternal use of pseudoephedrine, an over-the-counter oral decongestant. J Am Board Fam Pract 1992;5(5):527–528.
127. Onuigbo M, Alikhan M. Over-the-counter sympathomimetics: a risk factor for cardiac arrhythmias in pregnancy. South Med J 1998;91(12):1153–1155.
128. Weesner KM, Denison M, Roberts RJ. Cardiac arrhythmias in an adolescent following ingestion of an over-the-counter stimulant. Clin Pediatr (Phila) 1982;21(11):700–701.

129. Hanzlick R, Davis G. National Association of Medical Examiners Pediatric Toxicology Registry. Report 1: Phenylpropanolamine. Am J Forensic Med Pathol 1992;13(1):37–41.
130. Levisky JA, Karch SB, Bowerman DL, Jenkins WW, Johnson DG, Davies D. False-positive RIA for methamphetamine following ingestion of an ephedra-derived herbal product. J Anal Toxicol 2003;27:123–124.
131. Backer R, Tautman D, Lowry S, Harvey CM, Poklis A. Fatal ephedrine intoxication. J Forensic Sci 1997;42(1):157–159.
132. Levine B, Jones R, Klette K, Smith ML, Kilbane E. An intoxication involving BRON and verapamil. J Anal Toxicol 1993;17(6):381–383.
133. Baselt RC, Cravey RH, eds. *Disposition of toxic drugs and chemicals in man.* Foster City: Chemical Toxicology Institute, 1995.
134. Hedetoft C, Jensen CH, Christensen MR, Christensen O. [Fatal poisoning with Letigen]. Ugeskr Laeger 1999;161(50):6937–6938.
135. Blechman KM, Karch SB, Stephens BG. Demographic, pathologic, and toxicological profiles of 127 decedents testing positive for ephedrine alkaloids. Forensic Sci Int. 2004;139(1):61–9 as n.
136. Irvine G, Chen L, eds. The environmental impact and adverse health effects of clandestine manufacture of methamphetamine. NIDA Research Monograph 115. Miller MA, Kozel NJ, eds. N. Bethesda: National Institute on Drug Abuse, 1991, pp.33–46.
137. Dingemanse J, Guentert T, Gieschke R, Stabl M. Modification of the cardiovascular effects of ephedrine by the reversible monoamine oxidase A-inhibitor moclobemide. J Cardiovasc Pharmacol 1996;28(6):856–861.
138. Lefebvre H, Noblet C, Moore N, Wolf LM. Pseudo-phaeochromocytoma after multiple drug interactions involving the selective monoamine oxidase inhibitor selegiline. Clin Endocrinol (Oxf) 1995;42(1):95–98; discussion 98–99.
139. Dawson JK, Earnshaw SM, Graham CS. Dangerous monoamine oxidase inhibitor interactions are still occurring in the 1990s. J Accid Emerg Med 1995;12(1):49–51.
140. Skop BP, Finkelstein JA, Mareth TR, Magoon MR, Brown TM. The serotonin syndrome associated with paroxetine, an over-the-counter cold remedy, and vascular disease. Am J Emerg Med 1994;12(6):642–644.
141. Reeves RR, Pinkofsky HB. Postpartum psychosis induced by bromocriptine and pseudoephedrine. J Fam Pract 1997;45(2):164–166.
142. Inomata S, Nishikawa T, Kihara S, Akiyoshi Y. Enhancement of pressor response to intravenous phenylephrine following oral clonidine medication in awake and anaesthetized patients. Can J Anaesth 1995;42(2):119–125.
143. Tanaka M, Nishikawa T. Enhancement of pressor response to ephedrine following clonidine medication. Anaesthesia 1996;51(2):123–127.
144. Tekol Y, Tercan E, Esmaoglu A. Ephedrine enhances analgesic effect of morphine. Acta Anaesthesiol Scand 1994;38(4):396–397.
145. Walton R, Manos GH. Psychosis related to ephedra-containing herbal supplement use. South Med J 2003;96:718–720.

Chapter 2

Kava

Douglas D. Glover

Summary

Kava has a long history of traditional use for the treatment of symptoms related to anxiety, stress, and nervous restlessness and has demonstrated effectiveness for treatment of anxiety in double-blind, randomized, placebo-controlled trials. Lack of both dependence and documented adverse effects contributed to kava's popularity up through the 1990s. Typical adverse effects have been limited to reversible yellowing of the skin after chronic use and a temporary condition known as kava dermopathy. Subsequent reports of hepatotoxicity in Europe and less frequently in the United States have resulted in a decrease in its popularity as well as regulatory action by various government regulatory bodies against specific formulations. Although subsequent analysis has concluded that "there is no clear evidence that the liver damage reported in the United States and Europe was caused by the consumption of kava" much of the US market for the herb has been diminished.

Key Words: *Piper methysticum*; kava lactones; anxiolytic; hypnotic.

1. History

Kava is a term used to describe both *Piper methysticum* and the preparation made from its dried rhizome and root *(1)*. This South Pacific plant is a robust, branching, perennial shrub with heart-shaped, green, pointed leaves *(2)* that grows up to 28 cm long and flower spikes that grow up to 9 cm long *(1)*. The shrub grows best in warm, humid conditions with lots of sunlight, at altitudes of 150–300 m above sea level *(2)*, where it forms dense thickets *(3)*. Kava reproduces vegetatively, without fruit or seeds, usually under cultiva-

From Forensic Science and Medicine:
Herbal Products: Toxicology and Clinical Pharmacology, Second Edition
Edited by: T. S. Tracy and R. L. Kingston © Humana Press Inc., Totowa, NJ

tion *(3)*. There are reports of up to 72 varieties of the kava plant, which differ in appearance, and chemical analysis has shown differences in their composition as well, which may lead to differences in physiological activity *(2)*.

Kava has been described in the European literature since the early 1600s, when it was taken there by the Dutch explorers LeMaire and Schouten, who had acquired it while seeking new passages across the Pacific *(3)*. Captain James Cook was the first to describe the use of kava during the religious and cultural ceremonies of the people of the South Sea Islands, where it was, and still is, prepared as a beverage and consumed for its intoxicating, calming effects that promote sociability *(3)*. Thus, kava is used for the purposes that Western society uses alcohol, the Native American populations use peyote, and the people of the Middle or Far East use opium *(2)*. Events typically accompanied by kava ceremonies included weddings, funerals, births, religious occasions, seasonal feasts, reconciliations, welcoming of royalty or other guests, and the exchange of gifts *(3)*. Women and commoners seldom participated in these ceremonies because that was viewed as unacceptable; however, some cultures did permit use by commoners to relax after a hard day's work *(3)*.

The beverage was traditionally made by mixing grated, crushed, or chewed fresh or dried root with cool water or coconut milk and then straining the mixture through plant fibers to isolate the liquid, which was consumed *(3)*. However, all parts of the plant can be used *(4)*. Today the beverage is most often prepared by crushing dried roots with a large mortar and pestle, then straining the mixture in the traditional way or through cotton cloth *(3)*. Other folk uses of kava have included treatment of headaches, colds, rheumatism, sexually transmitted diseases, and inflammation of the uterus *(1)*. It has also been used as a sedative, aphrodisiac, urinary antiseptic *(5)*, wound healing agent, and a treatment for asthma *(1)*. Several substances extracted from the roots were also used briefly in Europe as diuretics *(3)*.

2. CURRENT PROMOTED USES

Kava is currently promoted for relief of anxiety, stress, and insomnia. Stress may be prolonged and difficult to cope with and affected individuals may suffer from insomnia. Kava has been promoted as an axiolytic agent with little risk for dependence or adverse reactions. An unblinded, comparative, crossover trial of kava (120 mg) and valerian (600 mg) was conducted, each agent administered for 6 weeks with a 2-week wash-out period between. This was followed with administration of a combination of the two compounds. Both stress and insomnia were measured regarding social, personal, and life events. Results: the severity of stress was equally relieved by each of the two compounds and there was further improvement of insomnia with combination therapy. With kava, 67% of the subjects reported no adverse events, 53% denied adverse events with valerian, and likewise with combination therapy. Vivid dreams were experienced by 21% of subjects taking combination therapy and 16% of those taking valerian alone. Dizziness or gastric discomfort was reported by 3%. The investigators concluded the results were extremely promising but recommended additional studies *(6)*.

3. SOURCES AND CHEMICAL COMPOSITION

P. methysticum, kava-kava, awa, kew, tonga *(1)*, kawa, yaqona, sakau *(3)*, ava, ava pepper, intoxicating pepper *(5)*.

4. PRODUCTS AVAILABLE

Kava is available from a variety of manufacturers in most health food stores under a variety of names. Kavatrol® is a popular brand found in retail outlets in the United States. Kava is marketed in Europe under a variety of names including Laitan® or Kavasporal® in Germany, Potter's Antigian Tablets in the United Kingdom, Viocava® in Switzerland, and Mosaro® in Austria *(7)*.

5. PHARMACOLOGICAL/TOXICOLOGICAL EFFECTS

5.1. Neurological Effects

The neurological effects of kava are attributed to a group of substituted dihydropyrones called kava lactones *(1)*. The main bioactive constituents include yangonin, desmethoxyyangonin, 11-methoxyyangonin, kavain (kawain), dihydrokavain, methysticin, dihydromethysticin, and 5,6-dehydromethysticin *(8)*. It is believed that the components present in the lipid-soluble kava extract, or kava resin, are responsible for the central nervous system (CNS) activities of kava including sedation, hypnosis, analgesia, and muscle relaxation *(9)*. Aqueous kava extract was not active orally in mice or rats.

A randomized, 25-week, placebo-controlled study by Volz and Kieser showed a significant benefit from the use of kava-kava extract WS 1490 over placebo in treating anxiety disorders of nonpsychotic origin. The study included 101 patients suffering from agoraphobia, specific phobia, generalized anxiety disorder, or adjustment disorder with anxiety—as per the *Diagnostic and Statistical Manual of Mental Disorders, Third Edition, Revised*—who were randomized to placebo or WS 1490 containing 90–100 mg dry extract per capsule three times daily. The main outcome criterion, the patients' score on the Hamilton Anxiety Scale, was significantly better ($p < 0.001$) for the WS 1490 patients compared to placebo at 24 weeks. Few adverse effects were judged to be related or possibly related to kava administration. Two patients in the WS 1490 group experienced stomach upset, two noted vertigo, and one experienced vertigo and palpitations. These results support use of kava as an alternative to antidepressants and benzodiazepines (10).

Pittler and Ernst (11) conducted a review of double-blind, randomized, placebo-controlled trials of kava extract monotherapy for treatment of anxiety. They reviewed 14 such studies and three were determined suitable for metaanalysis. They concluded that kava extract was not only relatively safe but superior to placebo in the treatment of anxiety.

Another study compared the cognitive effects of this same kava extract at a dose of 200 mg three times daily for 5 days to oxazepam 15 mg, followed by 75 mg on the experimental day (12). The results suggest that kava is less likely to affect cognitive function than oxazepam, but the oxazepam dosing regimen used was not typical of that seen in practice. Nevertheless, kava is purported to promote relaxation and sleep without dampening alertness, causing heavy sedation, or causing a "hangover" effect the morning after consumption (13). The limbic structures of the brain might represent the site of action of kava, explaining its ability to promote relaxation and sleep without cognitive effects (14).

The mechanism of the anxiolytic effect of kava is unclear. Studies of kava's effects in vitro, in vivo, and ex vivo report conflicting results in regard to kava's effects on benzodiazepine or γ-aminobutyric acid (GABA) receptors (5,14,15). This disparity may be explained by differences in GABA receptor subtypes among the different regions of the brain studied (14). It is thought that kavapyrones elicit a tranquilizing effect by enhancing GABA binding in the amygdala, but do not act directly as agonists at GABA receptors (14).

One study has suggested that a nonstereoselective inhibition of [3H]noradrenaline uptake may be responsible for, or at least contribute to, kava's anxiolytic effect (16). This investigation tested the effects of naturally occurring (+)-kavain, (+)-methysticin, and a synthetic racemic mixture of kavain on synaptosomes from the cerebral cortex and hippocampus of rat brains.

Both forms of kavain inhibited [3H]noradrenaline uptake more than methysticin, but the concentrations necessary to achieve this effect were approx 10 times higher than those in mouse brains after a dose of kavain high enough to cause significant sedation. This indicates that inhibition of noradrenaline uptake is probably only part of the psychotropic effects of kava. No effects were seen on the uptake of [3H]serotonin. A subsequent study *(17)* in rats showed that (+)-kavain and other kavapyrones affect serotonin levels in the mesolimbic area. The authors postulated that this effect could explain kava's hypnotic action. Dopamine levels in the nucleus accumbens were decreased by yangonin and low-dose (+)-kavain, but were increased by higher doses of (+)-kavain and desmethoxyyangonin. The investigators attributed kava's anxiolytic and euphoric effects to its action on mesolimbic dopaminergic pathways.

A study conducted in Germany indicates that kava may have neuroprotective properties, primarily owing to its constituent methysticum and dihydromethysticum *(18)*. The investigators studied the effects of kava extract WS 1490 and the individual pyrones kavain, dihydrokavain, methysticin, dihydromethysticin, and yangonin on the size of infarction in mouse brains. The extract as well as the individual pyrones methysticin and dihydromethysticin showed significant reductions in infarct area similar to those produced by memantine, an anticonvulsive agent known to have neuroprotective qualities *(18)*.

Kava lactones are also centrally acting skeletal muscle relaxants *(19)*. A study by Kretzschmar et al. compared the antagonistic effects of kavain, dihydrokavain, methysticin, and dihydromethysticin to those of mephenesin and phenobarbital in preventing convulsions and death caused by strychnine. All the kava pyrones showed an antagonistic effect, with methysticin being the most potent; however, kavain and dihydrokavain doses required to produce an effect approached the toxic range *(20)*. In contrast to mephenesin and phenobarbital, all the pyrones tested protected against strychnine at doses up to 5 mg/kg without causing impairment of motor function. Gleitz et al., in their studies on the antiseizure properties of kavain, conclude that the inhibition of voltage-dependent Ca^{++} and Na^+ channels by kavain resembles that of local anesthetics. They suggest kava pyrone accumulation in neuronal cell membranes may explain the antieleptic affects of kava *(21)*.

Kava also produces analgesic effects that appear to be mediated through a nonopiate pathway. A study conducted by Jamieson and Duffield compared the activity of an aqueous and a lipid extract of kava as well as eight purified pyrones on two tests for antinociception in mice. Both the aqueous and lipid extracts were effective analgesics, as were four of the eight purified pyrones (lactones): methysticin, dihydromethysticin, kavain, and dihydrokavain *(22)*. In hopes of discovering the mechanism of analgesia, the investigators attempted to antagonize the effects of kava with naloxone, a known inhibitor of opiate-mediated pathways of analgesia. Naloxone failed to inhibit kava's effects at doses high enough to inhibit the action of morphine, indicating that kava works through a nonopiate pathway to produce analgesia.

In humans, kava is reported to produce a mild euphoria characterized by happiness, fluent and lively speech, and increased sensibility to sounds *(1)*. It has also been reported to cause visual changes such as reduced near-point accommodation and convergence, increase in pupil diameter, and oculomotor balance disturbances *(23)*. It might even have an antipyretic effect *(19)*.

Tolerance and development of physical dependence by laboratory animals has been investigated for both the aqueous kava extract and kava resin, which contains the pharmacologically active pyrones. Duffield and Jamieson reported tolerance to be evident in mice only after parenteral administration of the aqueous kava extract, but not when given orally. Likewise, tolerance was not seen after daily dosing with kava resin over a 7-week period of time. They concluded tolerance to kava resin was not readily demonstrable *(24)*.

Kava's effects on the peripheral nervous system are limited to a local anesthetic effect, resulting in numbness in the mouth if kava is chewed *(1)*. Lipid-soluble kava extract, or resin, is also capable of causing anesthesia of the oral mucosa, whereas the water-soluble fraction is not *(9)*.

5.2. Dermatological Effects

There have been many reports of skin disturbances associated with the use of kava that date as far back as the 1700s *(3)*. Chronic ingestion of kava may cause a temporary yellowing of the skin, hair, and nails *(25)*. Two yellow pigments, flavokawains A and B, have been isolated from the kava plant *(8)* and may be responsible for this discoloration *(1)*. Chronic ingestion may also lead to a temporary condition known as kava dermopathy *(3)* or kawaism, characterized by dry, flaking, discolored skin and reddened eyes, which is reversible with discontinuation *(26)*. In the early 19th century, Peter Corney, a lieutenant on a fur-trading vessel, described this phenomenon in great detail as it applied to the use of this side effect in treating other skin disturbances:

"When a man first commences taking it, he begins to break out in scales about the head, and it makes the eyes very sore and red, then the neck and breasts, working downwards, till it approaches the feet, when the dose is reduced. At this time the body is covered all over with white scruff, or scale, resembling the dry scurvy. These scales drop off in the order of their formation, from the head, neck, and body, and finally leave a beautiful, smooth, clear skin, and the frame clear of all disease" *(3)*.

The exact mechanism for this dermopathy is unknown, but it has been speculated that kava may interfere with cholesterol metabolism, leading to a reversible, acquired ichthyosis similar to that seen with the use of lipid-lowering agents such as triparanol *(3)*. Skin biopsies of two recent cases associated with use of the commercially available product have revealed lymphocytic attacks on sebaceous glands, with subsequent destruction and necrosis caused by CD8+ cells *(see* Section 5) *(26)*. Yet another theory involves interference with B vitamin metabolism or action *(27)*.

5.3. Musculoskeletal Effects

As mentioned in Section 4.1, kava is a centrally acting skeletal muscle relaxant. The kava lactones kavain, dihydrokavain, methysticin, and dihydromethysticin isolated from kava rootstock were shown to antagonize strychnine-induced convulsions in mice *(20)*.

5.4. Antimicrobial Activity

Kava has been used traditionally as an antibacterial agent in the treatment of urinary tract infections *(28)*; however, no clinical trials have established its efficacy. Locher et al. investigated antiviral, antibacterial, and antifungal activity of extracts of Kava leaves, stems, and roots. They concluded that *P. methysticum* exerted no antiviral or antibacterial activity, although weak antifungal activity against *Epidermophyton fluccosum* was demonstrated by extracts made from the stems of the plant.

5.5. Hepatotoxicity

See Section 7.

5.6. Antiplatelet Effects

Racemic kavain, a component of kava, has been shown to have antiplatelet effects, presumably owing to inhibition of cyclooxygenase, and thus inhibition of thromboxane synthesis *(29)*. Antiplatelet effects have not been observed in vivo.

5.7. Cancer Prevention

Following the establishment of the South Pacific Commission Cancer Registry in 1977, interest abounded regarding the apparent dichotomy of high tobacco consumption and low cancer incidence of a number of South Pacific nations. Fugi, for example, has an age-standardized incidence rate of 75 cases per 100,000 males (112.2/100,000 females) compared to 237/100,000 males and 220/100,000 females in the United States. Both tobacco and kava are indigenous crops of these island nations. Men traditionally stop by kava bars to enjoy a bowl of kava on the way home from work. On the other hand, female consumption is highly variable. Data analysis reveals an inverse relationship between cancer incidence and kava consumption. A large difference exists between the age-standardized cancer incidence of these South Pacific island nations and that of the United States *(30)*.

6. PHARMACOKINETICS

6.1. Absorption

In mice and rats, the aqueous kava extract is inactive when administered orally *(9)*.

6.2. Metabolism/Elimination

Several kava lactones have been identified in human urine samples after ingestion of a kava beverage prepared from a commercial 450-g sample of *P. methysticin* extracted with 3 L of room temperature water *(31)*. Observed metabolic transformations include reduction of the 3,4 double bond and/or demethylation of the 4-methoxyl group on the α-pyrone ring system. Demethylation of the 12-methoxy substituent in yangonin and hydroxylation at carbon 12 of desmethoxyyangonin have also been observed. Chemical structures for these kava components and metabolites can be seen in the cited reference.

7. CASE REPORTS OF TOXICITY CAUSED BY COMMERCIAL KAVA PRODUCTS

Kava dermatopathy in association with traditional use of kava is well described in the literature *(3)*. In addition, two cases of dermopathy have recently been associated with commercially available kava products *(26)*. A 70-year-old man who had been using kava as an antidepressant for 2–3 weeks experienced itching, and later erythematous, infiltrated plaques on his chest, back, and face after several hours of sun exposure. Skin biopsy revealed CD8 lymphocytic infiltration with destruction of the sebaceous glands and lower infundibula. A 52-year-old woman presented with papules and plaques on her face, chest, back, and arms after taking a kava extract for 3 weeks. Skin biopsy revealed an infiltrate in the reticular dermis with disruption and necrosis of the sebaceous gland lobules. A kava extract patch test was strongly positive after 24 hours.

There have also been four cases of extrapyramidal effects associated with kava use *(7)*. A 28-year-old man with a history of antipsychotic-induced extrapyramidal effects experienced torticollis and oculogyric crisis 90 minutes after a single 100-mg dose of Laitan (kava extract). These effects resolved spontaneously after 40 minutes. A 22-year-old woman experienced oral and lingual dyskinesia, painful twisting movements of the trunk, and torticollis 4 hours after a 100-mg dose of the same product taken by the previously described male. The symptoms did not resolve spontaneously, so after 45 minutes, a 2.5-mg intravenous dose of beperiden was given, with immediate relief. A third patient, a 63-year-old female, also presented with oral and lingual dyskinesia after taking Kavasporal Forte® (150 mg of kava extract) three times a day for 4 days. A single 5-mg intravenous dose of beperiden was immediately effective.

Recently, an association of kava and Parkinsonism has been noted. A 45-year-old healthy woman, absent any signs of Parkinson's disease but with a family history of essential tremor, was prescribed fluoxetine and benzodiazepines for depression. Approximately 3 months later, she developed severe Parkinson's disease after 10 days use of kava extract *(32)*.

Finally, a 76-year-old woman experienced worsening of Parkinson's disease symptoms after taking Kavasporal Forte for 10 days. Improvement was noted 2 days after discontinuation of the product. These extrapyramidal side effects suggest cautious use of kava in the elderly, in patients with Parkinson's disease, and in patients taking antipsychotics.

8. TOXICITY ASSOCIATED WITH TRADITIONAL USE BY NATIVE POPULATIONS

Chronic use of the kava beverage has been associated with a wide range of abnormalities. A study *(27)* of an Australian Aboriginal community revealed malnutrition and weight loss associated with kava use. Red blood cell volume increased in proportion to kava use, whereas bilirubin, plasma protein, platelet volume, B-lymphocyte count, and plasma urea were inversely proportional to kava consumption. Although these values were not outside the normal range, it was hypothesized that malnutrition or reduced hemoglobin turnover might explain these observations. Other findings included hematuria and difficulty acidifying and concentrating the urine, suggesting an effect on the renal tubules; and increased serum transaminases and increased high-density lipoprotein cholesterol, suggesting some effect on the liver. Transaminase elevations were greater in the kava-using Aboriginal community compared to those in a community where alcohol, but not kava, was consumed. This suggests that kava might be more hepatotoxic than alcohol. Shortness of breath and electrocardiograph abnormalities (tall P waves) consistent with pulmonary hypertension were seen and are interesting in that, like kava, the prescription anorexiants fenfluramine and dexfenfluramine withdrawn from the US market in 1998 were associated with pulmonary hypertension. It was also noted by the authors of this observational study that sudden death in relatively young men is more common in kava-using Aboriginal communities than in nonusing communities.

In the United States and Europe, evidence of hepatic failure following the use of kava extracts is accumulating. A 50-year-old male, who had previously been well, experienced liver failure after consuming kava extract for 2 months. The dosage of kava extracts was at or slightly exceeded the maximum three-capsule-a-day dose recommended on the label. A liver transplant was performed and the individual survived (33).

A healthy 14-year-old adolescent girl developed nausea, vomiting, general malaise, and weight loss. Several days later she became icteric and was admitted to hospital with acute hepatitis. Drug history revealed no alcohol use; two kava products and occasional ibuprofen had been used over the preceding 4 months.

The packaging of the kava products was unavailable for identification purposes. Liver biopsy revealed active fulminant hepatitis with extensive necrosis and tests for viral hepatitis were negative. She underwent a successful liver transplantation and was able to return to normal activity upon recovery (34). Unfortunately, no information was provided indicating that acetaminophen toxicity had been ruled out, and the observed toxic effect could also have been associated with a large, undiagnosed acetaminophen ingestion.

A 33-year-old woman took 210 mg of kava extract for 3 weeks and discontinued the product. After 2 months, she resumed taking the same product for an additional 3-week period. Symptoms of hepatotoxicity developed a day after ingesting 60 mL of alcohol. Tests for viral hepatitis were negative and liver biopsy revealed evidence of hepatic necrosis. Phenotyping of CYP4502D6 activity with debrisoquine was consistant with deficiency of this enzyme. Her liver function returned to normal 8 weeks after kava discontinuation (35).

A US Food and Drug Administration (FDA) advisory letter dated March 25, 2002 warned health care providers of a total of 11 patients who had used kava products and developed liver failure requiring liver transplantation (34).

Additionally, there have been 25 reports of severe liver toxicity in Germany, Switzerland, and the United States *(36)*. The prevalence of CYP4502D6 deficiency in such cases has yet to be determined.

9. INTERACTIONS

Alcohol appears to at least add to the hypnotic effect of kava in mice, and was also observed to increase the lethality of kava *(37)*. These findings may be of importance because some Australian Aboriginal populations now frequently consume kava with alcohol. Concomitant use of barbiturates, melatonin, and other psychopharmacological agents might potentiate the effects of kava as well *(38)*. The hepatotoxic potential of kava *(27)* also raises concerns about concomitant alcohol use.

Although a Web site *(13)* promoting a kava product states that it is safe to use kava in combination with benzodiazepines, a case report *(39)* suggests otherwise. The combination of kava and alprazolam was believed to be responsible for hospitalizing a 54-year-old man. The patient's semicomatose (lethargic and disoriented) state improved after several hours. He had been taking an undisclosed brand of kava purchased in a health food store in combination with alprazolam for 3 days. Other medications taken included cimetidine and terazosin.

In vitro evidence suggests that kava components may inhibit the metabolism of drugs by cytochrome P450 1A2, 2C9, 2C19, 2D6, 2E1, and 3A4 *(40)*. However, in vivo studies were not confirmatory except for the case of CYP2E1. Gurley and colleagues studied the ability of kava to inhibit in vivo metabolism by several of these enzymes and found that kava coadministration had no effect on CYP1A2, CYP2D6, or CYP3A4 activity but did significantly inhibit CYP2E1 activity *(41)*. Because very few drugs are metabolized by CYP2E1, the clinical significance of this interaction is lessened. It appears that pharmacokinetic interactions with kava are unlikely and any drug interactions will likely be related to an additive pharmacodynamic effect (e.g., sedation).

10. REPRODUCTION

No information is available concerning the potential effects of kava on reproduction.

11. REGULATORY STATUS

Kava is currently sold as a dietary supplement in the United States *(25)*, though an FDA advisory warning concerning the potential for liver toxicity was issued in 2002 *(42)*. Germany had banned the sale of kava in 2002 but the ban was lifted in 2005. A number of countries have either banned the sale of kava or issued warnings concerning its use.

REFERENCES

1. Anonymous. Kava-kava. In: *The Review of Natural Products*. St. Louis: Facts and Comparisons, 1996.
2. Singh YN. Kava: an overview. J Ethnopharmacol 1992;37:13–45.
3. Norton S, Ruze P. Kava dermopathy. J Am Acad Dermatol 1994;31:89–97.
4. Malsch U, Kieser M. Efficacy of kava-kava in the treatment of non-psychotic anxiety, following pretreatment with benzodiazepines. Psychopharmacology 2001;157:277–283.
5. Heiligenstein E, Guenther G. Over-the counter psychotropics: a review of melatonin, St. John's wort, valerian, and kava-kava. J Am Coll Health 1998;46:271–276.
6. Wheatley D. Stress-induced insomnia treated with kava and valerian: singly and in combination. Hum Psychopharmacol 2001;16:353–356.
7. Schelosky L, Raffauf C, Jendroska K, Poewe W. Kava and dopamine antagonism. J Neurol Nerosurg Psychiatry 1995;58:639–640.
8. Keller F, Klohs MW. A review of the chemistry and pharmacology of the constituents of Piper methysticum. Lloydia 1963;26:1–15.
9. Jamieson DD, Duffield PH, Cheng D, Duffield AM. Comparison of the central nervous system activity of the aqueous and lipid extract of kava (Piper methysticum). Arch Int Pharmacodyn 1989;301:66–80.
10. Volz HP, Kieser M. Kava-kava extract WS 1490 versus placebo in anxiety disorders—a randomized placebo-controlled 25-week outpatient trial. Pharmacopsychiatry 1997;30:1–5.
11. Pittler MH, Ernst E. Efficacy of kava extract for treating anxiety: systematic review and meta-analysis. J Clin Psychopharmacol 2000;20(1):84–89.
12. Heinze HJ, Munthe TF, Steitz J, Matzke M. Pharmacopsychological effects of oxazepam and kava-extract in a visual search paradigm assessed with event-related potentials. Pharmacopsychiatry 1994;27:224–230.
13. Topic of the month. www.ownhealth.com/topic.html. Date accessed: Oct. 29, 1998.
14. Jussofie A, Schmiz A, Hiemke C. Kavapyrone enriched extract from Piper methysticum as modulator of the GABA binding site in different regions of rat brain. Psychopharmacology 1994;116:469–474.
15. Davies LP, Drew CA, Duffield P, Johnston GAR, Jamieson DD. Kava pyrones and resin: studies on GABAA, GABAB, and benzodiazepine binding sites in rodent brain. Pharmacol Toxicol 1992;71:120–126.
16. Seitz U, Schule A, Gleitz J. [3H]-Monoamine uptake inhibition properties of kava pyrones. Planta Med 1997;63:548–549.
17. Baum SS, Hill R, Rommelspacher H. Effect of kava extract and individual kavapyrones on neurotransmitter levels in the nucleus accumbens of rats. Prog Neuropsychopharmacol Biol Psychiatry 1998;22:1105–1120.

18. Backhauss C, Krieglstein J. Extract of kava (Piper methysticum) and its methysticin constituents protect brain tissue against ischemic damage in rodents. Eur J Pharmacol 1992;215:265–269.
19. Tyler VE, Brady LR, Robbers JE, eds. *Pharamcognosy, 8th ed.* Philadelphia: Lea and Febiger, 1981.
20. Kretzschmar R, Meyer HJ, Teschendorf HJ. Strychnine antagonistic potency of pyrone compounds of the kavaroot (Piper methysticum Forst). Experientia 1970;26:283–284.
21. Gleitz J, Friese J, Belle A, Ameri A, Peters T. Anti-convulsive action of (±)kavain estimated from its properties on stimulated synaptosomes and Na^+ channel receptor sites. Eur J Pharmacol 1996;315:89–97.
22. Jamieson DD, Duffield PH. The antinociceptive actions of kava components in mice. Clin Exp Pharmacol Physiol 1990;17:495–508.
23. Garner LF, Klinger JD. Some visual effects caused by the beverage kava. J Ethnopharamacol 1985;13:307–311.
24. Duffield PH, Jamieson D. Development of tolerance to kava in mice. Clin Exp Pharmacol Physiol 1991;18:571–578.
25. Blumenthal M. Kava-kava rhizome. In: *Popular Herbs in the U.S. Market.* Austin: American Botanical Council, 1997.
26. Jappe U, Franke I, Reinhold D, Gollnick HPM. Sebotropic drug interaction resulting from kava-kava extract therapy: a new entity? J Am Acad Dermatol 1998;38:104–106.
27. Mathews JD, Riley MD, Fejo L, et al. Effects of the heavy usage of kava on physical health: summary of a pilot survey in an Aboriginal community. Med J Aust 1988;148:548–555.
28. Locher CP, Burch MT, Mower HF, et al. Anti-microbial activity and anti comple ment activity of extracts obtained from selected Hawaiian medicinal plants. J Ethnopharmacol 1995;49:23 32.
29. Gleitz J, Beile A, Wilkens P, Ameri A, Peters T. Antithrombotic action of kava pyrone (+)-kavain prepared from Piper methysticum on human platelets. Planta Med 1997;63:27–30.
30. Steiner GG. The correlation between cancer incidence and kava consumption. Hawaii Med J. 2000;59:420–422.
31. Duffield AM, Jamieson DD, Lidgard RO, Duffield PH, Bourne DJ. Identification of some human urinary metabolites of the intoxicating beverage kava. J Chromatogr 1989;475:273–281.
32. Meseguer E, Sanchez V, Campos V. Life threatening parkinsonism induced by kava-kava. Mov Disord 2002;17(1):195–196.
33. Escher M, Desmeules J. Hepatitis associated with kava, a herbal remedy for anxiety. Br Med J 2001;322:139.
34. Centers for Disease Control and Prevention. Hepatic toxicity possibly associated with kava-containing products–United States, Germany, and Switzerland, 1999–2002. MMWR Morb Mortal Wkly Rep 2002;51(47):1065–1067.
35. Russmann S, Helbling A. Kava hepatotoxicity. Ann Intern Med 2001;135(1):68–69.
36. Anonymous. Herbal kava: reports of liver toxicity. Canadian Med J 2002;166(6):777.

37. Jamieson DD, Duffield PH. Positive interaction of ethanol and kava resin in mice. Clin Exp Pharmacol Physiol 1990;17:509–514.
38. Thorndyke A, Rhyne H. Kava. www.unc.edu/~cebradsh/kava.html. Date accessed: Oct. 29, 1998.
39. Almedia JC, Grimsley EW. Coma from the health food store: interaction between kava and alprazolam. Ann Intern Med 1996;125:940–941.
40. Mathews JM, Etheridge AS, Black SR. Inhibition of human cytochrome P450 activities by kava extract and kavalactones. Drug Metab Dispos 2002;30:1153–1157.
41. Gurley BJ, Gardner SF, Hubbard MA, et al. In vivo effects of goldenseal, kava kava, black cohosh, and valerian on human cytochrome P450 1A2, 2D6, 2E1, and 3A4/5 phenotypes. Clin Pharmacol Ther 2005;77:415–426.
42. FDA Advisory on Kava potential for liver toxicity from FDA Center for Food Safety and Applied Nutrition. http://www.cfsan.fda.gov/~dms/addskava.html. Date accessed: June 8, 2006.

Chapter 3

Ginkgo biloba

Timothy S. Tracy

SUMMARY

Controlled studies suggest that administration of *Ginkgo biloba* (GB) extract has limited effectiveness in improving memory and cognition, either in elderly subjects with dementia or healthy subjects. GB administration does seem to reverse sudden hearing loss in patients with mild cases of this disorder. Additionally, GB administration may blunt the rise in blood pressure in response to stress and may blunt the glycemic response after an oral glucose tolerance test. Despite the lack of evidence of effects on coagulation in vivo, a number of case reports of excessive bleeding in patients taking GB have been reported. Finally, GB does not appear to be prone to causing drug interactions, except for agents metabolized by cytochrome P450 2C19 (in which case, induction is observed).

Key Words: Ginkgolides; dementia; memory; diabetes; bleeding disorders.

1. HISTORY

The ginkgo tree, *Ginkgo biloba* (GB) L., is the last remaining member of the Ginkgoaceae family, which once included many species *(1)*. It has survived unchanged in China for more than 200 million years, and was brought to Europe in 1730 and to America in 1784. Since then, it has become a popular ornamental tree worldwide. Individual trees may live as long as 1000 years, and grow to a height of approx 125 feet *(2)*. GB fruits and seeds have been used in China for their medicinal properties since 2800 BCE *(1)*. Traditional Chinese physicians used GB leaves to treat asthma and chilblains (swelling of the hands and feet from exposure to damp cold) *(2)*. The ancient Chinese

From Forensic Science and Medicine:
Herbal Products: Toxicology and Clinical Pharmacology, Second Edition
Edited by: T. S. Tracy and R. L. Kingston © Humana Press Inc., Totowa, NJ

and Japanese ate roasted GB seeds as a digestive aid and to prevent drunkenness *(2)*. GB use had spread to Europe by the 1960s.

2. CURRENT PROMOTED USE

GB is sold as a dietary supplement in the United States. It is purported to improve blood flow to the brain and to improve peripheral circulation. It is promoted mainly to sharpen mental focus in otherwise healthy adults as well as in those with dementia. Other conditions for which it is currently used are diabetes-related circulatory disorders, impotence, and vertigo.

3. PRODUCTS AVAILABLE

An acetone-water mixture is used to extract the dried and milled leaves *(1)*. After the solvent is removed, the *Ginkgo biloba* extract (GBE) is dried and standardized. Most commercially prepared dosage forms contain 40 mg of GBE *(1)*, and are standardized to contain approx 24% flavonoids (mostly flavone glycosides, or ginkgoflavone glycosides) and 6% terpenes (ginkgolides and bilobalide) *(3–5)*. There are a more than 500 GB preparations on the market, in a number of dosage forms.

4. PHARMACOLOGICAL/TOXICOLOGICAL EFFECTS

The effects of GB are attributed to several chemical constituents of the whole plant rather than to any one individual component. These chemicals include many flavonoids (also called flavonol, flavone, or flavonoid glycosides, ginkgo flavone glycosides, dimeric bioflavones), and the terpene lactones (also called terpenoids, diterpenes, terpenes), including the ginkgolides and bilobalide *(2,5–7)*.

4.1. Nervous System Effects

The pharmacological basis of the effects of GBE on brain function has been addressed in a number of studies. One study *(6)* showed that dietary GBE 761 (prepared by the Henri Baeufour Institute) protected striatal dopaminergic neurons of male Sprague-Dawley rats from damage caused by N-methyl-4-phenyl-1,2,3,6-tetrahydropyridine (MPTP). MPTP, which has caused Parkinsonism in young drug abusers, is thought to damage these neurons through formation of free radicals. The mechanism of GBE's protective effect was attributed to an antioxidant action, rather than to prevention of neuronal uptake of MPTP. Whether chronic GBE ingestion could prevent development of idiopathic Parkinson's disease in humans remains to be seen.

The effectiveness of GB in improving memory and cognition remains controversial, with most studies demonstrating no effect or only very modest improvement. Readers of the original papers listed below are encouraged to closely examine the reported results and conclusions in addition to the abstract, as the claims are not always justified by the results. In one study *(8)*, 40-mg EGb 761 (Murdock, Springville, UT) tablets taken three times daily before meals was compared to placebo in a double-blind, randomized trial in patients with mild to severe Alzheimer type or multiinfarct dementia, diagnosed according to the *Diagnostic and Statistical Manual of Mental Disorders, Third Edition, Revised* and *International Statistical Classification of Diseases, Tenth Revision* criteria. The study lasted 52 weeks, and patients were assessed at weeks 3, 26, and 52 using the cognitive subscale of the Alzheimer's Disease Assessment Scale (ADAS-Cog), the Geriatric Evaluation by Relative's Rating Instrument (GERRI), and the Clinical Global Impression of Change (CGIC), three validated rating instruments. Thus, participants' cognitive impairment, daily living and social behavior, and general psychopathology were objectively evaluated. Modest improvement was appreciated using ADAS-Cog and GERRI, but the CGIC score did not reveal improvement compared to placebo. Adverse effects did not differ from those of placebo. The relatively large number of dropouts (only 202 of 309 patients were assessed at week 52) raises questions about the validity of the results. In addition, a metaanalysis of four double-blind, placebo-controlled studies including a total of 424 patients with Alzheimer's found a small (3%) but clinically significant improvement on the ADAS-Cog with 120–240 mg of GB administered for 3–6 months *(9)*. A number of double-blind, placebo-controlled studies have been conducted recently to assess the effect of GB on cognition and memory, particularly in elderly subjects or those with dementia. Four of these studies have reported no beneficial effect of GB administration on tests of memory and cognitive function *(10–13)*. Other studies have reported only modest beneficial effect of GB on some measures of memory and cognition, mostly related to measures of attention *(14–18)*. Thus, it appears that GB administration has little benefit on improvement of cognition and memory.

Anxiolytic effects have been demonstrated in animal models. The effect of Zingicomb® (Mattern et Partner, Starnberg, Germany), a combination product containing 24% ginkgo flavonoids and 23.5% gingerols, administered orally to rats at a dose of 0.5–100 mg/kg was compared with the effects of placebo and diazepam administered intraperitoneally at a dose of 1 mg/kg on anxiety-associated behaviors *(4)*. The rats were subjected to an elevated plus-maze consisting of enclosed and open arms. The 0.5 mg/kg dose of Zingicomb was associated with rats spending more time in the open arms and with more excursions toward the ends of the open arms as compared to placebo. At a dose of 100 mg/kg, excursions to the ends of the open arms and scanning (protruding the head over the edge of an open arm and looking around) were fewer. These results were interpreted to mean that the preparation exhibited anxiolytic effects at a dose of 0.5 mg/kg, but anxiogenic effects at 100 mg/kg. Both the herbal product at a 0.5 mg/kg dose and diazepam increased the number of entries into the open arms, but unlike diazepam, Zingicomb did not increase open-arm scanning, nor did it attenuate risk assessment (protruding the forepaws and head from an enclosed arm). These effects of the herbal preparation were attributed to blockade of 5-hydroxytryptamine3 (5-HT3; serotonin) receptors, which has been shown in previous studies to produce similar results in the elevated plus-maze. In addition, components of both ginger and GB have been shown in several animal studies to exert 5-HT3 receptor-blocking effects.

Though early reports had suggested that vertigo and tinnitus could be successfully relieved with GB treatment at doses of 16–160 mg/day for 3 months *(19)*, this effect has not been borne out in double-blind, placebo-controlled trials. Rejali and colleagues studied the effectiveness of GB administration in 66 adult subjects with tinnitus *(20)*. Using the Tinnitus Handicap Inventory as the primary outcome measure, these investigators found that administration of GB was of no benefit to patients with tinnitus. They also conducted a metaanalysis of five additional studies (plus theirs) and confirmed the lack of effect of GB in treating tinnitus. In a similar study of ginkgo treatment (Vertigoheel®), Issing and colleagues found that both placebo and the ginkgo preparation improved vertigo equally, suggesting no additional benefit of ginkgo treatment *(21)*. It should be noted that this study was conducted in a randomized, placebo-controlled, double-blind fashion.

Acute mountain sickness can occur when unacclimatized individuals ascend to altitudes above 2000 meters. Chow and colleagues compared the efficacy of either acetazolamide or GB prophylaxis for preventing acute mountain sickness *(22)*. The trial was completed by 57 subjects, with 20 receiving acetazolamide, 17 receiving GB, and 20 receiving placebo. GB had no effect on either the symptoms or incidence of acute mountain sickness. Acetazolamide reduced the symptoms of acute mountain sickness but did not reduce the incidence. A similar lack of effect on acute mountain sickness among Himalayan trekkers taking GB was noted by Gertsch et al. *(23)*.

Interestingly, GB treatment has been demonstrated to be effective in reversing the symptoms of sudden hearing loss. In one study of 106 patients receiving either GBE (EGb 761) or placebo and suffering from idiopathic sudden sensorineural hearing loss, the GBE treatment appeared to speed up recovery and improve the chances of complete recovery as compared to placebo *(24)*. Similarly, Reisser and Weidauer *(25)* observed that GBE (EGb 761) and pentoxifylline were equally effective in reversing hearing loss and reducing tinnitus. Though the study was randomized and double blinded, no placebo control was utilized.

4.2. Cardiovascular Effects

EGb 761 at a dose of 200 mg administered to 60 patients intravenously for 4 days improved skin perfusion and decreased blood viscosity without affecting plasma viscosity *(19)*. Another GB extract, LI 1730, increased blood flow in nailfold capillaries and decreased erythrocyte aggregation compared to placebo in 10 volunteers at a dose of 112.5 mg *(26)*. Blood pressure, heart rate, packed cell volume, and plasma viscosity were unchanged. A study in subjects with type 2 diabetes mellitus complicated with retinopathy evaluated the effects of administration of GB (EGb 761) for 3 months on erythrocyte hemorrheology *(27)*. At the end of the treatment period, it was observed that blood viscosity was significantly reduced, fibrinogen levels were decreased, and erythrocytes were more deformable. Finally, retinal capillary blood flow was improved. However, in a double-blind, placebo-controlled trial of GB in the treatment of Raynaud's disease, after 10 weeks of treatment there was no improvement in hemorrheology between the two groups *(28)*. There was a significant decrease in the number of attacks per day.

Studies have been conducted to examine the effects on GB administration on blood pressure and blood flow. In one study, either GB or placebo was administered in a double-blind, placebo-controlled crossover design to healthy volunteers and forearm blood flow was measured *(29)*. Forearm blood flow was significantly higher during GB therapy than with placebo and mean arterial pressure remained unchanged, thus rendering the forearm vascular resistance significantly lower during active treatment. In a related study, Jezova and colleagues studied the effect of GB (EGb 761) treatment on changes in blood pressure and cortisol release following exposure to stress stimuli *(30)*. The rise in systolic blood pressure following stress stimuli was significantly lower (~20 mmHg rise in subjects receiving EGb 761 vs an ~30 mmHg rise in subjects receiving placebo) in the GB group. Differences in diastolic pressure rise were similar (~10 mmHg difference between GB and placebo) between the two groups. GB did inhibit the stress-induced increase in cortisol release in the male subjects but had no effect in the female subjects.

Because of effects noted with in vitro studies demonstrating that ginkgolides are capable of inhibiting platelet-activating factor (PAF), which is involved in platelet aggregation and inflammatory processes such as those seen in asthma, ulcerative colitis, and allergies (reviewed in *5,19,31*), it has been suggested that bleeding parameters might be affected also. Several case reports of bleeding disorders among people receiving GB have been described (*see* Subheading 7.1.). However, at least in healthy volunteers, changes in platelet function or coagulation have not been substantiated. In a double-blind, placebo-controlled study of 32 healthy male volunteers receiving EGb 761 at three doses (120, 240, and 480 mg/day) for 14 days, no changes in platelet function or coagulation were noted (*32*). Similarly, Kohler and colleagues studied the influence of the same GBE (EGb 761) on bleeding time and coagulation in healthy volunteers (*33*). This double-blind, placebo-controlled study was carried out for 7 days in 50 healthy volunteers. No differences in bleeding time, coagulation parameters, or platelet activity were noted between the placebo and GB treatment groups. In a study of patients on chronic peritoneal dialysis, Kim and colleagues randomized the 66 patients into two groups; those receiving GB (160 mg/day) and those receiving no treatment (*34*). There was no placebo control. Except for a small but statistically significant change in the plasma D-dimer concentration, the administration of GB had no effect on any bleeding parameters. Finally, Kudolo and colleagues studied the effect of GBE on platelet aggregation and urinary prostanoid excretion in healthy subjects and patients with type 2 diabetes (*35*). Administration of GB had no effect on any parameter of coagulation or prostanoid excretion in the patients with type 2 diabetes. In the healthy volunteers, a modest but statistically significant decrease in thromboxane B_2, PGI_2, and prostanoid metabolite ratio was noted following GB treatment. It is of note that a placebo control was not included for either group.

4.3. Carcinogenicity/Mutagenicity/Teratogenicity

No mutagenic, carcinogenic, or teratogenic effects have been noted in studies performed using commercially available GB products containing 22–27% flavone glycosides and 5–7% terpene lactones (*36*).

4.4. Endocrine Effects

Kudolo (*37*) studied the effect of 3-month ingestion of a GBE on pancreatic β-cell function. Having taken a 3-month course of GB, 20 normal, healthy subjects were given an oral glucose tolerance test, and fasting plasma insulin and C-peptide were measured. Fasting plasma insulin area under the curve (AUC) was increased approx 20% whereas C-peptide AUC was increased approx 70%. In a follow-up study in noninsulin-dependent patients with diabetes mellitus who received the same 3-month GB therapy, following an oral glucose tolerance test, a blunted plasma insulin response was noted, leading to a reduction in insulin AUC *(38)*. Conversely, C-peptide levels were increased, leading to a dissimilar insulin/C-peptide ratio. The author suggested this indicated an increased hepatic extraction of insulin relative to C-peptide, potentially resulting in reduced insulin-mediated glucose metabolism and elevated blood glucose.

Administration of GB has also been studied for the treatment of sexual dysfunction. Kang and colleagues evaluated the efficacy of GB administered for 2 months to subjects with antidepressant-induced sexual dysfunction in a placebo-controlled, double-blind design *(39)*. Compared with baseline, both placebo and GB showed improvement in some aspects of sexual function, but there was no difference in effect between placebo and GB treatment. Similarly, Wheatley used a triple-blind, placebo-controlled design to again study the effect of GB on antidepressant-induced sexual dysfunction *(40)*. Again, though some individual subjects experienced improvement, no statistically significant improvement in sexual function was noted.

5. DRUG INTERACTIONS

In two of the five spontaneous bleeding episodes described in Heading 4, medications that can affect platelet function or prothrombin time (PT) (i.e., aspirin and warfarin) were involved. Because GB is known to be an inhibitor of PAF *(41)*, in theory GB could interact with antiplatelet drugs (e.g., aspirin, nonsteroidal anti-inflammatory drugs, clopidogrel, ticlopidine, dipyridamole) or anticoagulants (e.g., warfarin, heparin). EGb 761 was shown to potentiate the antiplatelet effect of ticlopidine in rats *(42)*. However, in two studies in humans, the coadministration of GB with warfarin had no effect on either international normalized ratio or warfarin metabolism *(43,44)*.

With respect to the effect of GB on cytochrome P450 drug metabolizing enzymes, in vitro studies have demonstrated that GBE and components have minimal effect on CYP2C9, CYP3A4, CYP1A2, and CYP2D6 *(45–47)*. This lack of effect on these particular cytochrome P450 enzymes has been confirmed by in vivo studies using probe drug substrates *(48)*. However, coadministration of GB with omeprazole (a CYP2C19 substrate) demonstrated significant induction of omeprazole metabolism, resulting in reduced AUC *(49)*. Finally, GB coadministration did not have an effect on donepezil pharmacokinetics *(50)*. A study in which 400 mg of EGb was administered to 24 healthy volunteers for 13 days demonstrated that GB is not an inducer of other hepatic microsomal enzymes *(51)*.

6. PHARMACOKINETICS/TOXICOKINETICS

6.1. Absorption

In humans, absolute bioavailability is 98–100% for ginkgolide A, 79–93% for ginkgolide B, and at least 70% for bilobalide *(36)*. In two healthy volunteers, flavonol glycosides administered as the product LI 1370 at doses of 50, 100, and 300 mg were absorbed in the small intestine with peak plasma concentration attained within 2–3 hours *(19)*. Additional data from human experiments from the manufacturer of 80-mg EGb 761 solution show that the absolute bioavailabilities of ginkgolides A and B were greater than 80%, whereas that of ginkgolide C was very low. Bioavailability of bilobalide was 70% after administration of 120 mg of the extract. Corroborating these results was a later pharmacokinetic study *(52)* that found mean bioavailabilities of 80, 88, and 79% for ginkgolide A, ginkgolide B, and bilobalide, respectively. Food intake did increase the time to peak concentration, but did not affect bioavailability.

A study in rats using radiolabeled EGb 761 revealed a bioavailability of at least 60% *(19)*. Peak blood concentrations occurred at 1.5 hours. At 3 hours, the highest radioactivity was measured in the stomach and small intestine, indicating that these are the sites of absorption.

6.2. Distribution

Rat studies using radiolabeled EGb 761 have revealed that the extract follows a two-compartment model of distribution *(19)*. The radiolabeled extract was distributed into glandular and neuronal tissues, as well as the eyes.

The volumes of distribution of ginkgolide A, ginkgolide B, and bilobalide are 40–60 L, 60–100 L, and 170 L, respectively *(19)*.

6.3. Metabolism/Elimination

The half-life of the flavonol glycosides administered as the product LI 1370 is 2–4 hours *(19)*. Similar results wereobtained using 80 mg of the product EGb 761; half-lives of ginkgolides A and B were 4 and 6 hours, respectively. The half-life of bilobalide was 3 hours after administration of 120 mg of this extract. Similar results were reported in another study *(52)* using this same product; mean half-lives of ginkgolide A, ginkgolide B, and bilobalide were 4.5, 10.57, and 3.21 hours, respectively.

A study in rats using radiolabeled EGb 761 revealed a half-life of 4.5 hours, with elimination following first-order (linear) kinetics *(19)*.

Approximately 70% of ginkgolide A, 50% of ginkgolide B, and 30% of bilobalide is excreted unchanged in the urine *(19)*. Metabolites isolated from human urine after administration of EGb include a 4-hydroxybenzoic acid conjugate, 4-hydroxyhippuric acid, 3-methoxy-4-hydroxyhippuric acid, 3,4-dihydroxybenzoic acid, 4-hydroxybenzoic acid, hippuric acid, and 3-methoxy-4-hydroxybenzoic acid (vanillic acid) *(53)*. In accord with previous data, these metabolites accounted for less than 30% of the administered EGb dose. Metabolites were not detectable in blood samples.

7. ADVERSE EFFECTS AND TOXICITY

7.1. Case Reports of Toxicity Caused By Commercially Available Products

Spontaneous intracerebral hemorrhage occurred in a 72-year-old woman who had been taking GB 50 mg three times daily for 6 months *(54)*. Bilateral subdural hematomas were discovered in a 33-year-old woman who had been taking 60 mg of GB twice daily for 2 year, acetaminophen, and occasionally an ergotamine/caffeine preparation *(55)*. Bleeding time was elevated, but had normalized when checked approx 1 month after discontinuation of the product.

In a similar case, a 61-year-old man presented with a subarachnoid hemorrhage after taking 40-mg GB tablets three or four times daily for more than 6 months *(56)*. Bleeding time was elevated (6 minutes, normal 1–3), but normalized with discontinuation of the product.

A 78-year-old woman suffered a left parietal hemorrhage after taking a GB preparation for 2 months *(57)*. Other medications included warfarin, which she had been taking for 5 years after undergoing coronary bypass. PT was unchanged.

A 70-year-old man experienced bleeding from the iris into the anterior chamber after self-medicating with 40 mg of Ginkoba® twice daily for 1 week *(58)*. Other medications included 325 mg of aspirin daily for 3 years post-coronary bypass. GB, but not aspirin, was discontinued, and no further bleeding problems occurred.

A 75-year-old woman had undergone outpatient surgery and developed bleeding complications on the first postoperative night *(59)*. PT, activated partial thrombin time, and platelets were normal but platelet aggregation was diminished. The patient had been taking no other medications except a GB preparation (Gingium®) that she had been taking for the past 2 years. The GB product was discontinued and her platelet aggregation returned to normal after 10 days.

In a similar case, a 77-year-old woman experienced persistent bleeding after total hip arthroplasty while taking GB therapy *(60)*. This bleeding persisted for 4 weeks, at which time the ginkgo was discontinued. After the GB had been discontinued for 6 weeks, the bleeding stopped.

"Gin-nan" food poisoning, a toxic syndrome associated with ingestion of 50 or more GB seeds, can result in loss of consciousness, tonic/clonic seizures, and death *(2)*. Between 1930 and 1960, 70 cases were reported, with a 27% mortality rate. Infants were at greatest risk. Although ginkgotoxin (4-*O*-methylpyridoxine), which is found mostly in the seeds, has been implicated as the responsible neurotoxin, its concentrations in several commercially available GB products tested were deemed too low to have a toxic effect *(61)*. If used as directed, the maximum daily intake of 4-*O*-methylpyridoxine would be approx 60 μg; however, the presence of this neurotoxin raises questions about the herb's ability to lower the seizure threshold in patients with seizure disorders *(61)*. The authors of this study cite evidence that bilobalide present in the formulations may decrease the severity of convulsions, thus counteracting any neurotoxic effects of 4-*O*-methylpyridoxine.

Adverse effects listed in the German Commission E GB leaf extract monograph include gastrointestinal upset, headache, and rash *(36)*.

8. REGULATORY STATUS

GB leaf extract is approved by the German Commission E for memory deficits, disturbances in concentration, depression, dizziness, vertigo, headache, dementia, and intermittent claudication *(36)*. It is regulated as a dietary supplement in the United States.

REFERENCES

1. Tyler VE, ed. *The Honest Herbal, 3rd edition*. Binghamton: Pharmaceutical Products Press, 1993.
2. Anonymous. Ginkgo. In: *The Review of Natural Products*. St. Louis: Facts and Comparisons, 1998.
3. Amri H, Ogwuegbu SO, Boujrad N, Drieu K, Papadopoulos V. In vivo regulation of peripheral type benzodiazepine receptor and glucocorticoid synthesis by Ginkgo Biloba extract EGb 761 and isolated ginkgolides. Endocrinology 1996;137:5707–5718.
4. Hasenohrl RU, Nichau CH, Frisch CH, et al. Anxiolytic-like effect of combined extracts of Zingiber officinale and Ginkgo biloba in the elevated plus-maze. Pharmacol Biochem Behav 1996;53:271–275.

5. Nemecz G, Combest WL. Ginkgo biloba. US Pharm 1997;144:147–148, 151.
6. Ramassamy C, Clostre F, Christen Y, Costentin J. Prevention by a ginkgo extract (GBE 761) of the dopaminergic neurotoxicity of MPTP. J Pharm Pharmacol 1990;42:785–789.
7. Houghton P. Ginkgo. Pharm J 1994;253:122–123.
8. Le Bars PL, Katz MM, Berman N, et al. A placebo controlled, double-blind, randomized trial of an extract of Ginkgo biloba for dementia. JAMA 1997;278:1327–1332.
9. Oken BS, Storzbach DM, Kaye JA. The efficacy of Ginkgo biloba on cognitive function in Alzheimer's disease. Arch Neurol 1998;55:1409–1415.
10. van Dongen M, van Rossum E, Kessels A, Sielhorst H, Knipschild P. *Ginkgo* for elderly people with dementia and age-associated memory impairment: a randomized clinical trial. J Epidemiol 2003;56:367–376.
11. van Dongen M, van Rossum E, Kessels AGH, Sielhorst HJG, Knipschild PG. The efficacy of *Ginkgo* for elderly people with dementia and age-associated memory impairment: new results of a randomized clinical trial. J Am Geriatr Soc 2000;48:1183–1194.
12. Mattes RD, Pawlik MK. Effects of *Ginkgo biloba* on alertness and chemosensory function in healthy adults. Hum Psychopharmacol Clin Exp 2004;19:81–90.
13. Mix JA, Crews WD Jr. An examination of the efficacy of *Ginkgo biloba* extract EGb 761 on the neuropsychologic functioning of cognitively intact older adults. J Altern Complement Med 2000;6(3):219–229.
14. Allain H, Raoul P, Lieury A, LeCoz F, Gandon J. Effect of two doses of Ginkgo biloba extract (EGb 761) on the dual-coding test in elderly subjects. Clin Ther 1993;15:549–557.
15. Mix JA, Crews WD. A double-blind, placebo-controlled, randomized trial of *Ginkgo biloba* extract EGb 761® in a sample of cognitively intact older adults: neuropsychological findings. Hum Psychopharmacol Clin Exp 2002;17:267–277.
16. Kennedy DO, Scholey AB, Wesnes KA. The dose-dependent cognitive effects of acute administration of *Ginkgo biloba* to healthy young volunteers. Psychopharmacology 2000;151:416–423.
17. Hartley DE, Heinze L, Elsabagh S, File SE. Effects on cognition and mood in postmenopausal women of 1-week treatment with *Ginkgo biloba*. Pharmacol Biochem Behav 2003;75:711–720.
18. Kanowski S, Hoerr R. Ginkgo biloba extract EGb 761® in dementia: intent-to-treat analyses of a 24-week, multi-center, double-blind, placebo-controlled, randomized trial. Pharmacopsychiatry 2003;36:297–303.
19. Kleijnen J, Knipschild P. Ginkgo biloba. Lancet 1992;340:1136–1139.
20. Rejali D, Sivakumar A, Balaji N. *Ginkgo biloba* does not benefit patients with tinnitus: a randomized placebo-controlled double-blind trial and meta-analysis of randomized trials. Clin Otolaryngol 2004;29:226–231.
21. Issing W, Klein P, Weiser M. The homeopathic preparation Vertigoheel® versus *Ginkgo biloba* in the treatment of vertigo in an elderly population: a double-blinded, randomized, controlled clinical trial. J Altern Complement Med 2005;11(1):155–160.

22. Chow T, Browne V, Heileson HL, Wallace D, Anholm J, Green SM. *Ginkgo biloba* and acetazolamide prophylaxis for acute mountain sickness. Arch Intern Med 2005;165:296–301.

23. Gertsch JH, Basnyat B, Johnson EW, Onopa J, Holck PS, on behalf of the Prevention of High Altitude Illness Trial Research Group. Randomised, double blind, placebo controlled comparison of *ginkgo biloba* and acetazolamide for prevention of acute mountain sickness among Himalayan trekkers: the prevention of high altitude illness trial (PHAIT). Br Med J 2004;328:797–801.

24. Burschka MA, Abdel-Hady Hassan H, Reineke T, van Bebber L, Caird DM, Mosges R. Effect of treatment with *Ginkgo biloba* extract EGb 761 (oral) on unilateral idiopathic sudden hearing loss in a prospective randomized double-blind study of 106 outpatients. Eur Arch Otorhinolaryngol 2001;258:213–219.

25. Reisseer C, Weidauer H. *Ginkgo biloba* extract EGb 761® or pentoxifylline for the treatment of sudden deafness: a randomized, reference-controlled, double-blind study. Acta Otolaryngol 2001;121:579–584.

26. Jung F, Mroweitz C, Kiesewetter H, Wenzel E. Effect of Ginkgo biloba on fluidity of blood and peripheral microcirculation in volunteers. Arzneim Forsch 1990;40:589–593.

27. Huang SY, Jeng C, Kao SC, Yu, JJH, Liu DZ. Improved haemorrheological properties by *Gingko biloba* extract (Egb 761) in type 2 diabetes mellitus complicated with retinopathy. Clin Nutr 2004;23:615–621.

28. Muir AH, Robb R, McLaren M, Daly F, Belch JJF. The use of *Ginkgo biloba* in Raynaud's disease: a double-blind placebo-controlled trial. Vasc Med 2002;7:265–267.

29. Mehlsen J, Drabaek H, Wiinberg N, Winther K. Effects of a *Gingko biloba* extract on forearm haemodynamics in healthy volunteers. Clin Physiol Funct Imaging 2002;22:375–378.

30. Jezova D, Duncko R, Lassanova M, Kriska M, Moncek F. Reduction of rise in blood pressure and cortisol release during stress by *Ginkgo biloba* extract (EBg 761) in healthy volunteers. J Physiol Pharmacol 2002;53:337–348.

31. Chavez ML, Chavez PI. Ginkgo (Part I): History, use, and pharmacologic properties. Hosp Pharm 1998;33:658–672.

32. Sollier CBD, Caplain H, Drouet L. No alteration in platelet function or coagulation induced by EGb761 in a controlled study. Clin Lab Haem 2003;25:251–253.

33. Kohler S, Funk P, Kieser M. Influence of a 7-day treatment with *Ginkgo biloba* special extract EGb 761 on bleeding time and coagulation: a randomized, placebo-controlled, double-blind study in healthy volunteers. Blood Coagul Fibrinolysis 2004;15:303–309.

34. Kim SH, Lee EK, Chang JW, Min WK, Chi HS, Kim SB. Effectsw of *Ginkgo biloba* on haemostatic factors and inflammation in chronic peritoneal dialysis patients. Phytother Res 2005;19:546–548.

35. Kudolo GB, Dorsey S, Blodgett J. Effect of the ingestion of *Ginkgo biloba* extract on platelet aggregation and urinary prostanoid excretion in healthy and Type 2 diabetic subjects. Thromb Res 2003;108:151–160.

36. Blumenthal M, ed. Ginkgo biloba. *The complete German commission E monographs*. Austin: American Botanical Council, 1998.

37. Kudolo GB. The effect of 3-month ingestion of *Ginkgo biloba* extract on pancreatic β-cell function in response to glucose loading in normal glucose tolerant individuals. J Clin Pharmacol 2000;40:647–654.
38. Kudolo GB. The effect of 3-month ingestion of *Ginkgo biloba* extract (EGb 761) on pancreatic β-cell function in response to glucose loading in individuals with non-insulin-dependent diabetes mellitus. J Clin Pharmacol 2001;41:600–611.
39. Kang BJ, Lee SJ, Kim MD, Cho MJ. A placebo-controlled, double-blind trial of *Ginkgo biloba* for antidepressant-induced sexual dysfunction. Hum Psychopharmacol Clin Exp 2002;17:279–284.
40. Wheatley D. Triple-blind, placebo-controlled trial of *Ginkgo biloba* in sexual dysfunction due to antidepressant drugs. Hum Psychopharmacol Clin Exp 2004;19:545–548.
41. Chung KF, McCusker M, Page CP, Dent G, Guinot P, Barnes PJ. Effect of ginkgolide mixture (BN 52063) in antagonizing skin and platelet responses to platelet activating factor in man. Lancet 1987;1:248–251.
42. Kim YS, Pyo MK, Park PH, Hahn BS, Wu SJ, Yun-Choi HS. Antiplatelet and antithrombotic effects of a combination of ticlopidine and Ginkgo biloba extract (EGb 761). Thrombos Res 1998;91:33–38.
43. Jiang X, Williams K, Liauw W, et al. Effect of *ginkgo* and ginger on the pharmacokinetics and pharmacodynamics of warfarin in healthy subjects. Brit J Clin Pharmacol 2005;59(4):425–432.
44. Engelsen J, Nielsen JD, Winthere K. Effect of coenzyme Q10 and *Ginkgo biloba* on warfarin dosage in stable, long-term warfarin treated outpatients. A randomized, double-blind, placebo-crossover trial. Thromb Haemost 2002;87:1075–1076.
45. Yale SH, Glurich I. Analysis of the inhibitory potential of *Ginkgo biloba*, *Echinacea purpurea*, and *Serenoa repens* on the metabolic activity of cytochrome P450 3A4, 2D6, and 2C9.
46. von Moltke LL, Weemhoff JL, Bedir E, et al. Inhibition of human cytochromes P450 by components of *Ginkgo biloba*. J Plant Pathol 2004;56:1039–1044.
47. He N, Edeki T. The inhibitory effects of herbal components on CYP2C9 and and CYP3A4 catalytic activities in human liver microsomes. Am J Therapeutics 2004;11:206–212.
48. Markowitz JS, Donovan JL, DeVane CL, Sipkes L, Chavin KD. Multiple-dose administration of Ginkgo biloba did not affect cytochrome P-450 2D6 or 3A4 activity in normal volunteers. J Clin Psychopharmacol 2003;23:576–581.
49. Yin OQP, Tomlinson B, Waye MMY, Chow AHL, Chow MSS. Pharmacokinetics and herb-drug interactions: experience with *Ginkgo biloba* and omeprazole. Pharmacogenetics 2004;14:841–850.
50. Yasui-Furukori N, Furukori H, Kaneda A, Kaneko S, Tateishi T. The effects of *Ginkgo biloba* extracts on the pharmacokinetics and pharmacodynamics of donepezil. J Clin Pharmacol 2004;44:538–542.
51. Duche JC, Barre J, Guinot P, Duchier J, Cournot A, Tillement JP. Effect of Ginkgo biloba extract on microsomal enzyme induction. Int J Clin Pharmacol Res 1989;9:165–168.
52. Fourtillan JB, Brisson AM, Girault J, et al. Pharmacokinetic properties of bilobalide and ginkgolides A and B in healthy subjects after intravenous and oral administration of Ginkgo biloba extract (EGb 761). Therapie 1995;50:137–144.

53. Pietta PG, Gardana C, Mauri PL. Identification of Ginkgo flavonol metabolites after oral administration to humans. J Chromatogr Biomed Sci Appl 1997;693:249–255.
54. Gilbert GJ. Ginkgo biloba [letter]. Neurology 1997;48:1137.
55. Rowin J, Lewis SL. Spontaneous bilateral subdural hematomas associated with chronic Ginkgo biloba ingestion [letter]. Neurology 1996;46:1775–1776.
56. Vale S. Subarachnoid hemorrhage associated with Ginkgo biloba [letter]. Lancet 1998;352:36.
57. Matthews MK. Association of Ginkgo biloba with intracranial hemorrhage [letter]. Neurology 1998;50:1933–1934.
58. Rosenblatt M, Mindel J. Spontaneous hyphema associated with   ingestion   of Ginkgo biloba extract [letter]. N Engl J Med 1997;336:1108.
59. Yagmur E, Piatkowski A, Groger A, Pallua N, Gressner AM, Kiefer P. Bleeding complication under *Gingko biloba* medication. Letter to the Editor. Am J Hematol 2005;79:343–345.
60. Bebbington A, Kulkarni R, Roberts P. *Ginkgo biloba*. Persistent bleeding after total hip arthroplasty caused by herbal self-medication. J Arthroplasty 2005;20(1):125–126.
61. Arenz A, Klein M, Fiehe K, et al. Occurrence of neurotoxic 4'-O-mehtylpyridoxine in Ginkgo biloba leaves, Ginkgo medications, and Japanese ginkgo food. Planta Med 1996;62:548–551.

Chapter 4

Valerian

Brian J. Isetts

SUMMARY

Valerian is a unique herb with a long history of use through western Europe as a sedative and hypnotic. A variety of pharmacologically active components are likely responsible for its clinical effects including volatile oils, monoterpenes, valepotriates, and sesquiterpenes. Valerenic acid, a sesquiterpene component of valerian, is postulated to produce sedation through inhibition of the breakdown of gamma-amino butyric acid. The herb is well tolerated, and side effects have been mild and self-limiting in most cases. Isolated reports of liver damage have occurred with valerian being a concomitantly consumed agent, yet anecdotal cases of attempted intentional self-poisoning with the herb have not resulted in fatality and long-term follow-up for subsequent hepatotoxicity in a number of these patients has not revealed liver abnormalities. The herb's postitive safety profile and demonstrated effectiveness in treating insomnia contributes to its popularity.

Key Words: Valerenic acid; valepotriates monoterpenes; sesquiterpenes; anxiolytic; hypnotic.

*1. HISTORY**

Valerian is a perennial herb comprised of grooved hollow stems and saw-toothed green leaves. White, pale pink, or reddish flowers appear from June to August. Valerian grows to heights of 3–5 feet in the temperate climates of North America, western Asia, and Europe, often in moist soil along

---

*Based, in part, on Chapter 4 from *Herbal Products: Toxicology and Clinical Pharmocology, First Edition*, edited by Morlea Givens and Melanie Johns Cupp.

From Forensic Science and Medicine:
Herbal Products: Toxicology and Clinical Pharmacology, Second Edition
Edited by: T. S. Tracy and R. L. Kingston © Humana Press Inc., Totowa, NJ

riverbanks. The vertical rhizome and attached roots of valerian are parts used medicinally, and are best harvested in the autumn of the second year *(1)*. Although the fresh drug has no distinctive odor, over time hydrolysis of compounds present in the volatile oil produces isovaleric acid, which has an offensive, somewhat putrid odor *(2)*. Fortunately, the smell can be removed from the skin and utensils by washing with sodium bicarbonate *(3)*. Even though valerian has a disagreeable odor, people in the 16th century considered it a fragrant perfume *(2)*. Traditional uses include treatment of insomnia, migraine headache, anxiety, fatigue, and seizures *(4)*. It has also been applied externally on cuts, sores, and acne. Traditional Chinese uses include treatment of headache, numbness caused by rheumatic conditions, colds, menstrual difficulties, and bruises. The pharmacological effects of valerian have been attributed to the constituents of volatile oils, monoterpenes, valepotriates, and sesquiterpenes (valerenic acid) *(5)*. Some of these constituents have been shown to have a direct action on the brain, and valerenic acid inhibits enzyme-induced breakdown of γ-amino butyric acid (GABA) in the brain resulting in sedation *(6)*.

2. CURRENT PROMOTED USES

Valerian is promoted in the United States primarily as a sedative-hypnotic for treatment of insomnia, and as an anxiolytic for restlessness and sleeping disorders associated with anxiety *(4,7)*.

3. SOURCES AND CHEMICAL COMPOSITION

Also referred to as: *Valeriana officinalis* (L.), *Valeriana wallichii* DC. (Indian valerian), *Valeriana alliariifolia Vahl*, *Valeriana sambucifolia Mik*, *Radix valerianae*, red valerian (*Centranthus ruber* [L.] DC) *(2)*, valerian root, *Valerianae radix (4)*, garden heliotrope, all heal, amantilla, and setwall *(8)*.

4. PRODUCTS AVAILABLE

Crude valerian root, rhizome, or stolon is dried and used either "as is" or to prepare an extract. Valerian is available as a capsule, tablet, oral solution, or tea *(4)*. Valerian is also administered externally as a bath additive *(7,9)*.

5. PHARMACOLOGICAL/TOXICOLOGICAL EFFECTS

5.1. Insomnia

Several studies have examined the effects of valerian on sleep *(10–15)*.

Donath and colleagues performed a randomized, double-blind, placebo-controlled, cross-over study assessing the short-term (single dose) and long-term (14-day multiple dosage) effects of valerian extract on sleep structure and sleep quality. There were significant differences between valerian and placebo for parameters describing slow-wave sleep (SWS) and shorter sleep latency, with very low adverse events. Leathwood and colleagues demonstrated valerian's effect on sleep quality *(11)*. A freeze-dried aqueous extract of valerian root (*Rhizoma valeriana officinalis* [L.]) 400 mg was compared to two Hova® (valerian 60 mg and hop flower extract 30 mg per tablet) tablets and placebo (finely ground brown sugar) in this crossover study involving 128 volunteers. Study participants took the study medication 1 hour before retiring, and filled out a questionnaire the following morning. This was repeated on nonconsecutive nights, such that each of the three treatments, identified only by a code number, was administered in random order three times to each patient. Valerian caused a significant improvement in subjectively evaluated sleep quality and a significant decrease in perceived sleep latency. The self-reported improvement in sleep quality was especially notable in smokers, those patients who considered themselves poor or irregular sleepers, and those who reported having difficulty falling asleep on a prestudy questionnaire. Hova® did not demonstrate any beneficial effect, but it was reported to cause a "hangover effect" the next morning. Because subjective sleep questionnaires may not correlate with sleep electroencephalogram (EEG) results, a parallel EEG sleep study was performed comparing valerian to placebo in 10 young men. There was not a statistically significant difference between valerian and placebo in this small study. The authors hypothesized that the results of this experiment might have differed from the questionnaire-assessed study because of small sample size and differences in study populations. The larger study involved young and older individuals, men and women, and good and poor sleepers, whereas the EEG study involved young men with no reported sleep abnormalities. Rather than place more credence on the objective study, the investigators concluded that the questionnaire provides a more sensitive means of detecting mild sedative effects.

A double-blind, placebo-controlled study *(12)* was performed in eight volunteers recruited from among the research staff at Nestle Products and their families who reported that they "usually have problems getting to sleep." Sleep latency was measured using an activity monitor and questionnaire. The investigators documented a small (7 minute) but statistically significant decrease in sleep latency with 450 mg of an extract of valerian (*V. officinalis* [L.]). No further improvement was demonstrated with a 900-mg valerian dose; however, patients receiving the higher dose were more likely to feel sleepy the next morning. Sleep quality, sleep latency, and sleep depth also improved according to a nine-point subjective rating scale. However, the appropriateness of the statistical analysis used to interpret the results of the subjective portion of the study is unclear.

A more objective double-blind, placebo-controlled trial *(13)* evaluated the effect of 450- and 900-mg doses of an aqueous valerian extract (*V. officinalis* [L.]) on two groups of healthy, young (21–44 years of age) volunteers at home and in a laboratory setting. The effect of valerian on sleep was measured using a questionnaire and night-time motor activity recordings in both settings. The effects of valerian on the volunteers in the sleep laboratory were also measured using polysomnography and spectral analysis of the sleep EEG. Both groups demonstrated the mild hypnotic effects of valerian; however, the benefits of valerian were statistically significant only under home conditions.

Another double-blind, placebo-controlled crossover study *(14)* evaluated Valerina Natt®, a preparation equivalent to 400 mg of valerian root composed mainly of sesquiterpenes from *V. officinalis* [L.], on subjective sleep quality assessed using a three-point rating scale. Study subjects were 27 consecutive patients seen in a medical clinic for evaluation of sleep difficulty and fatigue who were willing to participate in the investigation. Statistically significant improvement in sleep quality was noted with the valerian preparation. Valerian was rated as better than placebo by 21 subjects, two rated the preparations equally, and four preferred placebo. No adverse effects were reported. Although some study subjects had experienced nightmares when using conventional hypnotics, nightmares were not reported in the study.

The effects of repeated doses (three tablets three times daily) for 8 days of Valdispert Forte® (135 mg of dried extract of *V. officinalis* [L.]) in 14 elderly women with sleeping difficulties was assessed using polysomnography in a particularly well-designed study *(15)*. Inclusion criteria were well defined: sleep latency longer than 30 minutes, more than three nocturnal awakenings per night with inability to go back to sleep within 5 minutes, and total sleep time less than 5 hours. Subjects could not have medical, psychological, or weight-related causes of sleep difficulty, and had to have normal health status for their age. Sedatives, hypnotics, and other central nervous system (CNS)-active drugs were discontinued 2 weeks prior to the study, and drug screening for morphine, benzodiazepines, barbiturates, and amphetamine was done prior to study commencement. Results showed an increase in SWS, and a decrease in sleep stage 1. There was no effect on rapid eye movement (REM) sleep, sleep latency, time awake after sleep onset, or self-rated sleep quality.

In aggregate, the results of these clinical studies suggest that at doses of approx 450 mg of the aqueous extract, valerian has mild hypnotic effects, possibly by affecting non-REM sleep in patients with reduced SWS. Unlike benzodiazepines, valerian appears not to adversely affect SWS or REM sleep, and does not appear to cause nightmares or hangover. Further well-designed studies are needed to objectively evaluate valerian. Results of animal studies reflect the clinical data. Sedative properties of Valdispert® (dried aqueous extract of *V. officinalis* [L.]) in mice were documented based on reduced spontaneous movement and an increase in thiopental-induced sleep time; however, these effects were slightly less than those of diazepam and chlorpromazine. No significant anticonvulsant effect was observed *(16)*.

Hendriks and colleagues tested several components of the volatile oil, obtained by steam distillation of *V. officinalis* [L.], on mice. The essential oil, its hydrocarbon fraction, its oxygen fraction, valeranone, valerenal, valerenic acid, and isoeugenyl-isovalerate were injected intraperitoneally at various doses ranging from 50 to 1600 mg/kg, with three mice receiving each dose. The mice were observed between 15 and 30 minutes postinjection for various symptoms suggestive of CNS stimulation or depression, analgesia, sympathomimetic or sympatholytic activity, vasodilation, or vasoconstriction. It was concluded that components of the essential oil, particularly valerenic acid and valerenal, which are present in the oxygen fraction, have a sedative and/ or muscle relaxant effect. The authors *(17)* tested the effect of intraperitoneal valerenic acid compared to diazepam, chlorpromazine, and pentobarbital on ability to walk on a rotating rod and grip strength in mice. The effects of valerinic acid on spontaneous motor activity and on pentobarbital-induced sleeping time were also assessed. Diazepam, a muscle relaxant, affected the grip test but not the rotarod test, whereas chlorpromazine, a neuroleptic, affected the rotarod test but not the grip test. Valerenic acid, like pentobarbital, decreased performance in both the rotarod and grip tests. The authors concluded that valerenic acid, like pentobarbital, has general CNS depressant activity. Valerenic acid also decreased spontaneous motor activity and prolonged pentobarbital-induced sleeping time. Dose–response effects of valerenic acid were also observed by the investigators. At a dose of 50 mg/kg, a decrease in spontaneous motor activity occurred. At 100 mg/kg, mice exhibited ataxia, then remained motionless. Muscle spasms occurred at 150–200 mg/kg and convulsions at 400 mg/kg, followed by death in six of seven mice within 24 hours *(17)*.

Sedation is mediated predominantly through the inhibitory neurotransmitter GABA. Although the mechanism of action of valerian as a sleep aid is not fully understood, it may involve inhibition of the enzyme that breaks down

GABA. Dihydrovaltrate, hydroxyvalerenic acid, a hydroalcoholic extract containing 0.8% valerenic acid; a lipid extract; an aqueous extract of the hydroalcoholic extract, and another aqueous extract of *V. officinalis* (L.) were assessed for in vitro binding to rat GABA, benzodiazepine, and barbiturate receptors *(18)*. The results indicated that an interaction of some component of the hydroalcoholic extract, the aqueous extract derived from the hydroalcoholic extract, and the other aqueous extract had affinity for the $GABA_A$ receptor. Because hydroxyvalerenic acid (a volatile oil sesquiterpene) and dihydrovaltrate (a valepotriate) did not show any notable activity, the investigators could not identify the specific constituents responsible for this activity. The lipophilic extract derived from the hydroalcoholic extract, as well as dihydrovaltrate, showed affinity for barbiturate receptors, and some affinity for peripheral benzodiazepine receptors.

Other in vitro studies have also yielded results that suggest GABA-mediated activity; however, the active constituent was unidentified. Cavadas and colleagues verified that valerenic acid (0.1 mmol/L) was not able to displace [^{3}H] muscimol from the $GABA_A$ receptor, although both an aqueous and a hydroalcoholic extract were able to do so. The investigators then attempted to identify other compounds in the extracts capable of displacing [^{3}H] muscinol. Both glutamate and glutamine, amino acids present in the aqueous extract, had little inhibitory effect on [^{3}H] muscinol binding. However, glutamine can cross the blood-brain barrier (BBB) and can be taken up by nerve terminals and converted to GABA inside GABA-nergic neurons. Thus, glutamine could be responsible for the sedative effect of the aqueous extract, but not the hydroalcoholic extract, in which it is not present. GABA is found in both extracts, but GABA itself cannot explain the sedative effects of valerian because it is unlikely to cross the BBB in amounts significant enough to cause sedation *(19)*. However, the amount of GABA present in the aqueous extract is sufficient to have effects on peripheral GABA receptors, perhaps resulting in muscle relaxation *(20)*. Another study *(21)* suggests a different mechanism of action involving inhibition of neuronal GABA uptake and stimulation of GABA release from synaptosomes. These investigators did not attempt to elucidate which constituent of the aqueous extract was responsible for these effects.

The CNS-depressant component of valerian is still unknown. Thus far, three major constituents of valerian have been identified: the volatile or essential oil, containing sesquiterpenes and monoterpenes, nonglycosidic iridoid esters (valepotriates), and a small number of alkaloids *(2)*. Valepotriates are unstable compounds and are easily hydrolyzed by heat and moisture *(22)*. In addition, valepotriates are not water soluble, and aqueous extracts contain small amounts *(22)*. For example, the aqueous extract used in the study by Balderer and Borbely, described previously *(13)* was analyzed using thin-layer chromatography, and no valepotriates were detectable. Furthermore, valepotriates are not well absorbed orally *(23)*. Therefore, the likelihood that valepotriates are a major contributor to valerian's effects is questionable. Because of the low amount of alkaloid present in preparations, their contribution is also questionable *(24)*. It is postulated that a combination of volatile oils, valepotriates, and possibly certain water-soluble constituents that have not yet been identified are responsible for valerian's sedative effects *(23)*.

Antidepressant effects of valerian were identified by Oshima and associates using a methanol extract of *V. fauriei* roots *(25)*. They found a strong antidepressant activity in mice as measured by the forced swimming test. One active component isolated was α-kessyl alcohol, a volatile oil component. At 30 mg/kg intraperitoneally, α-kessyl alcohol exhibited an effect similar to imipramine, a commonly used antidepressant. Kessanol and cyclokessyl acetate, guaiane-type sesquiterpenoids, also exhibited antidepressant activity. Kanokonol, kessyl glycol, and kessyl glycol diacetate, valerane-type sesquiterpenoids, did not exhibit an effect.

A 30% ethanol extract of the Japanese valerian root ("Hokkai-Kisso") extract (4.1 g/kg and 5.7 g/kg) and imipramine (20 mg/kg) also demonstrated statistically significant antidepressant effects compared to placebo as measured by the forced swimming test in rats *(26)*. As in the Oshima study, kessyl glycol diacetate exhibited no antidepressant activity in the forced swimming test. Because the forced swimming test can be affected by stimulants, anticholinergics, and antihistamines as well as antidepressants, the effect of the valerian extract on reserpine-induced hypothermia, a test for antidepressant activity and inhibition of neuronal reuptake of monoamines, was measured. Both valerian (11.2 g/kg) and imipramine (20 mg/kg) reversed reserpine-induced hypothermia, suggesting that the antidepressant effect of valerian is caused by reuptake of monoamine neurotransmitters, as with conventional antidepressants.

More evidence is needed to evaluate the use of valerian in children. One study using a combination product of valerian root extract and lemon balm leaf extract found that symptoms of dyssomnia or pathological restlessness might decrease in children under age 12 *(27)*.

5.2. Anxiety

A few studies have examined the effects of valerian on anxiety *(28–30)*. Cropley and colleagues investigated whether kava or valerian could moderate physiological stress induced under laboratory conditions in healthy volunteers. Subject (n = 18-kava, and n = 18-valerian) and comparison group (n = 36) volunteers performed a standardized mental stress task 1 week apart. Cases had their blood pressure, heart rate, and subjective ratings of pressure assessed at rest and during the mental stress task (time 1 = *T1*). The valerian subjects took a standard dose for 7 days (time 2 = *T2*). In the valerian group, heart rate reaction to mental stress was found to decline, systolic blood pressure decreased significantly, and subjects reported less pressure during mental stress test tasks at *T2* relative to *T1*. Behavioral performance on the standardized mental stress test task did not change between the groups over the two time points. There were no significant differences in blood pressure, heart rate, or subjective reports of pressure between *T1* and *T2* in the control group.

Kohnen and Oswald conducted a study on the effects of valerian, propranolol, and combinations on activation, performance, and mood of healthy volunteers under social stressor conditions. The results of this study were equivocal and published over 15 years ago; however, it is mentioned here for historical reference *(29)*.

Andreatini and colleagues examined the effect of valerian extract (valepotriates) using a randomized, parallel, double-blind placebo-controlled pilot study design in patients with generalized anxiety disorder (GAD). After a 2-week wash-out period, 36 patients with GAD as defined by the *Diagnostic and Statistical Manual of Mental Disorders, Third Edition, Revised* were randomized to one of the following three treatment groups for 4 weeks: valepotriates, mean daily dose of 81.3 mg; diazepam, mean daily dose of 6.5 mg; or placebo. There was a significant reduction in the psychic factor of the Hamilton anxiety scale in the valepotriates group; however, the principal study analysis using between group comparisons on total Hamilton anxiety scale scores found negative results. The conclusion of this study suggests that there may be a potential anxiolytic effect of valepotriates on the psychic symptoms of anxiety, but the total number of subjects per group (n = 12) was very small and results must be viewed as preliminary *(30)*.

5.3. Musculoskeletal Relaxation

Isovaltrate and valtrate (valepotriates) and valeronone, an essential oil component, isolated from *V. edulis* ssp. procera Meyer (Valeriana "mexicana") caused suppression of rhythmic contractions in guinea pig ileum in vivo at a dose of 20 mg/kg administered intravenously via the jugular vein. The investigators also demonstrated that the same compounds as well as dihydrovaltrate isolated from the same valerian species produced relaxation of carbacholstimulated guinea pig ileum preparations in vitro. They concluded that these compounds have a musculotropic action in concentrations from 10^{-5} to 10^{-4} *M (31)*.

6. PHARMACOKINETICS

One study has evaluated the pharmacokinetics of valerian following administration to humans *(32)*. Following administration of a single 600-mg dose of valerian, the pharmacokinetics of valerenic acid were measured. The T_{max} occurred between 1 and 2 hours and the C_{max} was between 0.9 and 2.3 ng/mL. Concentrations of valerenic acid were measurable for at least 5 hours following the dose. The elimination half-life was approx 1 hour. The authors suggest that based on the expected use of valerian (sedative effects), dosing 30 minutes to 2 hours prior to bedtime would be appropriate based on the previously mentioned pharmacokinetics.

7. ADVERSE EFFECTS AND TOXICITY

7.1. Reproductive System

There has been a theoretical concern with regard to pregnant women taking valerian because of possible effects on uterine contractions *(1)*, but no problems were noted in three cases of intentional overdose with 2–5 g of valerian during weeks 3–10 of pregnancy *(33)*. A mentally retarded child was born to a woman who overdosed on valerian 3 g, phenobarbital, glutethamide, amobarbital, and promethazine at 20 weeks of gestation, but this same woman delivered a mentally retarded child 2 years later after an overdose attempt with glutethamide, amobarbital, and promethazine *(34)*.

V. officinalis (L.) was tested on rats and their offspring. A mixture, containing three valepotriates (80% dihydrovaltrate, 15% valtrate, and 5% acevaltrate), was orally administered to female rats for 30 days at 6-, 12-, and 24-mg/kg doses. Each dose was given to 10 rats, and placebo was given to another 10. No changes were noted in the average length of the estrus cycle, or the number of estrus phases during the 30-day observation period. The valepotriate mixture or placebo were also administered to 40 pregnant rats in the manner described previously from the day 1 through day 19 of pregnancy. Valerian did not increase the risk of fetotoxicity or external malformation. However, internal examination revealed a significant increase in the number of fetuses with retarded ossification with the 12- and 24-mg/kg doses. No developmental changes were detected in the offspring after treatment during pregnancy *(35)*.

7.2. Cardiovascular System

Pharmacological investigations using a particular valepotriate fraction called Vpt2 extracted from the roots of *V. officinalis* (L.) have shown antiar-rhythmic activity and ability to dilate coronary arteries in experimental animals. Moderate positive inotropic and a negative chronotropic effect were also observed. Vpt2 contains valtratum (50%), valeridine (25%), and valechlorin (3%), with trace amounts of acevaltrate, dihydrovaltratum, and epi-7-desacetyl-isovaltrate *(36)*.

Alcoholic extracts of *V. officinalis* (L.) root (labeled V103 and V115) demonstrated hypotensive effects in rats, cats, and dogs. The V115 fraction showed greater potency and was extracted by a countercurrent distribution to yield three fractions. The first two fractions demonstrated hypotensive effects in rats, with the first fraction showing a hypotensive effect at 30 mg/kg. The third fraction produced hypertensive effects at a dose of 200 mg/kg. The authors noted that, apparently, with each succeeding extraction, less of the hypotensive principle was extracted. The hypotensive effect of the V103 fraction in rats was demonstrated at a dose of 500 mg/kg, and was hypothesized to act via a parasympathomimetic effect, blockade of the carotid sinus reflex, and CNS depression *(37)*.

7.3. Cytotoxicity

The valepotriates valtrate/isovaltrate and dihydrovaltrate were isolated from *V. mexicana* and *V. wallichii*, respectively. The valepotriates tested were cytotoxic to granulocyte/macrophage colony-forming units (GM-CFCUs), lymphocytes, and erythrocyte colony-forming units (E-CFCUs). Valtrate was found to be a more potent inhibitor of GM-CFCUs (ID50 ~3.7×10^{-6} *M* vs ~1.7×10^{-5} *M*) and T-lymphocytes (ID50 ~2.8×10^{-6} *M* vs ~3×10^{-5} *M*) than dihydrovaltrate. Valtrate and dihydrovaltrate were similar in their activity against E-CFCUs (ID50 ~2.3×10^{-8} *M* vs ~4.2×10^{-8} *M*). Because pharmaceutical products containing valepotriates are orally administered, their cytotoxicity to gastrointestinal mucosal cells is of concern *(38)*.

The effects of valtrate, dihydrovaltrate, and deoxido-dihydrovaltrate, valepotriates extracted from *V. wallichii* (DC.), on cultured rat hepatoma cells have been studied. Valtrate killed 50% of the cell population at a concentration of 5 µM, Deoxido-dihydrovaltrate and dihydrovaltae demonstrated this same toxicity at double the dose. Valtrate was also the most potent inhibitor of DNA and protein synthesis *(39)*. These results suggest a mechanism by which valerian may cause hepatotoxicity.

7.4. Case Reports of Toxicity

Four cases of women who sustained liver damage after taking valerian-containing herbal medicines to relieve stress have been described *(40)*. In addition, valerian was used by a patient who exhibited hepatotoxicity attributed to Chaparral.

Hospitals admitted 23 patients for treatment of intentional overdose with Sleep-Qik® (75 mg of valerian dry extract, 0.25 mg of hyoscine hydrobromide 2 mg of cyproheptadine hydrochloride) between 1988 and 1991. Of these 23, 9 were men and 14 were women, with a mean age of 23.8 years (range 15–37 years). They were previously healthy, except for two patients with histories of psychiatric illness. The mean number of Sleep-Qik tablets taken per patient history was 33 (range 6–166), for an average of 2.5 g (range 0.5–12 g) of valerian. Four patients were asymptomatic. The other 19 patients reported drowsiness ($n = 11$), dilated pupils ($n = 11$), tachycardia ($n = 6$), nausea ($n = 4$), confusion ($n = 3$), urinary retention ($n = 3$), visual hallucination ($n = 2$), flushing ($n = 2$), dry mouth ($n = 1$), and dizziness ($n = 1$). Coingestants were alcohol ($n = 2$), a pesticide ($n = 1$), and Pansedan® ($n = 1$) (*Passiflora* extract, *Viscum album* extract, *Uncariarhyncophylla* extract, and *Humulus lupulus*). One patient who was drowsy had also taken Panseden, and one who was confused had ingested alcohol.

Most patients received gastric lavage ($n = 14$), and one received syrup of ipecac. The patient who took 60 tablets of Sleep-Qik required ventilatory support. Liver function tests were performed on 12 patients approx 6–12 hours after ingestion with normal results. Drowsiness and confusion resolved within 24 hours. All patients recovered completely and were discharged after an average of 1.7 days (range 1–6 days). At an average of 43 months (range 27–65 months) after presentation, 10 patients were contacted by telephone. They had all remained well after discharge and none continued taking Sleep-Qik. Delayed onset of severe liver damage was ruled out via telephone interview, but subclinical disease could not be ruled out *(41)*.

Subsequently, Chan reported on 24 cases of overdose of a product containing valerian dry extract 75 mg, hyoscine hydrobromide 0.25 mg, and cyproheptadine hydrochloride 2 mg. Six patients developed vomiting, and 15 underwent gastric lavage. Co-ingestants included alcohol ($n = 10$), cold products ($n = 3$), hypnotics ($n = 2$), unknown drugs ($n = 2$), and gasoline ($n = 1$). Symptoms were mainly CNS depression and anticholinergic symptoms. One patient required ventilatory support. Liver function tests were performed in 17 cases, and all were normal. Over the next 22–48 months postingestion, none of the patients returned to the hospital or clinic for any reason, suggesting that serious hepatotoxicity did not occur. The author points out that gastric lavage and spontaneous vomiting may have limited the amount of valerian absorbed in these patients, thus decreasing the risk of any delayed adverse effects *(42)*. Other adverse effects attributed to overdose or chronic use of valerian include headaches, excitability, restlessness, uneasiness, blurred vision, and cardiac disturbances *(4)*.

In another reported suicide attempt, an 18-year-old female ingested between 40 and 50, 470-mg capsules (18.8–23.5 g valerian) of 100% powered valerian root (Nature's Way®, Springville, UT). The patient complained of fatigue, crampy abdominal pain, chest tightness, tremor of the hands and feet, and lightheadedness 30 minutes after ingestion. She presented to the emergency room 3 hours postingestion. Her vital signs were: blood pressure 111/64 mmHg, pulse 72 beats/minute, respiratory rate 14 breaths/minute, and temperature 37.6°C. Physical exam was unremarkable except for mydriasis (6 mm bilaterally). Electrocardiograph, complete blood count, and chemistry profile including liver function tests were normal. Toxicology screen was positive for marijuana, which she admitted using 2 weeks previously. She denied ingesting anything else. After two doses of activated charcoal, her symptoms resolved within 24 hours *(43)*.

A withdrawal syndrome was described after abrupt discontinuation of valerian root extract in a 58-year-old man who had taken 530–2000 mg/dose five times daily as an anxiolytic and hypnotic for many years. Withdrawal symptoms included sinus tachycardia of up to 150 beats/minute, tremulousness, and delirium after recovery from general anesthesia (propofol, nitrous oxide, isoflurane, and thiopental) for open biopsy of a lung nodule. Medical history included coronary artery disease, hypertension, and congestive heart failure with an ejection fraction of 30–35%. Medications included isosorbide dinitrate, digoxin, furosemide, benazepril, aspirin, lovastatin, ibuprofen, potassium, zinc supplement, and vitamins. The biopsy was complicated by multiple episodes of oxygen desaturation, and after extubation, the patient experienced tacycardia, oliguria, and increasing oxygen requirement. Despite naloxone administration, symptoms worsened. Swan-Ganz catheterization revealed high-output heart failure. At this time, interview with family members revealed the patient's long-standing valerian use. Because valerian withdrawal was suspected, midazolam 1 mg/hour (total dose 11 mg in 17 hours) was administered. Signs and symptoms improved, and stabilized by the third postoperative day. He was switched to lorazepam 1 mg/hour as needed (total dose 5 mg in 24 hours), and then to a tapering dose of clonazepam. He was discharged on postoperative day 7, and was stable at 5-month follow-up. Other causes of high-output heart failure were ruled out, but because of the patient's multiple medical problems, postsurgical status, and medications administered, the cause of the patient's symptoms is unclear *(44)*. The authors of this case report note that valerian has been reported to attenuate benzodiazepine withdrawal in rats *(45)*.

8. INTERACTIONS

Two alcoholic valerian extracts were found to potentiate pentobarbital sleeping time in mice *(37)*, and Valdispert, an aqueous extract prepared from *V. officinalis* (L.), increased the thiopental sleeping time in a dose-dependent manner in rats *(16)*. Based on these animal studies, in vitro studies of valerian's effect on GABAnergic transmission, as well as the case series reported by Chan and colleagues, valerian would be expected to have at least an additive effect with barbiturates, alcohol, benzodiazepines, and other CNS depressants.

Valerian may have the potential to increase the level of drugs metabolized by the cytochrome P-450 3A4 (CYP3A4) enzyme. In vitro studies have found that valerian may have an inhibitory effect on CYP3A4. *(46,47)*. A clinical research study suggested that low to moderate doses of valerian did not significantly inhibit CYP3A4, although taking valerian extract 1000 mg/day increased alprazolam levels by 19% *(48)*. Therefore, it may be wise to use valerian cautiously in patients taking medications that are CYP3A4 substrates such as lovastatin, ketoconazole, itraconazole, fexofenadine, alprazolam, triazolam, and various chemotherapeutic agents.

See also Chapter 5, St. John's Wort, Section 8, Drug Interactions.

9. REPRODUCTION

No information is available concerning any potential effects of valerian on female reproductive function. However, Mkrtchyan and colleagues reported that valerian had no effect on human male sterility *(49)*.

10. REGULATORY STATUS

Valerian was included as an official drug in the US Pharmacopeia until 1936 and in the National Formulary until 1946. Currently, the USP advisory panel does not recommend valerian's use owing to lack of adequate scientific evidence and conflicting study results. They encourage further research *(4)*. Valerian is generally recognized as safe as a food and beverage flavoring by the FDA *(2)*. The German Commission Monograph E has approved valerian as a sleep-promoting and calmative agent to be used in the treatment of unrest and sleep disturbances caused by anxiety *(9)*. In Australia, valerian is acceptable as an active ingredient in the "listed products" category of the Therapeutic Goods Administration. In Belgium, subterranean parts, powder extract, and tincture are allowed for use as traditional tranquilizers. The Health Protection Branch of Health Canada allows products containing valerian as a single agent in the form of crude dried root in tablets, capsules, powders, extracts, tinctures, drops, or tea bags intended for use as sleeping aids and sedatives. In the United Kingdom, valerian is included on the General Sale Ledger.

REFERENCES

1. Combest WL. Valerian. US Pharmacist 1997;22:62–68.
2. Anonymous. Valerian. *Lawrence Review of Natural Products*. St. Louis: Facts and Comparisons, 1991.
3. Houghton P. Valerian. Pharm J 1994;253:95–96.
4. United States Pharmacopeial Convention (USP). Valerian. *Botanical Monograph Series*. Rockville: United States Pharmacopeial Convention, 1998.
5. Klepser TB, Klepser ME. Unsafe and potentially safe herbal therapies. Am J Health Syst Pharm 1999; 56:125–138.
6. Houghton PJ. The scientific basis for the reputed activity of valerian. J Pharm Pharmacol 1999;51:505–512.
7. Natural medicines comprehensive database online, 2004. http://www.naturaldatabase.com/. Date accessed: June 9, 2006.
8. Heiligenstein E, Guenther G. Over-the-counter psychotropics: a review of melatonin, St. Johns's wort, valerian, and kava-kava. J Am Col Health 1998;46:271–276.
9. Blumenthal M. Valerian root. *The complete German Commission E monographs*. Austin: American Botanical Council, 1998.
10. Donath F, Quispe S, Diefenbach K, et al. Critical evaluation of the effect of valerian on sleep structure and sleep quality. Pharmacopsychiatry 2000;33:47–53.
11. Leathwood PD, Chauffard F, Heck E, Munoz-Box R. Aqueous extract of valerian root (*Valeriana officinalis* L.) improves sleep quality in man. Pharmacol Biochem Behav 1982;17:65–71.
12. Leathwood PD, Chauffard F. Aqueous extract of valerian reduces latency to fall asleep in man. Planta Med 1985;2:144–148.
13. Balderer G, Borbely AA. Effect of valerian on human sleep. Psychopharmacology 1985;87:406–409.
14. Lindahl O, Lindwall L. Double blind study of a valerian preparation. Pharmacol Biochem Behav 1989;32:1065–1066.
15. Schulz H, Stolz C, Muller J. The effect of valerian extract on sleep polygraphy in poor sleepers: a pilot study. Pharmacopsychiatry 1994;27:147–151.
16. Leuschner J, Muller J, Rudmann M. Characterization of the central nervous depressant activity of a commercially available valerian root extract. Arzneim Forsch 1993;43:638–641.

17. Hendriks H, Bos R, Woerdenbag HJ, Koster AS. Central nervous depressant activity of valerenic acid in the mouse. Planta Med 1985;51:28–31.
18. Mennini P, Bernasconi P, Bombardelli E, Morazzoni P. In vitro study on the interaction of extracts and pure compounds from *Valeriana officinalis* roots with GABA, benzodiazepine and barbiturate receptors in rat brain. Fitoterapia 1993;64:291–300.
19. Cavadas C, Araujo I, Cotrim MD, et al. *In vitro* study on the interaction of *Valeriana officinalis* L. extracts and their amino acids on $GABA_A$ receptor in rat brain. Arzneim Forsch 1995;45:753–755.
20. Santos MS, Ferreira F, Faro C, et al. The amount of GABA present in aqueous extracts of valerian is sufficient to account for [^{3}H] GABA release in synaptosomes. Planta Med 1994;60:475–476.
21. Santos MS, Ferreira F, Cunha AP, Carvalho AP, Macedo T. An aqueous extract of valerian influences the transport of GABA in synaptosomes. Planta Med 1994;60:278–279.
22. Wagner J, Wagner ML, Hening WA, Beyond benzodiazepines: alternative pharmacologic agents for the treatment of insomnia. Ann Pharmacother 1998;32:680–691.
23. Tyler VE, ed. *The Honest Herbal, 3rd edition*, Binghamton: Pharmaceutical Products Press, 1993.
24. Houghton P. The biological activity of valerian and related plants. J Ethnopharmacol 1988;22:121–142.
25. Oshima Y, Matsuoka S, Ohizumi Y. Antidepressant principles of *Valeriana fauriei* roots. Chem Pharmacol Bull 1995;43:169–170.
26. Sakamoto T, Mitani Y, Nakajima K. Psychotropic effevts of Japanese valerian root extract. Chem Pharmacol Bull 1992;40:758–761.
27. Muller SF, Klement S. A combination of valerian and umun balm is effective in the treatment of restlessness and dyssomnia in children. Phytomedicine 2006;13:383–387.
28. Cropley M, Cave Z, Ellis J, Middleton RW. Effect of kava and valerian on human physiological and psychological responses to mental stress assessed under laboratory conditions. Phytother Res 2002;16:23–27.
29. Kohnen R, Oswald WD. Effects of valerian, propranolol and combinations on activation, performance, and mood of healthy volunteers under social stress conditions. Pharmacopsychiatry 1988;21:477–478.
30. Andreatini R, Sartori VA, Seabra ML, Leite JR. Effect of valepotriates (valerian extract) in generalized anxiety disorder: a randomized placebo-controlled pilot study. Phytother Res 2002;16:650–654.
31. Hazelhoff B, Malingre TM, Meijer DKF. Antispasmodic effects of valeriana compounds: and in vivo and in vitro study on the guinea-pig ileum. Arch Int Pharmacodyn 1982;257:274–287.
32. Anderson GD, Elmer GW, Kantor ED, Templeton IE, Vitiello MV. Pharmacokinetics of valerenic acid after administration of valerian in healthy subjects. Phytother Res 2005;19:801–803.
33. Czeizel AE, Tomcsik M, Timar L. Teratologic evaluation of 178 infants born to mothers who attempted suicide by drugs during pregnancy. Obstet Gynecol 1997;90:195–201.

34. Czeizel A, Szentesi I, Szekeres H, Molnar G, Glauber A, Bucski P. A study of adverse effects on the progeny after intoxication during pregnancy. Ach Toxicol 1988;62:1–7.
35. Tufik S, Fujita K, Seabra MDV, Leticia LL. Effects of a prolonged administration of valepotriates in rats on the mothers and their offspring. J Ethnopharmacol 1994;41:39–44.
36. Petkov V. Plants with hypotensive, antiatheromatous and coronarodilatating action. Am J Chin Med 1979;7:197–236.
37. Rosecrans Ja, Defeo JJ, Youngken HW. Pharmacological investigation of certain *Valeriana officinalis* L. extracts. J Pharmaceut Sci 1961;50:240–244.
38. Tortarolo M, Braun R, Hubner GE, Maurer HR. In vitro effects of epoxide-bearing valepotriates on mouse early hematopoietic progenitor cells and human T-lymphocytes. Arch Toxicol 1982;51:37–42.
39. Bounthanh C, Richert L, Beck JP, Haag-Berrurier M, Anton R. The action of valepotriates on the synthesis of DNA and proteins of cultured hepatoma cells. Planta Med 1983;49:138–142.
40. MacGregor FB, Abernethy VE, Dahabra S, Cobden I, Hayes PC. Hepatotoxicity of herbal remedies. Br Med J 1989;299:1156–1157.
41. Chan TYK, Tang CH, Crichley J. Poisoning due to an over-the-counter hypnotic, Sleep-Qik (hyoscine, cyproheptadine, valerian). Postgrad Med J 1995;71:227–228.
42. Chan TYK. An assessment of the delayed effects associated with valerian overdose [letter]. Int J Clin Pharmacol Ther 1999;36:569.
43. Willey LB, Mady SP, Cobaugh DG, Wax PM. Valerian overdose: a case report. Vet Hum Toxicol 1995;37:364–365.
44. Garges HP, Varia I, Doraiswamy PM. Cardiac complications and delirium associated with valerian root withdrawal [letter]. JAMA 1998;280:1566–1567.
45. Andreatini R, Loire JR. Effect of valepotriates on the behavior of rats in the elevated plus-maze during diazepam withdrawal. Eur J Pharmacol 1994;260:233–235.
46. Budzinski JW, Foster BC, Vandenhoek S, et al. An in vitro evaluation of human cytochrome P450 3A4 inhibition by selected commerical herbal extracts and tinctures. Phytomedicine 2000; 7:237–282.
47. Lefebrve T, Foster BC, Drouin CE, Livesey JF, Jordan SA. In vitro activity of commerical valerian root extracts against cytochrome P450 3A4. J Pharm Pharmaceuti Sci 2004; 7:265–273.
48. Donovan JL, DeVane CL, Chavin KD, et al. Multiple night-times doses of valerian (Valeriana officianalis) had minimal effects of Cyp3A4 activity and no effect on CYP2D6 activity in healthy volunteers. Drug Metab Dispos 2004; 32:1333–1336.
49. Mkrtchyan A, Panosyan V, Panossian A, Wikman G, Wagner H. A phase I clinical study of Andrographis paniculata fixed combination Kan Jang versus ginseng and valerian on the semen quality of healthy male subjects. Phytomedicine 2005;12:403–409.

Chapter 5

St. John's Wort

Dean Filandrinos, Thomas R. Yentsch,
and Katie L. Meyers

SUMMARY

St. John's wort has demonstrated clinical efficacy for mild to moderate depression and compares favorably to other more potent or toxic antidepressants. Low side effects and potential benefits warrant its use as a first-line agent for select patients with mild to moderate depression or anxiety-related conditions. Benefits related to other reported uses such as an antimicrobial, agent to treat neuropathic pain, antiinflammatory, treatment alternative for atopic dermatitis, and antioxidant are either not well documented or evidence is encouraging but not conclusive and further study is needed. St. John's wort has an inherently wide margin of safety when taken by itself, with most reported adverse drug reactions (ADRs) being related to skin reactions. Isolated, but more significant ADRs have been reported in relation to neurological effects, impact on thyroid function, and increased prothrombin time. Of greatest concern is the potential for interactions between St. John's wort and mainstream pharmaceuticals through induction of cytochrome P450. Patients on concomitant treatment with drugs metabolized through this pathway should be monitored closely for altered drug effect.

Key Words: *Hypericum perforatum*; hypericin; hyperforin; antidepressant; anxiolytic; P450 enzyme induction.

1. HISTORY

Hypericum perforatum is Greek for "over an apparition." It was believed that evil spirits disliked the plant's odor and thus could be warded away *(1)*. *Hypericum* is a perennial aromatic shrub with bright yellow flowers that bloom from June to September *(2)*. The flowers are said to be at their brightest and most abundant around June 24th, the day traditionally believed to be the birth-

From Forensic Science and Medicine:
Herbal Products: Toxicology and Clinical Pharmacology, Second Edition
Edited by: T. S. Tracy and R. L. Kingston © Humana Press Inc., Totowa, NJ

day of John the Baptist *(3)*. Also, the red spots on the leaves are symbolic of the blood of St. John *(1)*. The plant is native to Europe, North Africa, and West Asia, and now also found in Australia, North and South America, and South Africa *(2)*. It grows in the dry ground of fields, roadsides, and woods. The commercial products are prepared from the dried flowering tops and leaves that are harvested just before or during the flowering period *(2)*.

St. John's wort has been described in medical literature for thousands of years, including the writings of Hippocrates *(4)*. Historically, St. John's wort has been used to treat neurological and psychiatric disturbances (anxiety, insomnia, bed-wetting, irritability, migraine, excitability, exhaustion, fibrositis, hysteria, neuralgia, and sciatica), gastritis, gout, hemorrhage, pulmonary disorders, and rheumatism, and has been used as a diuretic *(2)*. Some forms of the herb have been used topically as an astringent and to treat blisters, burns, cuts, hemorrhoids, vitiligo, neuralgias, inflammation, insect bites, itching, redness, sunburn, and wounds. Oral doses of 300 mg of *Hypericum* extract three times daily for periods of 4 to 6 weeks is a typical dosing regimen *(2)*.

2. CURRENT PROMOTED USES

St. John's wort is used most often for the treatment of mild to moderate depression. It is also used to treat anxiety, sleep disorders, seasonal affective disorders (SADs), and wound healing *(1,4,5)*.

3. SOURCES AND CHEMICAL COMPOSITION

The botanical name of St. John's wort is *H. perforatum*. Other common names by which it is known are goat weed, klamath weed, rosin rose, amber touch and heal, tipton weed *(1)*; blutdkraut, Johnswort, qian ceng lou, Sankt Hans urt, St. Jan's kraut, St. Johnswort, toutsaine, tupfelhartheu, walpurgiskraut, zweiroboij, amber, chassediable, corazoncillo, hardhay, hartheu, herbe de millepertuis, herrgottsblut, hexenkraut, hierba de San Juan, hipericon, hypericum, iperico, Johannesort, pelatro, perforata, Johannisblut, Johanniskraut *(2)*.

4. PRODUCTS AVAILABLE

Most commercially available preparations of hypericum in the United States are dried alcoholic extracts in a solid oral dosage form. Other preparations include the dried herb, teas, tinctures or liquid extracts *(2)*. The following is a list of a few of the available formulations:

- GNC Herbal Plus® — 1000-mg softgel capsule concentrated St. John's Wort
- GNC Herbal Plus — 300-mg tablet standardized St. John's Wort (0.3% hypericin, 0.9 mg)

- GNC Herbal Plus — 500-mg capsule fingerprinted St. John's Wort
- Nature's Resource® — 450-mg capsule time released St. John's Wort (1.35 mg hypericin)
- Nature's Way® — Capsule, Mood Aid, with 5-HTP and St. John's Wort
- Nature's Way — Tablet, Perika St. John's Wort (3% hyperforin)
- Nature's Way — Capsule, standardized St. John's Wort (0.3% hypericin)
- Enzymatic Therapy® — Capsule, St. John's Wort (0.3% hypericin, 3% hyperforin)
- Herbs for Kids® — St. John's Wort Blend (alcohol extract, evaporated)
- Nature's Answer® — St. John's Wort (organic alcohol)
- Natrol® — 300-mg tablet (0.3% hypericin)
- NSI® — 450-mg capsule (0.3% hypericin, 3% hyperforin)
- Only Natural® — 450 mg (0.3% hypericin)
- Smart Basic® — 300 mg St. John's Wort
- Source Naturals® — 300 mg, 450-mg tablet (0.3% hypericin)
- Karuna® — 300-mg capsule (0.3% hypericin)
- Remotiv® — Tablet (1.375 g hypercium perforatum herb, 500 µg hypericin)
- NOW® — 300-mg capsule St. John's Wort
- Kira® — 300-mg capsule St. John's Wort extract

5. PHARMACOLOGICAL/TOXICOLOGICAL EFFECTS

The active component of St. John's wort is not known. It is composed of many different compounds. The concentrations of these chemicals vary from brand to brand and batch to batch. Hyperforin, hypericin, and pseudohypericin are considered by most to be the major active ingredients. Hypericin, pseudohypericin, isohypericin, protohypericin, protopseudohypericin, and cyclopseudohypericin are all anthraquinone derivatives (naphthodianthrones) *(1–5)*. Hyperforin and adhyperforin are both prenylated phorolucinols *(2,3)*. The flavonoids that are present include kaempferol, quercetin, luteolin, hyperoside, isoquercitrin, quercitrin, rutin, hyperin, hyperoside, I3-II8-biapigenin, 1,3,6,7-tetrhydroxyxanthone, and amentoflavone *(2,3,5)*. The phenols consist of caffeic, chlorogenic, p-coumaric, ferulic, p-hydroxy-benzoic, and vanillic acids *(3)*. The volatile oils include methyl-2-octane, n-nonane, methyl-2-decane and n-undecane, α- and β-pinene, α-terpineol, geraniol, myrcene, limonene, caryophyllen, and humulene *(3)*. Other chemicals that are found in St. John's wort include tannins, organic acids (isovalerianic, nicotinic, myristic, palmitic, stearic), carotenoids, choline, nicotinamide, pectin, β-sitosterol, straight-chain saturated hydrocarbons, and alcohols *(3)*. Most of these agents are found in other plants that do not possess antidepressant activity, which has led to most of the research being concentrated on the naphthodianthrones and hyperforin, which are only found in a few species.

5.1. Neurological Effects

There have been many clinical trials studying the effectiveness of St. John's wort in the treatment of depression. By the spring of 2002, there were 34 controlled trials including more than 3000 patients. Most of these trials included patients with mild to moderate depression and used the Hamilton Rating Scale of Depression (HAMD) to measure efficacy *(6)*. Schulz compared the results of all of the trials since 1990. Nine of the 11 placebo-controlled trials showed a significant difference in the HAMD scores favoring hypericum, and a trend favoring hypericum was demonstrated in the other two. Linde also compared clinical trials with hypericum, and came to the conclusion that hypericum was superior to placebo in mild to moderate depression *(7)*. When compared with the synthetic antidepressants, there was one trial with amitriptyline, four with imiprimine, two with fluoxetine, two with sertraline, one with bromazepam, and one with maprotiline. Of these trials, hypericum was equal to or superior to all of them except amitriptyline *(6)*. In two trials comparing hypericum in major depression (participants with HAMD scores of at least 20), hypericum failed to show any improvement over placebo or other antidepressants *(8,9)*. One randomized, controlled, double-blind, noninferiority trial that has been published since the Schulz review compared 900 mg/day of St. John's wort vs 20 mg/day of paroxetine in adult patients with acute major depression (HAMD score ≥22). Using the HAMD to assess efficacy, it was found that St. John's wort was at least as effective as paroxetine and was better tolerated *(10)*.

Although the results from trials in patients with mild to moderate depression appear encouraging in their support of St. John's wort, there are limitations. First, the longest duration of these trials was 56 days and several of the trials were as short as 28 days. Also, most of the trials used relatively low doses of synthetic antidepressants. Finally, two of the trials did not state the exact number of responders, making the results somewhat questionable.

The exact mechanism of action responsible for St. John's wort's neurological effects is not known. Additionally, it is not known if any one chemical constituent is responsible for its activity or if it is a combination of multiple components. It is known that the extracts of *H. perforatum* appear to inhibit the synaptic uptake of several neurotransmitters including norepinephrine, serotonin (5-HT), and dopamine *(3,11–13)*. Rats that were fed high doses of hypericum extracts standardized to flavonoids (50%), hypericin (0.3%), and hyperforin (4.5%) were shown to have dose-dependent enhanced 5-HT levels in all brain regions. Norepinephrine levels were increased in the diencephalon and brain stem, but not in the cortex, and higher levels of hypericum were needed. Dopamine levels were only increased in the diencephalon region with doses similar to those required for increased levels of norepinephrine *(12)*. Cott demonstrated that hypericum extracts had affinity for adenosine, γ-aminobutyric acid (GABA)-A, GABA-B, benzodiazapine, and monoamine oxidase (MAO) types A and B receptors *(3)*. However, with the exception of GABA-A and GABA-B receptors, it is unlikely that the concentrations required to produce a physiological effect can be reached *(3)*. Other studies have shown that hypericum extracts do not have high affinity for GABA-A and -B *(5)*. Additionally, *H. perforatum* extracts downregulate β receptors and upregulate 5-HT2 receptors in the frontal cortex when given to rats *(3,11)*. *Hypericum* extract standardized to flavonoids (50%), hypericin (0.3%), and hyperforin (4.5%) was shown to inhibit the release of interleukin-6 (IL-6) in vitro *(14,15)*. IL-6 levels have been shown to be increased in patients with depression *(14)*. It is thought that IL-6 induced stimulation of corticotropin-releasing hormone, adrenocorticotropic hormone, or cortisol may be responsible for increased depression *(5)*. Also, using the rat forced-swimming model for depression, high doses of the extract was shown to improve depression in wild-type rats but had no effect in rats that were IL-6 knockouts (IL-6 –/–). The wild-type mice had a significantly greater increase in 5-HT in the diencephalon portion of the brain compared to the IL-6 –/– mice. This finding indicates that IL-6 may be necessary to have an antidepressant response to hypericum *(14)*. *Hypericum* extract was shown to inhibit the enzyme dopamine-β-hydroxylase (DβII), and its inhibition is 200 times stronger than the inhibition by pure hypericin, suggesting that hypericin is not the component responsible. DβH is the enzyme that catalyzes the conversion of dopamine into norepinephrine; thus St. John's wort may increase dopamine levels in the brain while lowering norepinephrine *(16)*. High concentrations of hypericum extracts inhibit catechol-O-methyltransferase (COMT) activity *(5)*. Consumption of a single dose of 2700 mg of St. John's wort extract was found to significantly increase plasma growth-hormone levels and decrease prolactin levels in human males *(3)*.

It was once thought that hypericin was the main active ingredient in St. John's wort. In 1994, it was reported that hypericin inhibited MAO-A *(11)*. Further studies have shown that hypericin and pseudohypericin do not inhibit MAO-A, and hypericum extracts only inhibit MAO at extremely high concentrations *(5)*. Furthermore, hypericin did not display a significant (>25%)

inhibition of norepinephrine, dopamine, or 5-HT uptake sites, nor did it display high affinity to 5-HT, adenosine, adrenergic, benzodiazepine, dopamine, or GABA receptors *(5,13)*. Hypericin was found to have high (>30%) levels of inhibition of nonselective muscarinic cholinergic receptors, 5-HT1A receptors and nonselective σ receptors *(5,17)*. Butterweck also has shown that hypericin and pseudohypericin have significant activity at D3- and D4-dopamine receptors, and that hypericin has significant activity at β-adrenergic receptors *(18)*.

Most evidence now implicates hyperforin as the main component responsible for the neurological activity of St. John's wort. Hyperforin has been shown to inhibit synaptic reuptake of 5-HT, dopamine, norepinephrine, GABA, l-glutamate, and acetylcholine *(3,11,13,19)*. It is a potent uptake inhibitor of 5-HT, dopamine, norepinephrine, and GABA with 50% inhibition concentrations (IC50) of approx 0.05–2 μg/mL *(13)*. It has been shown that hyperforin increases the intracellular level of sodium, which may be directly responsible for its effect on 5-HT reuptake *(11)*. Hyperforin was also shown to strongly inhibit D1- and D5-dopamine receptors and weakly inhibit binding to the opioid receptor hδ *(18)*. Adhyperforin exists as a component in St. John's wort in approx one-tenth the concentration of hyperforin; but it, too, was found to be a potent uptake inhibitor of 5-HT, dopamine, and norepinephrine at lower IC50 values. Another factor that supports hyperforin's role as the active ingredient is that it is the major lipophilic constituent in hypericum extract, allowing it to cross the blood–brain barrier more easily *(20)*.

Pseudohypericin has been shown to be a corticotropin-releasing factor (CRF)$_1$ receptor antagonist. CRF has been implicated as a pathogenic factor in affective disorders, with elevated levels that are normalized after treatment with antidepressants found in the cerebrospinal fluid of patients with depression. CRF acts on CRF$_1$ receptors in the pituitary gland to stimulate the release of adrenocoticotropic hormone, which stimulates the release of glucocorticoid stress hormones from the adrenal glands *(19)*. It is possible that St. John's wort's activity comes from pseudohypericin's ability to block the CRF$_1$ receptor.

Amentoflavone is a biflavonoid with some pharmacological activity that may contribute to the activity of St. John's wort. It was found to significantly inhibit binding at 5-HT1d, 5-HT2c, D3-dopamine, δ-opiate, and benzodiazepine receptors *(18)*.

Along with depression, hypericum has been studied in SAD. There have been two studies in which participants received 300 mg St. John's wort three times daily with or without bright-light therapy either for 4 or 8 weeks. In both studies there were significant reductions in HAMD scores or SAD scores, but no statistically significant difference in scores between the groups that received light therapy and those that did not *(3)*.

A study using rats to measure the anxiolytic activity of *Hypericum* was conducted with efficacy being measured by several means. In this study, rats were given either *Hypericum* extract 100 or 200 mg/kg or lorazepam 0.5 mg/kg. Using the open-field observation test and the maze test to measure anxiety, the *Hypericum* treatment groups showed anxiolytic efficacy and were superior to placebo, whereas the lorazepam was either equivalent to or superior to the *Hypericum* groups. With respect to social interaction, both treatment groups with *Hypericum* increased the amount of time the animals spent in social interactions with respect to control animals. Lorazepam-treated rats were comparable to the higher dose *Hypericum* group *(21)*.

Obsessive-compulsive disorder (OCD) is a neurological disorder that affects 1.2–2.4% of the population *(22)*. Drugs that inhibit 5-HT uptake are often used to treat OCD, with limited results. In a 12-week, open-label study, 13 people were treated with 450 mg of *Hypericum* standardized to 0.3% hypericin twice daily. Efficacy was measured by the Yale-Brown Obsessive Compulsive Scale (Y-BOCS). Of the 13 members, 12 completed the trial, with an average reduction in their Y-BOCS scores of 7.42 from baseline, which is comparable to the results in studies using antidepressants. Additionally, five out of 12 patients rated themselves as much or very much improved, six out of 12 were minimally improved, and one noted no change. Interestingly, although the patients' average HAMD scores were a subclinical 6.09 at baseline, they dropped significantly to 1.91 at the end of the study *(22)*.

Hypericum has also been studied for its effect on sleep. In a small trial, 14 females were given 300 mg *Hypericum* three times daily for 4 weeks, given a 2-week washout period, then given placebo for 4 weeks. The continuity of sleep, onset of sleep, intermittent wake-up phases, and total sleep were not improved. There was, however, a significant increase in deep sleep (stage 3 and 4, slow wave) that was shown by analysis of electroencephalogram activities *(23)*. Thus, *Hypericum* may be able to improve sleep quality.

Somatoform disorders are a group of diseases that include the complaint of physical pain, which lead the patient to believe they have a physical disease, though none can be found by medical investigation *(24)*. A study was conducted where 151 patients received either *Hypericum* 300 mg twice daily or placebo. Efficacy was measured using the Hamilton Anxiety Scale, subfactor somatic anxiety (HAMA-SOM). After 6 weeks, the average HAMA-SOM decreased from 15.39 to 6.64 in the *Hypericum* arm and from 15.55 to 11.97 in the placebo arm, which was statistically significant, demonstrating the superiority of St. John's wort over placebo *(24)*.

5.2. Antimicrobial Effects

St. John's wort has been used topically for wound healing for hundreds of years. Antibacterial properties have been reported as early as 1959, with hyperforin found to be the active component. Using multiple concentrations, it was discovered that no hyperforin dilutions had antimicrobial effects on Gram-negative bacteria or *Candida albicans*. There was, however, growth inhibition for all of the Gram-positive bacteria tested, some with the lowest dilution concentration of 0.1 µg/mL. Hyperforin was also shown to be effective at inhibiting methacillin-resistant *Staphylococcus aureus (25)*.

Along with antibacterial properties, it has also been reported that both the hypericin and pseudohypericin components of St. John's wort have antiviral properties *(2,26)*. In vitro studies showed antiviral activity against cytomegalovirus, *Herpes simplex*, human immunodeficiency virus (HIV) type I. Influenza virus A, moloney murine leukemia virus, and sindbis virus *(2,3)*. Hypericin and pseudohypericin are thought to work by inhibiting viral replication via disruption of the assembling and processing of intact virions from infected cells. Mice coinjected with 150 µg of hypericin/pseudohypericin and the Friend virus had a 100% survival rate at 240 days whereas all control mice that were only injected with the virus were dead by day 23. Animals treated with lower doses (10 and 50 µg) were also protected, but not to the same degree *(27)*. In an in vitro study, HeLa cells carrying HIVcat transcriptional units were incubated at concentrations of 25, 50, and 100 µg/mL of hypericin and 100 µg/mL of *Ginkgo biloba*, then exposed to ultraviolet (UV) light. It was found that hypericin inhibited the UV-induced HIV gene expression by 50, 81, and 88% correlating with the 25, 50 100 µg/mL concentrations when compared to control cells (g2). *G. biloba* inhibited the UV-induced HIV gene expression by 19% *(28)*. The first study with St. John's wort in people with HIV was halted because of phototoxicity; further studies are on the way *(2)*. When the first study in patients with AIDS was stopped, no significant improvements were seen in CD4 counts, HIV titer, HIV-RNA copies, or HIV p24 antigen levels *(29)*. Flavonoid and catechins in St. John's wort have been shown to have some activity against influenza virus *(3)*.

5.3. Mutagenicity

Quercetin, a flavonoid component of St. John's wort and several other medicinal plants, has been implicated as a mutagen. However, St. John's wort aqueous ethanolic extract showed no mutagenic effects in mammalian cells. Tests used included the HGPRT (hypoxanthine guanidine phosphoribosyl transferase) test, the UDS (unscheduled DNA synthesis) test, the cell transformation test using Syrian hamster embryo cells, the mouse-fur spot test, and the chromosome aberration test using Chinese hamster bone marrow cells *(30)*.

5.4. Neuropathic Pain

Neuropathic pain is commonly treated with tricyclic antidepressants. Although generally efficacious, these drugs do have the potential to cause serious side effects. A crossover trial was conducted in which participants received St. John's wort standardized to 2700 μg of hypericin per day or placebo for 5 weeks, with a 1-week washout period between treatments. Patients rated several types of pain on a scale of 1–10. A total of 47 patients completed the trial, which showed a trend toward lower total pain with the St. John's wort treatment; however, it was not statistically significant. There was also a trend toward people reporting moderate to complete pain relief during their treatment with St. John's wort. When the study population was further broken down into patients with and without diabetes, it was found that in the 18 participants with diabetes, there was still a trend toward lower total pain and a significant reduction in lancinating pain, whereas in the 29 participants without diabetes, there was no significant differences or trends in any pain scores. Interestingly, 25 participants preferred the St. John's wort treatment arm, 16 preferred placebo, and six did not have a preference *(31)*.

5.5. Inflammation/Asthma

There are many articles that address the role of St. John's wort in inflammation. As previously mentioned, St. John's wort is an inhibitor of IL-6, which is an important cytokine involved in inflammation *(14,15)*. Additionally, hyperforin was found to inhibit cyclooxygenase (COX)-1 and 5-lipoxygenase (5-LO), key enzymes in the formation of proinflammatory eicosanoids. Moreover, it inhibited both enzymes at IC50 concentrations of 0.09 to 3 μM, which is close to the plasma concentrations achieved with standard dosing. Hyperforin was three times more potent then aspirin in its ability to inhibit COX-1 and almost equipotent to zileuton in its ability to inhibit 5-LO. Hyperforin did not significantly inhibit COX-2, 12-LO, or 15-LO enzymes *(32)*. St. John's wort's ability to act as a 5-LO inhibitor could lead to a future role in asthma.

Another way that St. John's wort may reduce inflammation is by reducing inducible nitric oxide synthase (iNOS), which is increased in the early phases of inflammation. Nuclear factor-κB (NF-κB) and signal transducer and activator of transcription-1α (STAT-1α) are both implicated in inducing iNOS leading to the production of nitric oxide (NO), which is produced in large amounts near areas of inflammation. St. John's wort was found to inhibit STAT-1α, thereby reducing both iNOS and NO formation. Surprisingly, St. John's wort did not inhibit NF-κB, which was shown in earlier reports to be inhibited by quercetin, a component of St. John's wort *(33)*.

5.6. Atopic Dermatitis

After it was found that St. John's wort, and more specifically hyperforin, has an inhibitory effect on epidermal langerhan cells, there was speculation that it may treat atopic dermatitis. A 4-week trial was conducted in which 21 patients with mild to moderate atopic dermatitis were treated twice daily with a cream standardized to 1.5% hyperforin on one side of their body and placebo on the other side. The primary end point of the study was severity scoring of atopic dermatitis (SCORAD) index, based on extent and intensity of erythema, papulation, crust, excoriation, lichenification, and scaling. Among the 18 participants that completed the study, the SCORAD index fell from a baseline score of 44.9 to 23.9 in the hyperforin group. The SCORAD index also fell from 43.9 to 33.6 in the placebo group. These results show statistically significant superiority of hyperforin cream over placebo, with no difference in skin tolerance to the two treatments. Of note, a secondary end point of the study showed a reduction of skin colonization with *S. aureus* with both hyperforin and placebo, with a trend toward better antibacterial activity with hyperforin cream *(34)*. Although these results are positive, further studies should be conducted comparing hyperforin to corticosteroids in the treatment of atopic dermatitis.

5.7. Antioxidant

Free radicals are highly reactive molecules that have been implicated in cardiovascular and neurodegenerative disease. Hunt et al. generated superoxide radicals in both cell-free and human placental tissue to determine if St. John's wort has antioxidant qualities. They then tested St. John's wort samples that were standardized to either hypericin or hyperforin. In cell-free studies, both samples had a prooxidant effect at a 1:1 concentration. Both showed an inverse dose-related relationship in their antioxidant effect at concentrations from 1:2.5 to 1:20, with 1:20 having the greatest antioxidant effect in both groups. St. John's wort standardized to hypericin was superior in its antioxidant properties compared with hyperforin. Both were shown to be significant antioxidants in human placental vein tissue at a 1:20 dilution, the only concentration tested owing to results in the cell-free experiments.

5.8. Premenstrual Syndrome

There are numerous accounts of anecdotal evidence supporting the use of St. John's wort for premenstrual syndrome (PMS) *(36)*. One open, uncontrolled study was conducted to determine the efficacy of St. John's wort in treating PMS. The primary outcome was measured by a daily symptom checklist of 17 symptoms rated on a scale of 0 to 4 based on the Hospital Anxiety and Depression (HAD) scale and modified Social Adjustment Scale (SAS-M) broken down into four subscales: mood, behavior, pain, and physical. A total of 25 women were selected to participate in the study in which they received 300 mg hypericum standardized to 900 μg hypericin daily. The results from the daily symptoms survey after the first cycle show a statistically significant reduction from the baseline value of 128.42 to 70.11. After the second cycle, there was a further reduction to 42.74. Of the four subscales, St. John's wort had the greatest improvement on the mood subscale (57%) and the least improvement on the physical subscale (35%). Of the individual symptoms, crying (92%) and depression (85%) were improved the most with treatment, and food cravings and headaches were improved the least *(36)*.

5.9. Tumor Growth Inhibition

Hypericin, as mentioned earlier, is a fluorescent photosensitizer. When subjected to UV light, hypericin produces singlet oxygen, a nonradical oxygen species, which is highly reactive and cytoxic *(37)*. In vitro studies with hypericin have shown significant growth inhibition in various human malignant cells, and in vivo studies demonstrate that it accumulates in bladder tumor cells when injected intravesically *(37,38)*. In vivo studies were conducted in rats with transitional cell carcinoma of the bladder. Rats who were given hypericin IV and then photo-irradiated had their tumors eliminated in 15 days. Further studies could demonstrate success in human models with lower toxicity than current therapies such as bacillus Calmette-Guerin immunotherapy *(37)*. It is also thought that other agents present in St. John's wort have cytotoxic qualities. When human erythroleukemic cells (K562) (human chronic myelogenous leukemia) were incubated with purified hypericin in the dark, there was only a weak inhibitory effect on cell growth and no apoptotic effect *(39)*. When K562 cells were incubated with various extracts of *Hypericum*, there was significantly more growth inhibition and apoptosis *(38,39)*. Extracts of *Hypericum* that were high in flavonoid content and low in hyperforin content had significantly greater growth-inhibitory activity compared to extracts with similar hypericin content, but low flavonoid and high hyperforin content. This supports earlier work that flavonoids have antiproliferative effects on malignant cell lines. Additionally, when the same extracts that were incubated in the dark were incubated with 7.5 J/cm^2 light activation, the IC50 values were lowered by roughly half, further demonstrating the phototoxic effects of hypericin *(38)*. The mechanism of action of hypericin appears to be a combination of inhibition of protein kinase C, free-radical induction, release of mithochondrial cytochrome-c, and the activation of procaspase-3 *(39,40)*. The flavonoids also appear to increase caspase activity and release of cytochrome-c *(39)*.

6. PHARMACOKINETICS/TOXICOKINETICS

Two pharmacokinetic studies have examined the pharmacokinetics of hypericin and pseudohypericin *(41,42)*. Standardized hypericum extract LI 160 (Jarsin 300®, Lichtwer Pharma GmbH, Berlin) was used in both trials. In Part I of the studies, subjects in both trials were administered a single dose of either 300, 900, or 1800 mg of the extract (one, three, or six coated tablets) at 10- to 14-day intervals. Each dose contained 250, 750, or 1500 µg of hypericin and 526, 1578, or 3156 µg of pseudohypericin, respectively. The doses were administered on an empty stomach in the morning after a 12-hour fast. Subjects fasted for an additional 2 hours after administration. Multiple plasma levels of hypericin and pseudohypericin were measured for up to 120 hours after administration. In addition, urine samples were collected in the study performed by Kerb and colleagues. After a 4-week washout from Part I, subjects were given one coated tablet containing 300 mg of hypericum extract three times a day (8 AM, 1 PM, and 6 PM) before meals for 14 days. Blood samples were obtained over the 2-week dosing period.

6.1. Absorption

For single doses of 300, 900, or 1800 mg of dried hypericum extract in humans, the median time between administration of the dose and detectable plasma concentration (t_{lag}) in hours were as follows:

- Hypericin: 2.6, 2.0, and 2.6 *(41)*
- 2.1, 1.9, and 1.9 *(42)*
- Pseudohypericin: 0.6, 0.4, and 0.4 *(41)*
- 0.5, 0.4, and 0.4 *(42)*

A difference was observed between the t_{lag} of hypericin compared with pseudohypericin. These differences may be a function of the dosage form given. Pseudohypericin may be released from the dosage form more quickly than hypericin. Also, hypericin and pseudohypericin may be absorbed in different locations in the gastrointestinal tract. Another explanation may be that hypericin may undergo first-pass hepatic metabolism *(41)*.

The median maximum plasma concentrations (C_{max}) in g/L for the respective doses were as follows:

- Hypericin: 1.5, 7.5, and 14.2 *(41)*
- 1.3, 7.2, and 16.6 *(42)*
- Pseudohypericin: 2.7, 11.7, and 30.6 *(41)*
- 3.4, 12.1, and 29.7 *(42)*

The maximum plasma concentrations increased in a nonlinear fashion *(41,42)*.

The median time to peak plasma concentration (T_{max}) in hours for the corresponding doses were as follows:

- Hypericin: 5.2, 4.1, and 5.9 *(41)*
- 5.5, 6.0, and 5.7 *(42)*
- Pseudohypericin: 2.7, 3.0, and 3.2 *(41)*
- 3.0, 3.0, and 3.0 *(42)*

Overall no correlation was observed between dose and T_{max}. However, hypericin took longer to reach maximum plasma concentration. This corresponds with the lag time data *(41)*.

After multiple dosing of 300 mg of *Hypericum* extract three times daily, the data for median C_{max} and trough plasma concentration (C_{min}) were as follows:

- Hypericin: C_{max} 8.5 µg/L *(41)*
- C_{max} 8.8 µg/L *(42)*
- C_{min} 5.3 µg/L[a] *(41)*
- C_{min} 7.9 µg/L *(42)*
- Pseudohypericin: C_{max} 5.8 mg/L *(41)*
- C_{max} 8.5 mg/L *(42)*
- C_{min} 3.7 mg/L[a] *(41)*
- C_{min} 4.8 mg/L *(42)*

---

[a] mean.

6.2. Distribution

For oral doses, the volume of distribution appears to be approx 162 L for hypericin and 63 L for pseudohypericum *(42)*.

6.3. Metabolism/Elimination

The median half-lives in hours for single 300, 900, and 1800 mg oral doses were as follows:

- Hypericin: 24.8, 26.0, and 26.5 *(41)*
- 24.5, 43.1, and 48.2 *(42)*
- Pseudohypericin: 16.3, 36.0, and 22.8 *(41)*
- 18.2, 24.8, 19.5 *(42)*

After multiple doses of *Hypericum* extract 300 mg three times daily, median half-lives in hours were:

- Hypericin: 28.0 [a] *(41)*
- 41.3 *(42)*
- Pseudohypericin: 23.5 [a] *(41)*
- 18.8 *(42)*

---

[a]mean.

The data in these two studies differ in regard to the elimination half-life of hypericin. It is difficult to ascertain whether the half-life for either hypericin or pseudohypericin is dose related.

Neither hypericin, pseudohypericin, their glucuronic acid conjugates, nor their sulfate conjugates were detected in the urine *(42)*. The chemical structure and molecular size (>500 Da) of hypericin and pseudohypericin suggest metabolism via hepatic glucuronidization followed by biliary excretion *(42)*.

7. ADVERSE EFFECTS AND TOXICITY

Although St. John's wort has proven to be relatively safe, there is still risk associated with its use and the development of adverse drug reactions (ADRs) can occur. The exact percentage of patients taking St. John's wort and developing an ADR varies greatly between studies. Observational studies report an incidence of ADRs to be between 1 and 3% *(43)*. A German study with 3250 patients taking St. John's wort (Jarsin, 300 mg St. John's wort extract) found that 79 (2.43%) patients reported an ADR and 48 (1.43%) patients had to be treated for withdrawal symptoms. Of these ADRs, 18 (0.55%) were gastrointestinal effects, 17 (0.52%) were allergic/rash reactions, 13 (0.4%) were tiredness, eight (0.26%) were anxiety, five (0.15%) were confusion, and 18 (0.55%) were others, including two cases each of dry mouth, sleep disorders, palpitations, weakness, and worsening of concurrent disease, and one case each of heart flutter, circulatory complaints, irritability, visual disorders,

disorders of micturition, burning eyes, euphoria, and nervous tension. Over a period from 1991 to 1999, the German ADR recording system received 95 reports of ADRs out of an estimated 8.5 million patients taking Jarsin. Of these, skin reactions were the highest-reported ADR with 27 reports. Other reported ADRs include an increased prothrombin time (16 cases), gastrointestinal complaints (9 cases), breakthrough bleeding with oral contraception (8 cases), decreased cyclosporine plasma levels (7 cases), tingling paraesthesias (4 cases), and cardiovascular symptoms (3 cases). All other ADRs reported had two or fewer reported cases *(43)*. In a meta-analysis of placebo-controlled trials, the frequency of ADRs with St. John's wort is similar to those reported with placebo *(3,44)*. In a comparison of trials where St. John's wort was compared to antidepressants, 26.3% of patients taking St. John's wort reported an ADR, whereas 44.7% taking a synthetic antidepressant reported an ADR *(3)*.

Grazing animals that have consumed large amounts of St. John's wort have been reported to develop photosensitivity reactions *(3,29,45)*. There are numerous case reports linking the use of St. John's wort to the development of severe rashes, both associated with and without light exposure *(46–49)*. In the AIDS study previously mentioned, 11 out of 23 patients who were receiving 6–12 mg of hypericin IV developed severe phototoxic reactions *(29,43)*. A multidose study with 40 participants was performed to assess the phototoxic effect of St. John's wort. Participants took two 300-mg tablets of hypericum extract three times daily for 15 days and were irradiated with a Dermalight-2001 lamp on days 1 and 15. The results of this study concluded that taking hypericum did reduce the median time for both tanning and erythema by 21% *(45)*. Another study in cows found that hypericin in the presence of light induced photo-polymerization of the lens proteins crystallins α, β, and to a lesser degree, γ. These changes could potentially lead to the development of cataracts *(50)*.

St. John's wort has also been associated with numerous neurological adverse effects. An acute psychotic delirium episode in a 76 year old female was attributed to St. John's wort. She was taken to the hospital after having visual hallucinations of people in her home. She was taking no prescription or herbal medications besides one 75-mg capsule of St. John's wort daily for three weeks *(51)*. More seriously, there have been several reported cases of serotonin syndrome associated with St. John's wort *(52–54)*. A 33-year-old female on no other medications started taking St. John's wort. She took a single dose on the first day and two doses the next day. She awoke at 1:00 AM after the second day with extreme anxiety and nausea, and after going to the emergency department it was found that she had a blood pressure (BP) of

195/110 mmHg and a pulse of 122 beats per minute (BPM). Over the next 4 weeks, she had four additional episodes, although less severe *(53)*. A 41 year-old male had an episode of delirium, with a BP of 210/140 mmHg and a heart rate of 115 BPM. He was not taking any medications other than the St. John's wort, which he started 7 days prior to the episode. Ten hours before the episode, he had consumed aged cheeses and one glass of red wine, the ingestion of which have been associated with hypertensive crisis with MAO inhibitors *(54)*. Although the previous cases illustrate that St. John's wort alone can cause serotonin syndrome, there are even more reports of patients developing this syndrome when they are taking a synthetic antidepressant, such as a selective serotonin reuptake inhibitor (SSRI), and add St. John's wort *(55,56)*. One case report involved a female who was taking paroxetine 40 mg a day for 8 months, then discontinued the paroxetine and started taking St. John's wort 600 mg daily. On day 10, she had difficulty sleeping and took 40 mg of paroxetine. The next day she was found to be incoherent, groggy, slow-moving, and difficulty getting out of bed *(56)*. There are also numerous case reports of hypomania and mania associated with St. John's wort *(57,58)*.

There are several other case reports of different ADRs associated with St. John's wort consumption. A patient taking 1800 mg three times daily for 32 days discontinued her therapy because of a possible photosensitivity reaction. Within a day, she developed nausea, anorexia, retching, dizziness, dry mouth, chills, and extreme fatigue. Her symptoms peaked by the third day and gradually improved until they had completely resolved by the eighth day *(59)*. Several patients taking St. John's wort have been reported to have elevated thyroid-stimulating hormone (TSH) levels. In a study where 37 patients with an elevated TSH level and 37 patients with a normal TSH were interviewed to determine if they had used St. John's wort, it was found that there was a probable association with St. John's wort and an elevated TSH, but this was not statistically significant *(60)*. A man who had been taking St. John's wort for 9 months reported having a severely diminished libido, which resolved after he discontinued St. John's wort and began citalopram *(61)*. Hair loss has also been associated with the use of St. John's wort is hair loss. A 24 year-old female who took 300 mg of St. John's wort three times daily began experiencing hair loss of the scalp and eyebrows after 5 months of therapy, with the hair loss continuing for 12 months *(52)*.

8. INTERACTIONS

St. John's wort has been shown to have many interactions with other drugs. Although one study found that St. John's wort has no effect on the cytochrome P450 (CYP) enzyme system *(62)*, most studies have shown it is a potent inducer of CYP3A4, and some studies have shown it induces CYP1A2 and CYP2C9 *(3,63–67)*. Other studies have not supported the induction of CYP1A2 and CYP2C9 by St. John's wort *(68)*. Induction of CYP3A4 is of the greatest concern because it is an important enzyme involved with the metabolism of many prescription medications. Induction of CYP3A4 can lower serum concentrations of drugs taken in combination with St. John's wort, thus reducing the efficacy of the drug. In vitro studies have shown an inhibition of CYP2D6, 2C9, 3A4, 1A2, and 2C19. Hyperforin was shown to be a potent noncompetitive inhibitor of CYP2D6 and a competitive inhibitor of CYP2C9 and 3A4 *(3)*. Hypericum extracts and hyperforin have been shown to significantly induce activity of CYP3A4 in hepatocytes *(69)*. Another study performed on human hepatocytes found that the hyperforin component in St. John's wort is responsible for the drug interactions caused by CYP450 induction, and the hypericin constituent doesn't seem to affect drug metabolism *(70)*. Hyperforin has been shown to be a potent ligand for the pregnane X receptor, a nuclear receptor that regulates the expression of CYP3A4 and P-glycoprotein (Pgp) *(68,69,71,72)*. This significantly induces the activity of both systems. However, both hypericin and hyperforin can inhibit Pgp with acute treatment. Pgp is found in many tissues and actively pumps various drugs and natural products out of cells *(71)*. Acute use of St. John's wort can lead to increased initial serum concentrations of drugs that use this transport pump. Chronic use of St. John's wort has the opposite effect and induces Pgp *(71,73)*. Subjects treated with St. John's wort for 16 days had a 4.2-fold increase in Pgp expression *(73)*. Induction of Pgp is important because it decreases the bioavailability of drugs that use the transporter.

There are numerous case reports where patients on a calcineurin inhibitor, such as cyclosporine or tacrolimus, began taking St. John's wort and developed significant reductions in plasma concentrations of the drugs *(74–81)*. Both cyclosporine and tacrolimus are metabolized by the CYP3A4 enzyme system, and cyclosporine is also a substrate of Pgp *(74,81)*. There are reports of acute graft rejections caused by low cyclosporine or tacrolimus serum concentrations in heart, liver, and kidney transplant recipients who were taking St. John's wort *(75,76)*.

There are also reports of complications associated with the combined use of oral contraceptives and St. John's wort owing to enzyme induction. The most frequent complication is breakthrough bleeding, although there are also reports of unwanted pregnancies *(82,83)*.

Patients with AIDS who are taking protease inhibitors and nonnucleoside reverse transcriptase inhibitors are at risk of being subtherapeutically treated because these drugs are metabolized by CYP3A4. Studies have shown that combined use of St. John's wort and indinavir reduced the area under the curve (AUC) of indinavir by 57% *(84)*. The same held true for nevirapine with an increased oral clearance of 35%, thus significantly lowering the exposure to the drug *(85)*.

Patients taking voriconazole to treat a fungal infection may also be at risk of being subtherapeutically treated if they are concurrently taking St. John's wort. One study found that the administration of St. John's wort caused a brief, clinically insignificant increase in voriconazole blood levels followed by a significant long-term reduction in voriconazole concentrations. The AUC of voriconazole was reduced by 59% after 15 days of 900 mg St. John's wort extract taken daily. This was assumed to be caused by voriconazole being metabolized by CYP3A4 and 2C19 *(86)*.

Imatinib mesylate, a drug recently approved for the treatment of chronic myeloid leukemia (CML), can also be affected by St. John's wort. Because imatinib is primarily metabolized by CYP3A4 and is also a Pgp substrate, the usage of St. John's wort in combination with imatinib has resulted in a significant reduction in exposure to the drug compared to imatinib alone. This is potentially significant because therapeutic outcomes for patients with CML have been shown to correlate with the dose and drug concentrations of imatinib *(87)*.

Another enzyme system that St. John's wort has been found to affect is topoisomerase II (Topo II). Hypericin was found to be an inhibitor of cleavage complex stabilization by Topo II inhibitors, used in cancer chemotherapy. Hypericin seems to intercalate into or distort DNA structure, precluding Topo II binding and/or DNA cleavage. Because hypericin appears to antagonize Topo II-poisoning chemotherapy drugs, concomitant usage of these medications could inhibit the antitumor effects of these drugs *(67)*.

There have also been several reports of delayed emergence from anesthesia; decreased international normalized ratios (INRs) in patients taking warfarin; and decreased drug levels of digoxin, buspirone, methadone, mephenytoin, chlorzoxazone, and some benzodiazepines with concomitant use of St. John's wort *(88–91)*.

9. REPRODUCTION

There have been only limited studies in humans and only rare anecdotal evidence of St. John's wort's effects on reproduction and lactation. A study using hamster oocytes incubated in either 0.06 or 0.6 mg/mL of hypericum extract for 1 hour showed normal sperm penetration at the lower concentration, whereas no penetration occurred at the higher concentration. Sperm incubated in the same concentrations for 1 week demonstrated sperm DNA denaturation and decreased viability with both concentrations. None of these effects have been seen in vivo. In vitro testing using animal uterine tissue showed weak uterine tonus-enhancing activity, but there have been no reports of abortions in animals or humans taking St. John's wort. A study using female mice fed 180 mg/kg hypericum extract or placebo starting 2 weeks prior to pregnancy and lasting until delivery demonstrated that the birth weight of the male offspring was significantly lower in the hypericum arm compared to placebo (1.67 g vs 1.74 g), but the weights were equivalent by day 3. There was no difference in female weights. There were no differences in mice of both genders in any other areas measured, including body length, head circumference, sexual maturation, or attainment of developmental milestones *(92)*.

A case report involving a 38-year-old female who was in a major depressive episode began taking St. John's wort at 24 gestational weeks and continued with the therapy until delivery. The pregnancy was generally unremarkable, with a relatively mild case of late onset thrombocytopenia and neonatal jaundice that developed at day 5 and responded to treatment. The child was 7 lbs 8 oz, had Apgar scores of 9 at 1 and 5 minutes, normal physical and laboratory results, and normal behavioral assessments at 4 and 33 days (92).

Postpartum depression is a relatively common occurrence in women after childbirth. One female who started taking 300 mg of St. John's wort (Jarsin 300) three times daily after meeting the *Diagnostic and Statistical Manual of Mental Disorders, Fourth Edition* criteria for major depressive episode 5 months after delivery agreed to have milk samples tested. Hypericin was not detected in the milk samples, but hyperforin was detected at low concentrations, with higher levels in the hind-milk than the foremilk samples. The milk/plasma ratio was well below one for both hypericin and hyperforin. Both levels were undetectable in the infant's serum and the baby showed no negative side effects *(93)*. A larger study that involved 30 women who were taking St. John's wort and breastfeeding compared results to women who were not taking St. John's wort. There were no differences in maternal events, including duration of breastfeeding, decreased lactation, or maternal demographics. Women taking St. John's wort did report a significantly higher level of infant side effects, such as lethargy and colic, vs one case of infant colic in 97 women not taking St. John's wort. None of these infants required medical attention *(94)*.

10. REGULATORY STATUS

The proposed United States Pharmacopoeia National Formulary (USP-NF) monograph for hypericum requires that products contain a minimum of 0.04% of hypericins *(95)*.

The German E Commission has approved St. John's wort for internal consumption for psychogenic disturbances, depressive states, sleep disorders, and anxiety and nervous excitement, particularly that associated with menopause. Oily *Hypericum* preparations are approved for stomach and gastrointestinal complaints, including diarrhea. Oily *Hypericum* preparations are also approved by the Commission E for external use for the treatment of incised and contused wounds, muscle aches, and first degree burns *(96)*.

The USP advisory panel recognizes that St. John's wort has a long history of use. However, because of a lack of well-controlled clinical trials its use is not recommended *(2)*.

REFERENCES

1. Bradshaw C, Nguyen A, Surles J. www.unc.edu/~cebradsh/stjohn.html. Date accessed: Oct 28, 1998.
2. United States Pharmacopeial Convention (USP), ed. Hypericum (St. John's wort). *Botanical Monograph Series*. Rockville: United States Pharmacopeial Convention, 1998.
3. Barnes J, Anderson LA, Phillipson JD. St. John's wort (*Hypericum perforatum* L.): a review of its chemistry, pharmacology and clinical properties. J Pharm Pharmacol 2001;53:583–600.
4. Pepping J. Alternative therapies. St. John's wort: *Hypericum perforatum*. Am J Health Syst Pharm 1999;56:329–330.
5. Bennett DA Jr, Phun L, Polk JF, Voglino SA, Zlotnik V, Raffa RB. Neuropharmacology of St. John's Wort (*Hypericum*). Ann Pharmacotherapy 1998;32:1201–1208.
6. Schulz V. Clinical trials with Hypericum extracts in patients with depression—results, comparisons, conclusions for therapy with antidepressant drugs. Phytomedicine 2002;9:469–474.
7. Linde K, Ramirez G, Mulrow CD, Pauls A, Weidenhammer W, Melchart D. St. John's wort for depression—an overview and meta-analysis of randomized clinical trials. BMJ 1996;313:253–258.
8. Shelton RC, Keller MB, Gelenberg A, et al. Effectiveness of St. John's Wort in major depression. JAMA 2001;285(15):1978–1986.
9. Davidson JRT, Gadde KM, Fairbank JA, Krishnan KRR, Califf RM, Binanay C. Effect of *Hypericum perforatum* (St. John's Wort) in major depressive disorder. JAMA 2002;287(14):1807–1814.
10. Szegedi A, Kohnen R, Dienel A, Kieser M. Acute treatment of moderate to severe depression with hypericum extract WS 5570 (St. John's wort): randomized controlled double blind non-inferiority trial versus paroxetine. BMJ 2005;330:503–508.
11. Singer A, Wonnemann M, Muller WE. Hyperforin, a major antidepressant constituent of St. John's Wort, inhibits serotonin uptake by elevating free intracellular Na$^+$. J Pharmacol Exp Ther 1999;290(3):1363–1368.

12. Calapai G, Crupi A, Firenzuoli F, et al. Serotonin, norepinephrine, and dopamine involvement in the antidepressant action of *Hypericum Perforatum.* Pharmacopsychiatry 2001;34:45–49.

13. Muller WE, Singer A, Wonnemann M, Hafner U, Rolli M, Schafer C. Hyperforin represents the neurotransmitter reuptake inhibiting constituent of Hypericum Extract. Pharmacopsychiatry 1998;31(Suppl):16–21.

14. Calapai G, Crupi A, Firenzuoli F, et al. Interleukin-6 involvement in antidepressant action of *Hypericum Perforatum.* Pharmacopsychiatry 2001;34(Suppl 1):S8–S10.

15. Fiebich BL, Hollig A, Lieb K. Inhibition of substance P-induced cytokine synthesis by St. John's Wort extracts. Pharmacopsychiatry 2001;34(Suppl 1):S26–S28.

16. Kleber E, Obry T, Hippeli S, Schneider W, Elstner EF. Biochemical activities of extracts from Hypericum perforatum L. Drug Res 1999;2:106–109.

17. Raffa R. Screen of receptor and uptake-site activity of hypericin component of St. John's wort reveals receptor binding. Life Sci 1998;62:PL265–270.

18. Butterweck V, Nahrstedt A, Evans J, et al. In vitro receptor screening of pure constituents of St. John's wort reveals novel interactions with a number of GPCRs Psychopharmacology 2002;162:193–202.

19. Simmen U, Bobirnac I, Ullmer C, et al. Antagonist effect of pseudohypericin at CRF_1 receptors. Eur J Pharmacol 2003;458:251–256.

20. Jensen AG, Hansen SH, Nielsen EO. Adhyperforin as a contributor to the effect of *Hypericum* perforatum L. in biochemical models of antidepressant activity. Life Sci 2001;68:1593–1605.

21. Kumar V, Jaiswal AK, Singh PN, Bhattacharya SK. Anxiolytic activity of Indian *Hypericum perforatum* Linn: an experimental study. Indian J Exp Biol 2000;38:36–41.

22. Taylor L, Kobak KA. An open-label trial of St. John's Wort (*Hypericum perforatum*) in obsessive-compulsive disorder. J Clin Psychiatry 2000;61:575–578.

23. Schulz H, Jobert M. Effects of hypericum extract on the sleep EEG in older volunteers. J Geriatr Psychiatry Neurol 1994;7(Suppl 1):S39–S43.

24. Volz HP, Murck H, Kasper S, Moller HJ. St. John's wort extract (LI 160) in somatoform disorders: results of a placebo-controlled trial. Psychopharmacology 2002;164:294–300.

25. Schempp CM, Pelz K, Wittmer A, Schopf E, Simon JC. Antibacterial activity of hyperforin from St. John's wort, against multiresistant Staphylococcus aureus and gram-positive bacteria. Lancet 1999;353:2129.

26. Liebes L, Mazur Y, Freeman D, et al. A method for the quantification of hypericin, an antiviral agent, in biological fluids by high-performance liquid chromatography. Anal Biochem 1991;195:77–85.

27. Lavie G, Valentine F, Levin B, et al. Studies of the mechanisms of action of the antiretroviral agents hypericin and pseudohypericin. Proc Natl Acad Sci USA 1989;86:5963–5967.

28. Taher MM, Lammering GM, Hershey CM, Valerie KC. Mood-enhancing antidepressant St. John's Wort inhibits the activation of human immunodeficiency virus gene expression by ultraviolet light. Life 2002;54:357–364.

29. Gulick RM, McAuliffe V, Holden-Wiltse J, Crumpacker C, Liebes L, Stein DS, for the AIDS Clinical Trials Group 150 and 258 Protocol Teams. Phase I studies of hypericin, the active compound in St. John's wort, as an antiretroviral agent in HIV-infected adults. Ann Intern Med 1999;130:510–514.

30. Okpanyi SN, Lidzba H, Scholl BC, Miltenburger HG. Genotoxicity of a standardized hypericum extract. Arzneimittelforschung 1990;40:851–855.

31. Sindrup SH, Madsen C, Bach FW, Gram LF, Jensen TS. St. John's wort has no effect on pain in polyneuropathy. Pain 2000;91:361–365.

32. Albert D, Zundorf I, Dingermann T, Muller WE, Steinhilber D, Werz O. Hyperforin is a dual inhibitor of cyclooxygenase-1 and 5-lipoxygenase. Biochem Pharmacol 2002;64:1767–1775.

33. Tedeschi E, Menegazzi M, Margotto D, Suzuki H, Forstermann U, Kleinert H. Anti-inflammatory actions of St. John's Wort: inhibition of human inducible nitric-oxide synthase expression by down-regulating signal transducer and activator of transcription-α (STAT-1α) activation. J Pharmacol Exp Ther 2003;307:254–261.

34. Schempp CM, Windeck T, Hezel S, Simon JC. Topical treatment of atopic dermatitis with St. John's wort cream—a randomized, placebo controlled, double blind half-sided comparison. Phytomedicine 2003;10(Suppl IV):31–37.

35. Hunt EJ, Lester CE, Lester EA, Tackett RL. Effect of St. John's wort on free radical production. Life Sci 2001;69:181–190.

36. Stevinson C, Ernst E. A pilor study of *Hypericum perforatum* for the treatment of premenstrual syndrome. Br J Obstet Gynaecol 2000;107:870–876.

37. Orellana C. St. John's Wort helps to fight bladder cancer. Lancet Oncol 2001;2:399.

38. Hostanska K, Bommer S, Weber M, Krasniqi B, Saller R. Comparison of the growth-inhibitory effect of *Hypericum perforatum* L. extracts, differing in the concentration of phloroglucinols and flavonoids, on leukaemia cells. J Pharm Pharmacol 2003;55:973–980.

39. Roscetti G, Franzese O, Comandini A, Bonmassar E. Cytotoxic activity of *Hypericum perforatum* L. on K562 erythroleukemic cells: differential effects between methanolic extract and hypericin. Phytother Res 2004;18:66–72.

40. Kimura H, Harris MS, Sakamoto T, et al. Hypericin inhibits choroidal endothelial cell proliferation and cord formation in vitro. Curr Eye Res 1997;16:967–972.

41. Staffeldt B, Kerb R, Brockmoller J, Ploch M, Roots I. Pharmacokinetics of hypericin and pseudohypericin after oral intake of Hypericum perforatum extract LI 160 in healthy volunteers. J Geriatr Psychiatry Neurol 1994;7(Suppl 1):S47–S53.

42. Kerb R, Brockmoller J, Staffeldt B, Ploch M, Roots I. Single-dose and steady state pharmacokinetics of hypericin and pseudohypericin. Antimicrob Agents Chemother 1996;40:2087–2193.

43. Schulz V. Incidence and clinical relevance of the interactions and side effects of Hypericum preparations. Phytomedicine 2001;8(2):152–160.

44. Ernst E, Rand JI, Barnes J, Stevinson C. Adverse effects profile of the herbal anti-depressant St. John's wort (*Hypericum perforatum* L.). Eur J Clin Pharmacol 1998;54:589–594.

45. Brockmoller J, Reum T, Bauer S, Kerb R, Hubner WD, Roots I. Hypericin and Pseudohypericin: pharmacokinetics and effects on photosensitivity in humans. Pharmacopsychiatry 1997;30(Suppl):94–101.
46. Holme SA, Roberts DL. Erythroderma associated with St. John's wort. Brit J Dermatol 2000;143:1097–1131.
47. Sultana D, Peindl KS, Wisner KL. Rash associated with St. John's Wort treatment in premenstrual dysphoric disorder. Arch Womens Ment Health 2000;3:99–101.
48. Lane-Brown MM. Photosensitivity associated with herbal preparations of St. John's wort (*Hypericum perforatum*). Med J Aust 2000;172:302.
49. Schempp CM, Muller K. Single-dose and steady-state administration of *Hypericum perforatum* extract (St. John's Wort) does not influence skin sensitivity to UV radiation, visible light, and solar-simulated radiation. Arch Dermatol 2001;137:512–513.
50. Roberts JE, Wang RH, Tan IP, Datillo M, Chignell CF. Hypericin (active ingredient in St. John's Wort) photooxidation of lens proteins. Abstracts of the 27th Annual Meeting of the American Society of Photobiology 1999;42S.
51. Laird RD, Webb M. Psychotic episode during use of St. John's wort. J Herb Pharmacother 2001;1(2):81–87.
52. Parker V, Wong AHC, Boon HS, Seeman MV. Adverse reactions to St. John's Wort. Can J Psych 2001;46(1):77–79.
53. Brown TM. Acute St. John's Wort toxicity. Am J Emerg Med 2000;18(2):231–232.
54. Patel S, Robinson R, Burk M. Hypertensive crisis associated with St. John's Wort. Am J Med 2002;112:507–508.
55. Lantz MS, Buchalter E, Giambanco V. St. John's Wort and antidepressant drug interactions in the elderly. J Geriatr Psychiatry Neurol 1999;12:7–10.
56. Gordon JB. SSRIs and St. John's Wort: possible toxicity? Am Fam Physician 1998;57(5):952–953.
57. Nierenberg AA, Burt T, Matthews J, Weiss AP. Mania associated with St. John's Wort. Soc Biol Psych 1999;46:1707–1708.
58. Moses EL, Mallinger AG. St. John's Wort: three cases of possible mania induction. J Clin Psychopharmacol 2000;20(1):115–117.
59. Dean AJ. Suspected withdrawal syndrome after cessation of St. John's wort. Ann Pharmacother 2003;37:150.
60. Ferko N, Levine MAH. Evaluation of the association between St. John's Wort and elevated thyroid-stimulating hormone. Pharmacotherapy 2001;21(12):1574–1578.
61. Bhopal JS. St. John's Wort-induced sexual dysfunction. Can J Psych 2001;46(5):456–457.
62. Markowitz JS, DeVane CL, Boulton DW, Carson SW, Nahas Z, Risch SC. Effect of St. John's Wort (*Hypericum perforatum*) on cytochrome P-450 2D6 and 3A4 activity in healthy volunteers. Life Sci 2000;66(9):133–139.
63. Roby CA, Dryer DA, Burstein AH. St. John's Wort: effect on CYP2D6 activity using dextromethorphan-dextrorphan ratios. J Clin Psychopharmacol 2001;21(5):530–532.
64. Wenk M, Todesco L, Krahenbuhl S. Effect of St. John's Wort on the activities of CYP1A2, CYP3A4, CYP2D6, N-acetyltransferase 2, and xanthine oxidase in healthy males and females. Br J Clin Pharmacol 2004;57(4):495–499.

65. Wang Z, Gorski JC, Hamman MA, Huang SM, Lesko LJ, Hall SD. The effects of St. John's wort (*Hypericum perforatum*) on human cytochrome P450 activity. Clin Pharmacol Ther 2001:317–326.

66. Roby CA, Anderson GD, Kantor E, Dryer DA, Burstein AH. St. John's Wort: effect on CYP3A4 activity. Clin Pharmacol Ther 2000;67(5):451–457.

67. Peebles KA, Baker RK, Kurz EU, Schneider BJ, Kroll DJ. Catalytic inhibition of human DNA topoisomerase IIα by hypericin, a naphthodianthrone from St. John's wort (*Hypericum perforatum*). Biochem Pharmacol 2001;62:1059–1070.

68. Wentworth M, Agostini M, Love J, Schwabe JW, Chatterjee VKK. St. John's wort, a herbal antidepressant, activates the steroid X receptor. J Endocrinol 2000;166:R11–R16.

69. Moore LB, Goodwin B, Jones SA, et al. St. John's wort induces hepatic drug metabolism through activation of the pregnane X receptor. Proc Natl Acad Sci USA 2000;97(13):7500–7502.

70. Komoroski BJ, Zhang S, Cai H, et al. Induction and inhibition of cytochromes P450 by the St. John's wort constituent hyperforin in human hepatocyte cultures. Drug Metab Dispos 2004;32(5):512–518.

71. Wang E, Barecki-Roach M, Johnson WW. Quantitative characterization of direct P-glycoprotein inhibition by St. John's wort constituents hypericin and hyperforin. J Pharm Pharmacol 2004;56:123–128.

72. Izzo AA. Drug interactions with St. John's Wort (Hypericum perforatum): a review of the clinical evidence. Int J Clin Pharmacol Ther 2004;42(3):139–148.

73. Hennessy M, Kelleher D, Spiers JP, et al. St. John's wort increases expression of P-glycoprotein: implications for drug interactions. Br J Clin Pharmacol 2002;53:75–82.

74. Barone GW, Gurley BJ, Ketel BL, Lightfoot ML, Abul-Ezz SR. Drug interaction between St. John's Wort and cyclosporine. Ann Pharmacotherapy 2000;34:1013–1016.

75. Ruschitzka F, Meier PJ, Turina M, Luscher TF, Noll G. Acute heart transplant rejection due to Saint John's wort. Lancet 2000;355:548–549.

76. Breidenbach TH, Hoffmann MW, Becker TH, Schlitt H, Klempnauer J. Drug interaction of St. John's wort with ciclosporin. Lancet 2000;355:1912.

77. Mai I, Kruger H, Budde K, et al. Hazardous pharmacokinetic interaction of Saint John's wort (Hypericum perforatum) with the immunosuppressant cyclosporine. Int J Clin Pharmacol Ther 2000;38(10):500–502.

78. Moschella C, Jaber BL. Interaction between cyclosporine and *Hypericum perforatum* (St. John's Wort) after organ transplantation. Am J Kidney Dis 2001;38(5):1105–1107.

79. Mandelbaum A, Pertzborn F, Martin-Facklam M, Wiesel M. Unexplained decrease of cyclosporine trough levels in a compliant renal transplant patient. Nephrol Dial Transplant 2000;15:1473–1474.

80. Karliova M, Treichel U, Malago M, Frilling A, Gerken G, Broelsch CE. Interaction of *Hypericum perforatum* (St. John's wort) with cyclosporine A metabolism in a patient after liver transplantation. J Hepatol 2000;33:853–855.

81. Mai I, Stormer E, Bauer S, Kruger H, Budde K, Roots I. Impact of St. John's wort treatment on the pharmacokinetics of tacrolimus and mycophenolic acid in renal transplant patients. Nephrol Dial Transplant 2003;18(4):819–822.
82. Pfrunder A, Schiesser M, Gerber S, Haschke M, Bitzer J, Drewe J. Interaction of St. John's wort with low-dose oral contraceptive therapy: a randomized controlled trial. Br J Clin Pharmacol 2003;56(6):683–690.
83. Gorski JC, Hamman MA, Wang Z, Vasavada N, Huang S, Hall SD. The Effect of St. John's Wort on the efficacy of oral contraception. Clin Pharmacol Ther;71(2):P25.
84. Piscitelli SC, Burstein AH, Chaitt D, Alfaro RM, Falloon J. Indinavir concentrations and St. John's wort. Lancet 2000;355:547–549.
85. de Maat MMR, Hoetelmans RMW, Mathot RAA, et al. Drug interactions between St. John's wort and nevirapine. AIDS 2001;15(3):420–421.
86. Rengelshausen J, Banfield M, Riedel KD, et al. Opposite effects of short-term and long-term St. John's wort intake on voriconazole pharmacokinetics. Clin Pharmacol Ther 2005;78(1):25 33.
87. Smith PF, Bullock JM, Booker BM, Haas CE, Berenson CS, Jusko WJ. The influence of St. John's Wort on the pharmacokinetics and protein binding of imatinib mesylate. Pharmacotherapy 2004;24(11):1508–1514.
88. Crowe S, McKeating K. Delayed emergence and St. John's Wort. J Am Soc Anesth 2002;9(4):1025–1027.
89. Irefin S, Sprung J. A possible cause of cardiovascular collapse during anesthesia: long-term use of St. John's Wort. J Clin Anesth 2000;12:498–499.
90. Johne A, Brockmoller J, Bauer S, Maurer A, Langheinrich M, Roots I. Pharmacokinetic interaction of digoxin with an herbal extract from St. John's wort (*Hypericum perforatum*). Clin Pharmacol Ther 1999;66(4):338–345.
91. Xie HG, Kim RB. St. John's wort-associated drug interactions: short-term inhibition and long-term induction? Clin Pharmacol Ther 2005;78(1):19–24.
92. Briggs GG, Freeman RK, Yaffe SJ. Drugs in pregnancy and lactation. Briggs Update 1999;12(3):17–19.
93. Klier CM, Schafer MR, Schmid-Siegel B, Lenz G, Mannel M. St. John's Wort (Hypericum Perforatum)—Is it safe during breastfeeding? Pharmacopsychiatry 2002;35:29–30.
94. Lee A, Minhas R, Ito S. Safety of St. John's Wort during breastfeeding. Am Soc Clin Pharmacol Ther 2000;130.
95. United States Pharmacopoeia (USP), ed. National Formulary, 18th edition, Supplement 9. Rockville: United States Pharmacopeial Convention, 1998.
96. Blumenthal M. St. John's wort. The complete German Commission E monographs. Austin: American Botanical Council, 1998.

Chapter 6

Echinacea

Daniel Berkner and Leo Sioris

SUMMARY

Echinacea remains a popular supplement used as an immunostimulant in the prevention and treatment of infection. Despite inconsistent results from clinical trials attempting to assess effectiveness, its relatively wide margin of safety makes the herb an attractive alternative for prevention and treatment of common infections such as upper respiratory infections. Given the herb's inherent ability to inhibit various CYP450 enzymes, further studies to identify the clinical implications for herb–drug interactions are needed.

Key Words: *Echinacea purpurea*; *Echinacea augustifolia*; immune stimulation; arabino-galactans; P450 enzyme induction.

1. HISTORY

Echinacea is a group of American coneflowers in the Family Asteraceae/Compositae. There are nine species of the plant included in the genus. Three of these are typically seen in herbal preparations: *Echinacea purpurea*, *Echinacea angustifolia*, and *Echinacea pallida*. Common preparations consist of freshly pressed or ethanolic extracts of the roots, leaves, and flowers as well as dried portions of the plants. *E. purpurea* is the most commonly used species, although it is often seen in combination with *E. angustifolia (1)*.

Echinacea was first used by Native Americans for treatment of many conditions. These included pain relief, cough and sore throat, fever, smallpox, mumps, measles, rheumatism, arthritis, and as an antidote for poisons and venoms *(2)*. As early as 1762, Echinacea was mentioned for use on saddle

From Forensic Science and Medicine:
Herbal Products: Toxicology and Clinical Pharmacology, Second Edition
Edited by: T. S. Tracy and R. L. Kingston © Humana Press Inc., Totowa, NJ

sores on horses. Until 1885, no further study is documented in the literature. That year marked the beginning of the rise of Echinacea into mainstream medicine. H.C.F. Meyer, a Nebraska physician, began promoting the product for conditions such as syphilis, hemorrhoids, and rabies, among many other claims. This same year the Lloyd Brothers pharmaceutical company was finally persuaded to produce and market an Echinacea product. They began producing a number of products containing Echinacea and had a great deal of success with their line of products. In a survey conducted on preference and use of phyto-pharmaceuticals early in the 20th century, 6000 physicians ranked Echinacea 11th overall out of a list of several hundred products. Antibiotics and the push for patentable medicines led to the fall of Echinacea and herbal medicine in the United States and Europe *(3)*. In recent years, Echinacea has made a comeback in the United States and, in 2002, it was the second-best-selling herbal product *(4)*.

2. CURRENT PROMOTED USES

In the United States, Echinacea is marketed primarily in oral dosage forms (tablet, capsule, and liquid) as an immune stimulant used to help with the symptoms of upper respiratory infections (URIs). It has also been promoted as a general immune stimulant to help fight various other infections. Topical preparations are also available for treatment of wounds and inflammatory skin conditions.

3. SOURCES AND CHEMICAL COMPOSITION

E. purpurea (L.) Moench, *E. angustifolia* D.C., *E. pallida* (Nutt.); American Coneflower, Black Sampson, Black Susan, *Brauneria angustifolia*, *Brauneria pallida*, Cock-Up-Hat, Comb Flower, Coneflower, Echinaceawurzel, Hedgehog, Igelkopfwurzel, Indian Head Lyons, Kansas Snakeroot, Missouri Snakeroot, Narrow-Leaved Purple Cone Flower, Pale Coneflower, Purple Kansas Coneflower, Purpursonnenhutkraut, Purpursonnenhutwurzel, Racine d'echinacea, Red Sunflower, Rock-Up-Hat, Roter Sonnenhut, *Rudbeckia purpurea*, Schmallblaettriger Kegelblumenwurzel, Schmallblaettriger Sonnenhut, Scurvy Root, Snakeroot, and Sonnenhutwurzel

4. PRODUCTS AVAILABLE

Numerous forms of Echinacea are available in the United States. Dried herb and concentrated extracts in oral dosage forms make up the bulk of the products available. There are also available fresh, freeze-dried, and liquid alcoholic extracts, which come in a variety of forms including tablets, capsules, lozenges, liquids, teas, and salves. A good number of these products are combined with other herbs such as ginseng, goldenseal, and various other supplements to enhance the efficacy of Echinacea.

Consumer Reports analyzed 19 different Echinacea products in its February 2004 issue to determine content and potency. As a primary standard they used the phenolic content as a measure of potency. The products varied with their phenolic content, some from bottle to bottle of the same manufacturer. Some of the combination products that were tested were found to have unacceptable lead levels according to California standards. Only three of the products were deemed to have adequate labeling with regards to precautions *(5)*.

5. *PHARMACOLOGICAL/TOXICOLOGICAL EFFECTS*

5.1. *Immunological Effects*

The majority of literature published about Echinacea focuses on its activity as an immunostimulant. Many of the studies focus on the activity of macrophages and Echinacea's ability to activate and stimulate immune function.

One constituent of Echinacea, the polysaccharide arabinogalactan, has been identified as a macrophage activator in vitro, causing macrophages to attack tumor cells and microorganisms. When injected into mice intraperitoneally, arabinogalactan was able to activate macrophages. Macrophage production of tumor necrosis factor (TNF)-α, interleukin (IL)-1, and interferon-B_2 was increased in vitro, and production of oxygen free radicals was increased both in vitro and in vivo *(6)*.

Three of the main active components of Echinacea, cichoric acid, polysaccharides, and alkylamides, were separated and tested at various doses in rats for phagocytic activity in alveolar macrophages and splenocytes. The alveolar macrophages from the group of rats treated with the alkylamides were the only cells to show any significant increases in phagocytic activity, phagocytic index, TNF-α, and nitric oxide. None of the components tested had any activity on splenocytes *(7)*.

Echinacea was fed to aging male rats and was found to cause an increase in total white cell counts during the first 2 weeks of the administration and increases in IL-2 levels in the final 5 weeks. Differential white counts were altered during the entire 8-week study, with mononuclear cells significantly increased, whereas granulocytes decreased *(8)*.

5.2. Antimicrobial/Antiviral Effects

The focus of Echinacea research most recently has been for the treatment and prevention of URIs of varying causes. There are a number of studies using Echinacea products in the treatment of URI. Many of these show positive results with reduction of symptoms and duration of URI. Studies evaluating the preventative role of Echinacea in URI have shown less impressive results. A few reasons for the differences in efficacy may have to do with the quality of the Echinacea used in the study and study design. The treatment studies demonstrating effectiveness tend to start treatment early in the course of a URI *(9)*.

A double-blind, placebo-controlled, randomized trial evaluating Echinacea for the treatment of colds involved 559 adult patients and three different Echinacea products (Echinaforce® [*E. purpurea* 95% herb and 5% root], *E. purpurea* concentrate, and *E. purpurea* root preparation). In the patients that were treated, two of the Echinacea products produced a statistically significant reduction in symptoms compared to placebo and the *E. purpurea* root preparation *(10)*. In another study evaluating the prophylactic role of Echinacea, 302 patients were enrolled in a three-armed, randomized, double-blind, placebo-controlled trial. Each of the groups was given ethanolic extract of *E. purpurea* roots, *E. angustifolia* roots, or placebo for 12 weeks. They did not find a significant reduction in occurrence of URI in either treatment group; however, they speculated from their results and the results of two other similar studies that there was a 10 to 20% relative risk reduction for URI. It was concluded that larger sample sizes were needed to confirm their observation *(11)*.

A randomized, double-blind, placebo-controlled study was done in 48 healthy patients given *E. purpurea* extract or placebo for 7 days and then inoculated with rhinovirus type 39. Treatment was continued for 7 more days after the inoculation. They did not find a statistically significant decrease in the rate of infection. Because of the small sample size, power analysis did not detect any differences in the frequency and severity of the illness that ensued after inoculation. However, their findings did show a trend of reduced symptoms consistent with previous studies relative to prevention of URI *(12)*.

Eight different varieties of Echinacea were found to have antiviral activity against Herpes simplex virus (HSV) Type I in vitro. The two most potent inhibitors found were ethanol extracts of *E. pallida* var. *sanguinea* and n-hexane extracts of *E. purpurea (13)*.

In another study, to evaluate the prophylactic action of Echinacea on Influenza virus Type A, a mixture of four herbal extracts, which included

Thujae occidentalis herb, *Baptisiae tinctoriae* root, *E. purpurea* root, and *E. pallida* root, were given to mice. After 6 days the mice were inoculated with Influenza virus Type A. They found a statistically significant increase in survival rate, survival time, reduced lung consolidation, and virus titer *(14)*.

A single center, prospective, double-blind, placebo-controlled, crossover trial investigated the activity of Echinacea in humans for the treatment of recurrent genital herpes. The 1-year study involved 50 patients who were each given the product Echinaforce for 6 months and placebo for 6 months. The study found no statistically significant benefit in using Echinaforce vs placebo for frequently recurrent genital herpes *(15)*.

5.3. Antifungal Effects

Acetylenic isobutylamides and polyacetylenes occurring in Echinacea have been shown to inhibit the growth of yeast strains of *Saccharomyces cerevisiae, Candida shehata, Candida kefyr, Candida albicans, Candida steatulytica,* and *Candida tropicalis.* This growth inhibition occurred to a greater extent under ultraviolet irradiation than without it. There are other compounds in Echinacea that are suspected to be phototoxic to microbes, but this has yet to be demonstrated *(16)*.

Pretreatment with a polysaccharide *E. purpurea* extract was found to decrease morbidity and mortality in mice infected by *C. albicans* immunosuppresed with cyclophosphamide and cyclosporine A. They found that macrophages in the Echinacea group produced an increased amount of TNF-α. The authors state that this led to an increased resistance toward *Listeria monocytogenes*, *C. albicans*, and the intracellular parasite *Leishmania enrietti* *(17)*.

5.4. Antineoplastic Activity

See also Chapter 7, Section 4.1. An investigation details the isolation of (Z)-1,8-pentadecadiene from *E. angustifolia* and *E. pallida*. This root oil constituent has inhibitory effects against Walker carcinosarcoma 256 and P-388 lymphocytic leukemia in the mouse and rat, respectively *(18)*.

An examination of mature and precursor cells in the bone marrow and spleen was conducted to determine the activity of Echinacea in these cell's development. After an extract of Echinacea was given to mice for 1- and 2-week periods, populations of natural-killer cells and monocytes were increased in both organs, whereas other hemopoietic and immune cell populations remained at control levels. This study confirmed Echinacea's effectiveness as a nonspecific immune stimulant and suggested a prophylactic role for Echinacea in treating virus-based tumors and infection *(19)*.

An investigation into the effects of Echinacea in mice with leukemia was performed to determine the antineoplastic activity. Two groups of mice were used; one group of mice was given a vaccine of killed erytholeukemia cells and then given live tumor cells to induce leukemia, the other group received live tumor cells. It was found that the mice that had been given the vaccine and Echinacea survived longer than the control group and the group given the vaccine alone. They found significant elevations in natural killer cells in these mice. It was concluded that combination therapy of Echinacea and vaccine prolonged the life more than the group that had received the vaccine alone *(20)*. In an earlier study by the same investigator, it was shown that leukemic mice treated with Echinacea had a much higher survival rate than the control. The treatment group showed a 2.5-fold increase in natural killer cells in their spleens. All other major hemopoietic and immune cell lineages remained normal in the treatment group at 3 months after tumor onset. The authors concluded that the positive effects observed suggest a potential use of Echinacea in treatment of leukemia *(21)*.

5.5. Wound Healing

Echinacea has been used topically for wound-healing. The exact mechanism is unknown but is likely caused by antihyaluronidase activity of echinacoside. A study investigating this activity found that *E. pallida,* which is known to contain echinacoside, had more anti-inflammatory and wound-healing activity in rats after topical application. The effects were much greater with *E. pallida* compared with *E. purpurea* and control *(22)*.

5.6. Anti-Inflammatory Effects

Alkylamides from the roots of *E. purpurea* have been shown to have anti-inflammatory activity in vitro. In a study by Clifford et al., they demonstrated 36–60% and 15–46% of cyclooxygenase (COX)-I and COX-II, respectively *(23)*.

Two in vitro studies have demonstrated anti-inflammatory activity by various Echinacea preparations. Speroni et al. showed anti-inflammatory activity attributed to echinacosides in *E. pallida* in rats. Another in vitro study used *E. purpurea* in mice that had induced paw edema. Only the higher dose used in the study downregulated COX-2 expression. The authors suggested that the anti-inflammatory properties of Echinacea are related to this inhibition *(24)*.

5.7. Mutagenicity/Carcinogenicity

E. purpurea gave negative results in mammalian cells and bacteria in vitro and in vivo in mice mutagenicity tests. Hamster embryo cell carcinogenicity studies revealed no morphological transformations *(25)*.

5.8. Antioxidant Effects

In an analysis of the different components of the extracts of the roots and leaves of *E. purpurea, E. angustifolia,* and *E. pallida*, all possessed antioxidant properties in a free-radical scavenging assay and in a lipid peroxidation assay *(26)*.

6. PHARMACOKINETICS

To date, only one study has evaluated the pharmacokinetics of the alkamides contained in the Echinacea products administered to humans *(27)*. Subjects *(n* = 11) received a single oral 2.5-mL dose of the 60% ethanolic extract from *E. angustifolia* roots or placebo (60% ethanol). Six different alkamides were analyzed: (1) Undeca-2D/Z-ene-8,10-diynoic acid isobutylamides; (2) Dodeca-2D,4Z-diene-8,10-diynoic acid isobutylamide; (3) Dodeca-2E-ene-8,10-diynoic acid isobutylamide; (4) Dodeca-2E,4E,8Z,10E/ Z-tetraenoic acid isobutylamides; (5) Dodeca-2E,4E,8Z-trienoic acid isobutylamide; and (6) Dodeca-2E,4E-dienoic acid isobutylamide. The extract contained approx 2.5 mg of (4), and approx 0.5 mg of all other components. The C_{max} and area under the curve (AUC) for (4) were approx 10-fold that achieved with each of the other components. Thus, despite a fivefold higher amount per dose, the 10-fold greater C_{max} and AUC achieved with (4) suggest it exhibits a greater bioavailability than the other components.

7. ADVERSE EFFECTS AND TOXICITY

7.1. Common Adverse Reactions

Side effects that have been observed with administration of Echinacea are generally mild and uncommon. Infrequent adverse effects include abdominal upset, nausea, unpleasant taste, and dizziness. Rarely seen effects are anaphylaxis, exacerbation of asthma, and angioedema *(28)*.

There is a potential for interaction in patients with immunosuppression, especially those on medications designed to suppress an autoimmune disorder or prevent transplant rejection. In a study looking at the use of herbal medications in a population of patients that had liver transplants, the researchers found that of five patients that had taken Echinacea, two of them had elevated aminotransferase levels, which returned to normal after stopping the product *(29)*.

In Germany from 1989 to 1995, there were 13 adverse events possibly associated with the use of Echinacin® (pressed juice from *E. purpurea* herb). Only four of these, all allergic skin reactions, were thought to be related to use of the product. During this time period, several million people were thought to treat themselves or obtain a prescription for an Echinacea product *(9)*.

In a study of Australian adverse events thought to be caused by Echinacea, 26 separate cases were studied. Four of these cases developed anaphylaxis, 12 were acute asthma attacks, and 10 were urticaria/angioedema. Four of the 26 reacted after their first ever dose of Echinacea. More than 50% of the patients were found to have some form of atopic disease. In a study of 100 patients with atopic disease, 20 demonstrated positive skin-prick testing (SPT) to other plants in the Family Asteraceae (ragweed, daisies, and others), and specifically Echinacea. Only three of this group of 100 had ever taken an Echinacea supplement. Five of these patients were studied more closely and had what was believed to be immunoglobulin E-mediated reactions to Echinacea products. All had SPT and four of five had radioallergosorbent testing (RAST) performed. Three of the five had positive skin-reaction SPT and three of four had positive RAST results. The authors concluded that atopic patients should exercise caution when using Echinacea *(30)*.

7.2. Case Reports of Toxicity

Echinacea has been thought to have potential for liver toxicity because of the presence of pyrrolizidine alkaloids, and some authors have warned about its concurrent use with known hepatoxic drugs. The importance of this purported toxicity has been questioned, as Echinacea lacks the 1,2-unsaturated necrine ring system that is associated with the hepatoxicity of pyrrolizidine alkaloids *(31)*.

A case report describing a 41-year-old male who took Echinacea routinely at the start of influenza-like illness recalled taking it before each of four clinical episodes of erythema nodosum. These episodes lasted anywhere from a few days to 2 weeks and each time they resolved when the Echinacea was stopped. He was followed a year later and had not had any more recurrences of erythema nodosum. He had other bouts of intermittent influenza illness similar to the previous episodes, which he treated with Echinacea, but the patient was unwilling to rechallenge with Echinacea. The authors concluded that the erythema nodosum could have been caused by a number of pathways but were more convinced that Echinacea had brought on his symptoms because each recurrence of his disease resolved after stopping his Echinacea use *(32)*.

Another case report identified a 51-year-old woman who had been taking Echinacea for 2 months and was found to have a depressed white blood cell count (WBC). After stopping the Echinacea, she was retested and her WBC returned to normal. After 1 year, she returned for a routine check up and 2 months earlier started taking Echinacea again. Her WBC was found to be depressed again and similar increases into the normal range were found after discontinuing the Echinacea 2 months later. The authors could not be certain that Echinacea decreased the WBC in this patient, but suggested a type IV allergic response to Echinacea may be responsible *(33)*.

A 36-year-old patient started taking a combination of herbal products including Echinacea, and 2 weeks later she presented with generalized muscle weakness that limited her ambulation and ability to use her hands. She was found to have distal renal tubular acidosis and was extremely hypokalemic (K^+ of 1.3). Over 4 days she received 1200 mEq of sodium bicarbonate and 400 mEq of potassium chloride along with other electrolyte supplements to correct the imbalances. After her serum electrolytes were corrected, her muscle weakness improved rapidly. She was diagnosed and treated for Sjögren's syndrome and her condition rapidly improved. The researchers suggested that her use of the immunostimulant Echinacea could have contributed to the activation of her autoimmune disease, which ultimately caused her severe metabolic disturbances. Because she had remained symptom free for more than 3 years, the authors concluded that, after review, her disease was relatively mild and was exacerbated by Echinacea (34).

In a study to determine the LD_{50} of express juice of *E. purpurea* in rats and mice, one researcher was unable to kill either rat or mouse at oral doses greater than 15 g/kg body weight and intravenous doses greater than 5 g/kg body weight *(25)*. Injected concentrates of polysaccharide fractions produced an LD_{50} of 2.5 g/kg body weight in mice. Echinacea has a wide margin of safety considering a typical oral dose in a 50–80 kg human is 200–2000 mg of Echinacea or 2.5–40 mg/kg body weight *(9)*.

8. DRUG INTERACTIONS

Analysis of dilutions of extract of *E. angustifolia* and *E. purpurea* showed medium to high levels of inhibition of cytochrome P450 3A4 in vitro. Testing of chicoric acid and echinacoside alone showed low to very low inhibition.

The authors did not speculate on what other compounds may be present in either *Echinacea* sp. that caused the more significant enzyme inhibition *(35)*.

Caffeine, tolbutamide, dextromethorphan, and oral and intravenous midazolam were given to 12 healthy subjects. This was followed with a course of *E. purpurea*, 400 mg four times a day for 8 days, then these drugs were given again and the subjects were assessed for cytochrome P450 activity. They found that the Echinacea significantly reduced the activity of hepatic CYP 1A2 and intestinal CYP 3A and induced hepatic CYP 3A activity. They advised caution when giving drugs that are metabolized by these same enzyme systems *(36)*.

9. REPRODUCTION

In a prospective, controlled study, 206 women who reported use of Echinacea during their pregnancy were compared to a group of 206 women who were matched to the study group with regards to maternal age, alcohol, and cigarette use. In comparing the rates of major and minor malformation, it was found that there were no statistical differences in number of live births, spontaneous abortions, therapeutic abortions, or major malformations. This study suggests that use of Echinacea during organogenesis is not associated with any detectable increased risk for any major malformations *(37)*.

In a study of human sperm, Echinacea was found to inhibit the motility of the sperm only at high concentrations and after 24 hours. One potential effect of Echinacea is thought to be the inhibition of hyaluronidase activity. Hyaluronidase is localized on the sperm head and helps the sperm to penetrate the oocyte. This potential inhibition could prevent sperm from fertilizing oocytes, but further studies are needed to confirm this potential interaction *(38)*.

Another study of human sperm and oocytes showed that Echinacea at high concentrations had adverse effects on oocytes and suggested that Echinacea damages reproductive cells *(39)*.

10. REGULATORY STATUS

Echinacea is regulated as a dietary supplement in the United States *(40)*. The Homeopathic Mother tincture is a Class C over-the-counter drug official in the *Homeopathic Pharmacopoeia of the United States (41)*, Official Compendium (1992). *E. angustifolia* powdered and powdered extract, *E. pallida* powdered and powdered extract, *E. purpurea* root, powdered root extract, and powdered extract have monographs for their identity, quality, and other properties in the United States Pharmacopeia National Formulary (USP-NF) *(42)*. *E. purpurea* herb is being reviewed and will likely be added to the next USP-NF.

REFERENCES

1. Hobbs C. Echinacea: a literature review; botany, history, chemistry, pharmacology, toxicology, and clinical uses. HerbalGram 1994;30:33.
2. Borchers AT, Keen CL, Stern JS, Gershwin ME. Inflammation and Native American medicine: the role of botanicals. Am J Clin Nutr 2000;72:339–347.
3. Flannery, MA. From Rudbeckia to Echinacea: The emergence of the purple cone flower in modern therapeutics. Pharm Hist 1999;41(2):52–59.
4. Blumenthal M. Herb sales down in mainstream market, up in natural food stores. HerbalGram 2002;55:60.
5. Sandrof R. Echinacea: cold comfort. Consumer Reports 2004;69(2):30–32.
6. Luettig B, Steinmuller C, Gifford GE, Wagner H, Lohmann-Matthes ML. Macrophage activation by the polysaccharide arabinogalactan isolated from plant cell cultures of *Echinacea purpurea*. J Natl Cancer Inst 1989;81(9):669–675.
7. Goel V, Chang C, Slama JV, et al. Alkylamides of *Echinacea purpurea* stimulate alveolar macrophage function in normal rats. Int Immunopharmacol 2002;2:381–387.
8. Cundell DR, Matrone MA, Ratajczak P, Pierce JD Jr. The effect of aerial parts of *Echinacea* on the circulating white cell levels and selected immune functions of the aging male Sprague-Dawley rat. Int Immunopharmacol 2003;3:1041–1048.
9. Barrett B. Medicinal properties of Echinacea: a critical review. Phytomedicine 2003,10.66–86.
10. Brinkeborn RM, Shah DV, Degenring FH. Echinaforce® and other Echinacea fresh plant preparations in the treatment of the common cold. A randomized, placebo controlled, double-blind clinical trial. Phytomedicine 1999;6(1):1–6.
11. Melchart D, Walther E, Linde K, Brandmaier R, Lersch C. Echinacea root extracts for the prevention of upper respiratory tract infections: a double-blind, placebo-controlled randomized trial. Arch Fam Med 1998;7:541–545.
12. Sperber SJ, Shah LP, Gilbert RD, Ritchey TW, Monto AS. *Echinacea purpurea* for prevention of experimental rhinovirus colds. Clin Infect Dis 2004;38:1367–1371.
13. Binns SE, Hudson J, Merali S, Arnason JT. Antiviral activity of characterized extracts from *Echinacea* spp. (Heliantheae: Asteraceae) against *Herpes simplex v*irus (HSV-I). Planta Med 2002;68:780–783.
14. Bodinet C, Mentel R, Wegner U, Lindequist U, Teuscher E, Freudenstein J. Effect of oral application of an immunomodulating plant extract on influenza virus type A infection in mice. Planta Med 2002;68:896–900.
15. Vonau B, Chard S, Mandalia S, Wilkinson D, Barton SE. Does the extract of the plant *Echinacea purpurea* influence the clinical course of recurrent genital herpes? Int J STD AIDS 2001;12:154–158.
16. Binns SE, Purgina B, Bergeron C, et al. Light-mediated antifungal activity of *Echinacea* extracts. Planta Med 2000;66:241–244.

17. Steinmüller C, Roesler J, Grottrup E, Franke G, Wagner H, Lohmann-Matthes ML. Polysaccharides isolated from plant cell cultures of *Echinacea purpurea* enhance the resistance of immunosuppressed mice against systemic infections with *Candida albicans* and *Listeria monocytogenes*. Int J Immunopharmacol 1993;15:605–614.

18. Voaden DJ, Jacobsen M. Tumor Inhibitors. 3. Identification and synthesis of an oncolytic hydrocarbon from American coneflower roots. J Med Chem 1972;15:619–623.

19. Sun LZ, Currier NL, Miller SC. The American coneflower: a prophylactic role involving nonspecific immunity. J Altern Complement Med 1999;5:437–446.

20. Currier NL, Miller SC. The effect of immunization with killed tumor cells, with/without feeding of *Echinacea purpurea* in an erythroleukemic mouse model. J Altern Complement Med 2002;8(1):49–58.

21. Currier NL, Miller SC. *Echinacea purpurea* and melatonin augment natural-killer cells in leukemic mice and prolong life span. J Altern Complement Med 2001;7(3):241–251.

22. Speroni E, Govoni P, Guizzardi S, Renzulli C, Guerra MC. Anti-inflammatory and cicatrizing activity of *Echinacea pallida* Nutt. root extract. J Ethnopharmacol 2002;79:265–272.

23. Clifford LJ, Nair MG, Rana J, Dewitt DL. Bioactivity of alkamides isolated from *Echinacea purpurea* (L.) Moench. Phytomedicine 2002;9:249–253.

24. Raso GM, Pacilio M, Di Carlo G, Esposito E, Pinto L, Meli R. In-vivo and in-vitro anti-inflammatory effect of *Echinacea purpurea* and *Hypericum perforatum*. J Pharm Pharmacol 2002;54:1379–1383.

25. Mengs U, Clare CB, Poiley JA. Toxicity of *Echinacea purpurea*. Acute, subacute and genotoxicity studies. Arzneimittelforschung 1991;41(10):1076–1081.

26. Sloley BD, Urichuk LJ, Tywin C, Coutts RT, Pang PK, Shan JJ. Comparison of chemical components and antioxidant capacity of different *Echinacea* species. J Pharm Pharmacol 2001;53:849–857.

27. Woelkart K, Koidl C, Grisold A, et al. Bioavailability and pharmacokinetics of alkamides from the roots of *Echinacea angustifolia* in humans. J Clin Pharmacol 2005;45:683–689.

28. Kligler B. Echinacea. Am Fam Physician 2003;67:77–80,83.

29. Neff GW, O'Brien C, Montalbano M, et al. Consumption of dietary supplements in a liver transplant population. Liver Transpl 2004;10:881–885.

30. Mullins RJ, Heddle R. Adverse reactions associated with echinacea: the Australian experience. Ann Allergy Asthma Immunol 2002;88:42–51.

31. Miller LG. Herbal medicinals: selected clinical considerations focusing on known or potential drug-herb interactions. Arch Intern Med 1998;158:2200–2211.

32. Soon SL, Crawford RI. Recurrent erythema nodosum associated with echinacea herbal therapy. J Am Acad Dermatol 2001;44:298–299.

33. Kemp DE, Franco KN. Possible leukopenia associated with long-term use of echinacea. J Am Board Fam Pract 2002;15:417–419.

34. Logan JL, Ahmed J. Critical hypokalemic renal tubular acidosis due to Sjögren's syndrome: association with the purported immune stimulant echinacea. Clin Rheumatol 2003;22:158–159.

35. Budzinski JW, Foster BC, Vandenhoek S, Arnason JT. An *in vitro* evaluation of human cytochrome P450 3A4 inhibition by selected commercial herbal extracts and tinctures. Phytomedicine 2000;7(4):273–282.
36. Gorski JC, Huang SM, Pinto A, et al. The effect of echinacea (*Echinacea purpurea* root) on cytochrome P450 activity in vivo. Clin Pharmacol Ther 2004;75:89–100.
37. Gallo M, Sarkar M, Au W, et al. Pregnancy outcome following gestational exposure to echinacea: a prospective controlled study. Arch Intern Med 2000;160:3141–3143.
38. Ondrizek RR, Chan PJ, Patton WC, King A. Inhibition of human sperm motility by specific herbs used in alternative medicine. J Assist Reprod Genet 1999;16(2):87–91.
39. Ondrizek RR, Chan PJ, Patton WC, King A. An alternative medicine study of herbal effects on the penetration of zona-free hamster oocytes and the integrity of sperm deoxyribonucleic acid. Fertil Steril 1999;71:517–522.
40. United States Congress. Public Law 103-417: *Dietary Supplement Health and Education Act of 1994* (S 7840). Washington, DC: 103rd Congress of the United States, 1994.
41. *Homeopathic Pharmacopoeia of the United States—Revision service official compendium from July 1, 1992*. Falls Church: American Institute of Homeopathy, 1991, 3212, ECHN.
42. *United States Pharmacopeia (USP 27th Revision)—The national formulary (NF 22nd edition)*. Rockville: United States Pharmacopeial Convention, 2004.

Chapter 7

Feverfew

Richard L. Kingston

SUMMARY

Although feverfew has been demonstrated to provide therapeutic benefit to isolated patient groups and at varying dosage levels, its clinical effectiveness has not been consistently reported. A variety of caveats associated with study design, identification or concentration of appropriate substance, or duration of evaluation from previous clinical studies complicate the assessment of the overall benefit of the herb. Despite inconsistent reports of effectiveness, the favorable safety profile and low risk of use suggest a positive risk benefit for patients looking for migraine prophylaxis treatment alternatives.

Key Words: *Tanacetum partheium*; migraine; parthenolide; oral ulceration; sesquiterpene lactones.

1. HISTORY

Feverfew is a short perennial bush that grows along fields and roadsides. It reaches heights of 15–60 cm. With its yellow-green leaves and yellow flowers, it can be mistaken for chamomile (*Matricaria chamomilla*). The flowers bloom from July to October (*1*). Since the time of Dioscorides in the first century CE, feverfew has been used for the treatment of headache, menstrual irregularities, and fever. The common name is in fact a corruption of the Latin word *febrifugia (2)*. Other traditional uses include treatment of menstrual pain, asthma, arthritis (*1*), psoriasis, threatened miscarriage, toothache, opium abuse, vertigo, *(3)*, tinnitus, anemia, the common cold, and gastrointes-

From Forensic Science and Medicine:
Herbal Products: Toxicology and Clinical Pharmacology, Second Edition
Edited by: T. S. Tracy and R. L. Kingston © Humana Press Inc., Totowa, NJ

tinal disturbances *(4)*. It was also used to aid in expulsion of the placenta and stillbirths *(3)*, and in difficult labor *(4)*. Feverfew has been planted around houses to act as an insect repellant, as well as for use as a topical remedy for insect bites *(1)*.

2. CURRENT PROMOTED USES

In the 1970s, use of feverfew as an alternative to traditional medicines for relief from arthritis and migraine headache began gaining popularity *(2)*. Prevention of migraine headache and nausea and vomiting associated with migraine headache is the most commonly promoted indication for feverfew.

3. SOURCES AND CHEMICAL COMPOSITION

Tanacetum parthenium Schulz-Bip, formerly *Chrysanthemum parthenium* (L.) Bernh, *Leucanthemum parthenium* (L.) Gren and Gordon, *Pyrethum parthenium* (L.) Sm; also described as a member of the genus *Matricaria (1,4)*; featherfew, altamisa, bachelor's button, featherfoil, febrifuge plant, midsummer daisy, nosebleed, Santa Maria, wild chamomile, wild quinine *(1)*, amargosa, flirtwort, manzanilla, mutterkraut, varadika *(4)*.

4. PRODUCTS AVAILABLE

Feverfew is available as a fresh leaf; dried, powdered leaf; capsules; tablets; fluid extract; dry standardized extract; crystals; and oral drops *(4)*. Brand names include Migracare® (600 µg of parthenolide per capsule), Migracin® (feverfew extract 1:4 and white willow bark), MigraSpray®, MygraFew®, Lomigran, 125-mg Migrelief® (light green round tablet containing 600 µg of parthenolide), Partenelle, Phytofeverfew, and 125-mg Tanacet® (not <0.2% parthenolide). Feverfew contains flavonoid glycosides and sesquiterpene lactones. Parthenolide can constitute up to 85% of the sesquiterpene lactones in feverfew grown in Europe *(1,4)*, but is present in lesser amounts or is even totally absent from North American feverfew *(4)*. Parthenolide is concentrated in the flowers and leaves, as opposed to stems and roots, and parthenolide content of the leaves may decrease during storage *(4,5)*. The vegetative cycle also influences parthenolide content *(4)*. Although parthenolide is thought to be the active ingredient in feverfew, and preparations are often standardized based on parthenolide content *(6)*, there may be other active compounds, including the lipophilic flavonol tanetin, other methyl monoterpene ethers, and chrysanthenyl acetate, monoterpene *(4)*.

Most tablet and capsule formulations contain 300 mg of feverfew, and the recommended dose is usually two to six tablets or capsules per day. A dose of 250 µg of parthenolide is considered an adequate daily dose, and 0.2% parthnolide is considered the acceptable minimum parthenolide concentration; therefore, the manufacturer's recommended dose is probably in excess of what is considered therapeutic *(6)*. In the prevention of migraine headache, doses used in studies have been 50–100 mg of dried feverfew leaves daily *(7–9)* (60 mg of dried feverfew leaves = 2.5 leaves) *(4)*. However, the parthenolide content of the North American plant is low *(6)*, and parthenolide content in feverfew products varies widely *(5)*, and may be lower than stated on the label *(10)* or even absent from some preparations *(5,10)*. For example, no parthenolide could be detected in two-thirds of feverfew products purchased in Louisiana health food stores *(6)*. The parthenolide content of powdered feverfew leaves falls during storage *(5)*. Given that the active ingredients of feverfew have yet to be definitively determined, it is difficult to designate a therapeutic or toxic dosage range. (*See* Subheading 5.1. for discussion of melatonin content of feverfew products).

5. PHARMACOLOGICAL/TOXICOLOGICAL EFFECTS

5.1. Neurological Effects

Feverfew's mechanism of action in the prevention of migraine headaches is not known. It is speculated that feverfew affects platelet activity or inhibits vascular smooth-muscle contraction, perhaps by inhibiting prostaglandin synthesis *(4)*. Results of in vitro studies suggest that rather than acting as a cyclooxygenase inhibitor, feverfew inhibits phospholipase A2, thus inhibiting release of arachidonic acid from the cell membrane phospholipid bilayer *(11,12)*.

Drugs that are serotonin antagonists are used in migraine prevention (e.g., methysergide) *(9)*. During a migraine, serotonin is released from platelets *(9)*, and in vitro studies using a bovine platelet bioassay have shown that parthenolide, as well as other sesquiterpene lactones, inhibits platelet serotonin release *(13)*. Both parthenolide and a chloroform extract of dried, powdered leaves were also able to inhibit serotonin release and platelet aggregation in an in vitro study using human platelets and a variety of platelet-activating agents *(14)*. The effect of these substances on platelet aggregation caused by a variety of chemicals was tested, and was similar except that inhibition of platelet aggregation induced by the calcium ionophore A23187 by chloroform extract leveled off at a relatively low concentration and was not complete, whereas parthenolide inhibited aggregation in a dose-dependent manner, suggesting a different mechanism of action.

Similarly discrepant results were reported in a study comparing chloroform extracts of fresh and dried feverfew and parthenolide *(15)*. In this in vitro study, both the fresh extract and parthenolide were able to irreversibly inhibit contraction of rabbit aortic ring and rat anococcygeus muscle in a dose-dependent manner. In contrast, the extract from dried powdered feverfew leaves was spasmogenic, causing a slow, maintained, reversible contraction. The differences in pharmacological effect were explained by the differences in composition of the extracts; unlike the extract of fresh leaves, the extract of dried powdered leaves did not contain parthenolide or other sesquiterpene lactones. The specific functional group responsible for inhibition of smooth muscle contraction has been identified as the α-methylene moiety present on parthenolide and other sesquiterpene lactones *(16)*. It has been hypothesized that the irreversible inhibition of platelet aggregation and inhibition of smooth muscle contraction are caused by covalent binding of parthenolide and other lactones to sulfhydryl (SH-) groups on proteins *(15)*.

Another study using chloroform extract of fresh feverfew leaves demonstrated reversible blockade of open voltage-dependent potassium channels, but not of calcium-dependent potassium channels, in smooth muscle cells in vitro *(17)*. Inhibition of potassium channels would be expected to increase the excitability of smooth muscle cells, potentiate the effects of depolarizing stimuli, and open voltage-dependent calcium channels, thus leading to muscle contraction. In the study described previously *(17)*, the extract of dried, powdered feverfew had this very effect, which could be explained by potassium channel blockade; however, the fresh extract had the opposite effect (i.e., it irreversibly inhibited contractility). In addition, parthenolide, which was present in fresh but not dried extracts, did not appear to inhibit potassium channels. The substances in feverfew that cause potassium channel blockade and muscle contraction have not been identified, but because voltage-dependent potassium channels present in smooth muscle cells are similar to those present in neurons, it is possible that feverfew interferes with the neurogenic response in migraine *(17)*.

One of the first studies to attempt to objectively evaluate the efficacy of feverfew for migraine prophylaxis enrolled 17 patients with common or classical migraine who had been self-medicating with raw feverfew leaves (average 2.44 leaves [60 mg]) daily for at least 3 months *(7)*. Patients were randomized to receive either 50 mg of freeze-dried feverfew powder or placebo for 6 months. One patient in each group was taking conjugated equine estrogens (Premarin®) and one patient in the feverfew group was taking Orlest® 21, an oral contraceptive. Efficacy was assessed using patient diaries in which patients recorded the duration and severity of headache pain, severity and duration of nausea and vomiting, and analgesic use on an ordinal scale. The frequency of migraine, nausea, and vomiting, was significantly ($p < 0.02$) lower in the feverfew group, but analgesic use was similar. Two patients taking placebo withdrew from the study because of recurrent severe migraine. Patients taking feverfew reported a similar number of migraine attacks during the study compared to before the study, when they were self-medicating with feverfew. Conversely, placebo patients reported a frequency of headache that was greater than when they were self-medicating, and similar to the frequency of headache before beginning feverfew self-treatment. At the end of the study, the patients assessed the overall efficacy of the treatment; feverfew had a more favorable rating than placebo ($p < 0.01$). Because there was underreporting of headache in the placebo group, the difference between feverfew and placebo may have been even greater. Adverse events were not reported in the feverfew group, but patients did complain of the product's taste. A potential problem with this study was blinding; most patients guessed correctly which treatment they were receiving. Another criticism of this study is that because the participants were recruited from a population already taking feverfew and who presumably felt they were benefiting from feverfew, the investigators were in effect selecting known "feverfew responders" for their study. Such a selection process limits the extent to which these study results can be extrapolated to the general population.

In a subsequent study, efficacy of feverfew in migraine prophylaxis was further assessed in a double-blind, randomized, crossover design *(9)*. One capsule of dried feverfew leaves (70–114 mg, average 82 mg) was compared to placebo in 72 adult volunteers with classical or common migraine. All subjects had migraine of at least a 2-year duration, and suffered at least one attack per month. Patients were excluded if they were being treated for any other disease, but women taking oral contraceptives were eligible for the study if they had been on the same contraceptive for at least 3 months. Females of child-bearing potential were excluded unless they were using adequate contraception. All migraine-related drugs were stopped at the beginning of the trial, which commenced with a 1-month single-blind placebo run-in period. Patients were then randomized to placebo or feverfew for 4 months each. Efficacy was assessed based on a patient diary in which patients recorded the number, severity, and duration of any migraine attacks, as well as the presence of nausea and vomiting, on a scale from 0 to 3. In addition, every 2 months, the patient's overall impression of migraine control was assessed using a 10-cm visual analog scale. There was a significant difference ($p <$ 0.05) between placebo and feverfew in number of attacks only after month 4, but there wasa significant difference between the two groups in overall impression after month 4 ($p < 0.05$) and after month 6 ($p < 0.01$) when assessed via the visual analog scale. Feverfew decreased the number of classical migraine attacks by 32% (95% confidence interval [CI] 11–53%, $p <$ 0.05), but the effect on the number of common migraine attacks was not statistically significant ($p = 0.06$). When assessing the responses of patients who had never used feverfew before study enrollment ($n = 42/59$), the number of attacks was reduced by 23% (95% CI 10–33%, $p = 0.06$). This nonsignificant result gives credence to the concerns about selection bias in the study by Johnson and colleagues. The overall impression of both patients with common and classical migraines was favorable based on the visual analog scale ($p < 0.01$). Vomiting associated with attacks was also decreased with feverfew, and there was a trend toward reduction in migraine severity. Duration of attacks was unchanged. Incidence of adverse effects, including mouth ulceration, indigestion, heartburn, dizziness, lightheadedness, rash, and diarrhea was low and comparable to placebo.

Another randomized, double-blind, crossover study assessed the efficacy of 100 mg of feverfew (0.2% parthenolide) daily compared to placebo in 57 patients *(8)*. Efficacy was assessed using a questionnaire. Feverfew was superior to placebo in reducing intensity of migraine pain and other symptoms. Unfortunately, no results were reported for the actual number of headache attacks occurring during the study. An alcoholic extract of feverfew providing 0.5 mg of parthenolide daily for 4 months was not superior to placebo in the number of migraine attacks in a randomized, double-blind, crossover study in 44 evaluable patients *(18)*.

Surprisingly, melatonin, a human pineal hormone, has been identified in fresh green feverfew leaves at a concentration of 2.45 µg/g, and in a commercially available feverfew tablet (Tanacet, Ashbury Biologicals, Inc., Toronto, CA) at a concentration of 0.143 µg/g. Each Tanacet tablet contains 70–80 ng of melatonin, and the recommended dose is one or two tablets daily. Freeze-dried green leaf contains 2.19 µg/g of melatonin, fresh golden feverfew leaf contains 1.92 µg/g, oven-dried green leaf contains 1.69 µg/g, freeze-dried golden leaf conatins 1.61 µg/g, and oven-dried golden leaf contains 1.37 µg/g. Because chronic migraine headaches are associated with lower circulating melatonin levels, it is possible that melatonin plays a role in feverfew's purported efficacy in preventing migraine headache. This finding underscores the need to fully characterize the ingredients in herbs and medicinal preparations made from them *(19)*.

A concentrated CO_2 extract of *T. parthenium* (feverfew) indentified as MIG-99 was evaluated in a 12-week, double-blind, multicenter, randomized, placebo-controlled, dose-response study involving 147 patients *(20)*. The clinical effectiveness of three dosage levels of MIG-99 (2.08, 6.25, and 18.75 mg) administered three times daily was studied. In general, the compound failed to demonstrate a significant prophylactic effect in any treatment group. Only the maximum migraine intensity, severity, and the number of attacks with confinement to bed were reduced by MIG-99. In the intent-to-treat analysis, MIG-99 was shown to be effective only in a small, predefined subgroup of patients receiving the 6.25-mg dose. These patients were noted to have a total of four attacks reported in a baseline period. Regarding toxicity and safety, the incidence of adverse events was similar between all treatment groups as compared to placebo, and the incidence of patients reporting at least one adverse effect was lowest in the patients receiving the highest dose. Additionally there were no negative laboratory investigations or changes in vital signs during the treatment regimen in any patient group.

A randomized, double-blind, placebo-controlled trial comparing the effects of a compound containing a combination of riboflavin (4000 mg daily), magnesium (300 mg daily), and feverfew (100 mg standardized to 0.7 mg parthenolide daily) to placebo (25 mg riboflavin) showed a placebo effect comparable to the combination compound *(21)*. In this particular study, the placebo effect exceeded that reported for any other placebo in migraine prophylaxis trials. The trial was undertaken to study patient response to a "natural" multicombination product containing ingredients with previously demonstrated efficacy in at least one double-blind, placebo-controlled trial. Although there was no statistical difference between groups during the 3-month trial, both groups were superior to baseline in reduction of number of migraines, migraine days, and migraine index, but not superior to previously reported positive results for any of the agents alone. Possible reasons for the high placebo response (44%) included a potential therapeutic effect of the small dose of riboflavin in the placebo group, adverse interaction between the three agents used in the combination product, and a short duration of study (3 months).

In a study designed to evaluate the pharmacokinetics and toxicity of parthenolide, the active component of feverfew, doses of 1, 2, 3, and 4 mg were studied in a dose escalation fashion *(22)*. Administration of feverfew in escalating doses up to 4 mg showed no toxicity and a maximum tolerated dose was not reached. Despite a parthenolide detection level of 0.5ng/mL, no measurable concentrations of this component could be measured at any of the administered doses levels.

Larger studies are needed to definitively determine the efficacy of feverfew in the prevention of migraine and to identify the component or components responsible for its pharmacologic effects. Although parthenolide is considered the active constituent of feverfew, the pharmacokinetics of this constituent have not been characterized, and challenges remain in detecting this component analytically to allow evaluation of its metabolic fate.

5.2. Anti-Inflammatory Effects

Organic and aqueous feverfew powdered leaf extracts were found to inhibit IL-1-induced prostaglandin E2 release from synovial cells, IL-2-induced thymidine uptake by lymphoblasts, and mitogen-induced uptake of thymidine by peripheral blood mononuclear cells (PBMCs) *(23)*. Parthenolide also inhibited thymidine uptake by PBMCs. Both parthenolide and the extracts were cytotoxic to the PBMCs and synovial cells; thus, the anti-inflammatory effects of feverfew may be secondary to cytotoxicity. These results reflect those of previous researchers who found parthenolide and other sesquiterpene lactones to be cytotoxic to cultures of human fibroblasts, human laryngeal carcinoma cells, and human cells transformed with simian virus 40 *(24)*.

The anti-inflammatory effect of dried powdered feverfew leaf was compared to placebo in the treatment of rheumatoid arthritis (RA) *(25)*. This double-blind, randomized study used dried powdered feverfew leaf 70–86 mg (mean 76 mg), equivalent to 2–3 μmol of parthenolide. A total of 41 female patients with RA from a rheumatology clinic participated. Patients were allowed to continue their usual doses of nonsteroidal anti-inflammatory drugs and other analgesics. If a patient deteriorated acutely during the study, a single intraarticular dose of 20 mg of triamcinolone hexacetonide was allowed at week 3. Efficacy was determined by clinical assessments at weeks 3 and 6, and included duration of early morning stiffness in minutes, inactivity stiffness (present/absent), pain (10 cm visual analog scale), grip strength, and Richie articular index. Patients were also questioned about adverse effects. At weeks 0 and 6, hemoglobin, white blood cell count, platelet count, urea, creatinine, erythrocyte sedimentation rate, C reactive protein, immunoglobulin G (IgG), IgM, IgA, latex fixation test, Rose-Waaler titer, C3 degradation products, and Steinbrocker functional capacity were determined. At week 6, a global impression from both the patient and the clinician were recorded as better, same, or worse. One patient in the placebo group dropped out after the third day because of lightheadedness, but complete data were obtained for the remaining 40 patients. One patient receiving feverfew reported minor ulceration and soreness of the tongue. At baseline, hemoglobin and serum creatinine levels were lower in the placebo group than in the feverfew group. By week 3, urea levels had significantly increased ($p = 0.04$) in the feverfew group, but this was not apparent at week 6. At week 6, grip strength and IgG were increased in the feverfew group compared with baseline ($p = 0.47$ and 0.025, respectively). Overall, the results of this study do not support the efficacy of 76 mg of dried feverfew leaf in the treatment of RA.

5.3. Mutagenicity/Carcinogenicity/Teratogenicity

In 30 patients with migraine who had been taking feverfew leaves, tablets, or capsules for at least 11 months, there was no increase in chromosomal aberrations or sister chromatid exchange in circulating lymphocytes compared to patients with migraine not taking feverfew matched for age and sex. The Ames salmonella mutagenicity test was also performed on urine samples from 10 patients using feverfew and 10 matched nonusers, with no indication of mutagenicity *(26)*.

No problems have been reported in offspring of pregnant women who used feverfew, but feverfew has purportedly been associated with spontaneous abortion in cattle and uterine contractions in term human pregnancies *(4)*.

6. ADVERSE EFFECTS AND TOXICITY

6.1. Case Reports of Toxicity Caused By Commercially Available Products

Adverse effects associated with feverfew use include dizziness, lightheadedness, nausea, heartburn, indigestion, bloating, gas, constipation, diarrhea, inflammation, and ulceration of the oral mucosa, weight gain, palpitations, heavier menstrual flow, contact dermatitis, and rash *(4)*. Feverfew belongs to the *Compositae* family *(1)*, and persons allergic to other members of this family such as chamomile, ragweed, asters, chrysanthemums *(27)*, and echinacea could also be allergic to feverfew. Out of 300 feverfew users, 18% of those questioned reported adverse effects, with mouth ulceration reported in 11.3% *(7)*. Feverfew-induced mouth ulceration is not a manifestation of contact dermatitis; it is a systemic reaction. In contrast, inflammation of the tongue and oral mucosa accompanied by lip swelling and loss of taste is probably caused by direct contact with feverfew and is not associated with use of feverfew capsules or tablets *(28)*.

In the study by Johnson and colleagues, in which 10 patients who had been taking fresh feverfew leaves were switched to placebo, patients experienced recurrence of migraine, tension headaches, joint pain and stiffness, nervousness, insomnia and disrupted sleep, and tiredness. The investigators dubbed these symptoms the "postfeverfew syndrome." Dr. Johnson had documented this syndrome in a previous publication when approx 10% of 164 patients who discontinued feverfew reported anxiety, poor sleep, joint and muscle aches, pains, and stiffness *(7)*.

7. DRUG INTERACTIONS

In vitro studies suggest that feverfew may inhibit platelet aggregation, leading to recommendations that patients avoid use of feverfew with anticoagulants and medications with antiplatelet activity *(4)*. Platelets from 10 patients who had taken feverfew for at least 3.5 years responded normally to aggregation induced by adenosine diphosphate and thrombin compared to platelets from four control patients who had stopped taking feverfew at least 6 months earlier. In patients who had been taking feverfew for at least 4 years, the threshold for platelet aggregation in response to $11\alpha,9\alpha$;-epoxymethanoprostaglandin H2 (U46619) and serotonin was elevated *(29)*. Whether these results translate into the potential for drug interactions and bleeding diatheses remains to be documented.

8. REGULATION

In the United States, feverfew may be marketed as a dietary supplement, but is not approved as a drug. A United States Pharmacopeia advisory panel, although recognizing that feverfew has a long history of use and lack of documented adverse effects, does not recommend its use owing to the paucity of scientific evidence of safety and efficacy. The panel encourages further research, including at least one properly designed clinical trial *(4)*.

In Canada, the Health Protection branch allows sale of tablets and capsules made from feverfew crude dried leaves for decreasing the frequency and severity of migraine headaches. The products should be standardized to contain no less than 0.2% parthenolide. In France, feverfew has traditional use in the treatment of heavy menstrual flow and prevention of migraine headache *(4)*.

REFERENCES

1. Anonymous, ed. Feverfew. In: *The Lawrence Review of Natural Products*. St. Louis: Facts and Comparisons, 1994.
2. Tyler VE, ed. *The Honest Herbal, 3rd edition*. Binghamton: Pharmaceutical Products Press, 1993.
3. Knight DW. Feverfew: chemistry and biological activity. Nat Prod Rep 1995;12:271–276.

4. United States Pharmacopeial Convention (USP). Feverfew. *Botanical Monograph Series*. Rockville: United States Pharmacopeial Convention, 1998.
5. Heptinstall S, Awang DVC, Dawson BA, Kindack D, Knight DW, May J. Parthenolide content and bioactivity of feverfew (*Tanacetum parthenium* [L.] schultz-Bip.). Estimation of commercial and authenticated feverfew products. J Pharm Pharmacol 1992;44:391–395.
6. Tyler VE, ed. *Herbs of Choice: the Therapeutic Use of Phytomedicinals*. Binghamton: Pharmaceuticals Products Press, 1994.
7. Johnson ES, Kadam NP, Hylands DM, Hylands PJ. Efficacy of feverfew as prophylactic treatment of migraine. Br Med J 1985;291:569–573.
8. Palevitch D, Earon G, Carasso R. Feverfew (Tanacetum parthenium) as a prophylactic treatment for migraine: a double-blind placebo-controlled study. Phytother Res 1997;11:508–511.
9. Murphy JJ, Heptinstall S, Mitchell JRA. Randomized, double-blind placebo-controlled trial of feverfew in migraine prevention. Lancet 1988;2:189–192.
10. Groenewegen WA, Heptinstall S. Amounts of feverfew in commercial perparations of the herb. Lancet 1986;1:44–45.
11. Collier HOJ, Butt NM, McDonald-Gibson WJ, Saeed SA. Extract of feverfew inhibits prostaglandin biosynthesis. Lancet 1980;2:922–923.
12. Makheja AN, Bailey JM. A platelet phospholipase inhibitor from the medicinal herb feverfew (Tanacetum parthenium). Prostaglandins Leukot Med 1982;8:653–660.
13. Marles RJ, Kaminski J, Arnason JT. A bioassay for the inhibition of serotonin release from bovine platelets. J Nat Prod 1992;55:1044–1056.
14. Groenewegen WA, Heptinstall S. A comparison of the effects of an extract of feverfew and parthenolide, a component of feverfew, on human platelet activity in vitro. J Pharm Pharmacol 1990;42:553–557.
15. Barsby RWJ, Salan U, Knight DW, Hoult JRS. Feverfew and vascular smooth muscle: extracts from fresh and dried plants show opposing pharmacological profiles, dependent upon sesquiterpene lactone content. Planta Med 1993;59:20–25.
16. Hay AJB, Hamburger M, Hostettmann K, Hoult JRS. Toxic inhibition of smooth muscle contractility by plant-derived sesquiterpenes caused by their chemically reactive α;-methylenebutyrolactone functions. Br J Pharmacol 1994;112:9–12.
17. Barsby RWJ, Knight DW, McFadzean I. A chloroform extract of the herb feverfew blocks voltage-dependent potassium currents recorded from single smooth muscle cells. J Pharm Pharmacol 1993;45:641–645.
18. DeWeerdt CJ, Bootsma HPR, Hendriks H. Herbal medicine in migraine prevention: randomized, double-blind, placebo-controlled crossover trial of a feverfew preparation. Phytomedicine 1996;3:225–230.
19. Murch SJ, Simmons CB, Saxena PK. Melatonin in feverfew and other medicinal plants. Lancet 1997;350:1598–1599.
20. Pfaffenrath V, Diener HC, Fischeer M, Friede M, Henneicke-von Zepelin HH. The efficacy and safety of Tanacetum parthenium (feverfew) in migraine prophylaxix — a double-blind, multicentre, randomized placebo-controlled dose-response study. Cephalalgia 2002;22(7):523–532.

21. Maizels M, Blumenfeld A, Burchette MS. A combination of riboflavin, magnesium, and feverfew for migraine prophylaxis: a randomized trial. Headache 2004;44(9):885–890.
22. Curry EA III, Murray DJ, Yoder C, Fife K, Armstrong V. Phase I dose excalation trial of feverfew with standardized doses of parthenolide in patients with cancer. Invest New Drugs 2004;22(3):299–305.
23. O'Neill LAJ, Barrett ML, Lewis GP. Extracts of feverfew inhibit mitogen-induced human peripheral blood mononuclear cell proliferation and cytokine mediated responses: a cytotoxic effect. Br J Clin Pharmacol 1987;23:81–83.
24. Lee K, Huang E, Piantadosi C, Pagano JS, Geissman TA. Cytotoxicity of sesquiterpene lactones. Cancer Res 1971;31:1649–1654.
25. Pattrick M, Heptinstall S, Doherty M. Feverfew in rheumatoid arthritis: a double blind, placebo controlled study. Ann Rheum Dis 1989;48:547–549.
26. Anderson D, Jenkinson PC, Dewdney RS, Blowere SD, Johnson ES, Kadam NP. Chromosomal aberrations and sister chromatid exchanges in lymphocytes and urine mutagenicity of migraine patients: a comparison of chronic feverfew users and age matched nonusers. Hum Toxicol 1988;7:145–152.
27. Benner MH, Lee HJ. Anaphylactic reaction to chamomile tea. J Allergy Clin Immunol 1973;52:307–308.
28. Awang DVC. Feverfew fever. A headache for the consumer. HerbalGram 1993;29:34–36,66.
29. Biggs MJ, Johnson ES, Persaud NP, Ratcliffe DM. Platelet aggregation in patients using feverfew for migraine. Lancet 1982;2:776.

Chapter 8

Garlic

Leslie Helou and Ila M. Harris

SUMMARY

Garlic possesses a variety of beneficial pharmacological properties affecting most notably the cardiovascular system (lipid management, decreased blood pressure, platelet inhibition, and decreased fibrinolytic activity), and the immune system as an antineoplastic and immunostimulant agent. It is also a potent antioxidant. Although its effects are modest in some clinical applications, its inherent safety and culinary benefits usually support its use when a mild clinical effect is acceptable. There is the potential for interactions with drugs possessing antiplatelet and anticoagulant effects. Additionally, potential induction of various cytochrome P450 enzymes warrants closer monitoring of drugs metabolized through this pathway when garlic is concomitantly administered.

Key Words: *Allium sativum*; antilipemic; platelet inhibition; antioxidants; cancer prevention; P450 enzyme induction.

1. HISTORY

Garlic use dates back to Old Testament times, when it was a favored food. Drawings of garlic from 3700 BCE were uncovered in Egyptian tombs. Over the centuries, garlic has been used to ward off vampires, demons, witches, and evil beings and was thought to have magical properties. Medicinal uses date back to 1550 BCE, when it was used as a remedy for heart disease, headaches, and tumors. It has also been used as an aphrodisiac to improve sexual performance and desire, and as a cure-all for everything from hemorrhoids to snake bites *(1–4)*.

From Forensic Science and Medicine:
Herbal Products: Toxicology and Clinical Pharmacology, Second Edition
Edited by: T. S. Tracy and R. L. Kingston © Humana Press Inc., Totowa, NJ

2. CURRENT PROMOTED USES

In 1997, garlic was the most widely used natural supplement in US households. Garlic was shown to be used more than twice as much as any other natural supplement *(5)*. Garlic is promoted to lower cholesterol and blood pressure, delay the progression of atherosclerosis, prevent heart disease, improve circulation, prevent cancer, and is used topically for tinea infections *(3,4)*.

3. SOURCES AND CHEMICAL COMPOSITION

Allium sativum, *Allii sativi bulbus*, knoblauch, ail, ajo, allium, Camphor of the Poor, Garlic Clove, Nectar of the Gods, Poor Man's Treacle, Rust Treacle, Stinking Rose *(3,6)*.

4. PRODUCTS AVAILABLE

Four types of garlic preparations are currently available on the US market: garlic essential oils, garlic oil macerate, garlic powder, and aged garlic extract (AGE). Most garlic preparations report allicin yield potentials, whereas AGE products standardize to S-allylcysteine (SAC) amounts. Some products have been shown to release differing amounts of active components depending on when the product was made *(7)*.

4.1. Product Names

- Garlic-Gold, extract 600 mg (*A. sativum*), (7200-µg allicin yield)
- Garlic-Go!, 1000 mg AGE
- Garlic HP — Physiologics, 400-mg garlic bulb (1000 µg allicin/g yield)
- Garlic HP 650 — Physiologics, bulb powder (allicin yield 6500 µg, total thiosulfinates 6500 µg, alliin yield 14,500 µg, γ-glutamylcysteines 5200 µg)
- Garlife — Life Extension, 900-mg pure garlic extract (odor suppressed)
- Garlinase 4000 — Enzymatic Therapy, extract equivalent to 4 g fresh garlic (3.4% allicin)
- Garlique® — Chattem, 400-mg bulb powder (500 µg allicin yield)
- GNC Garlic Oil — Basic Nutrition, 0.65 mg
- Herbscience® Garlic, 600-mg caplets
- Kwai® Odor-Free Garlic, 150-mg tablets (900 µg allicin yield)
- Kyo-Chrome AGE Cholesterol Formula, 400-mg extract powder, niacin 20 mg, chromium 200 µg
- Kyolic® Aged Garlic Extract Kyolic HI-PO™, 600-mg tablets AGE powder
- Kyolic Liquid Aged Garlic Extract, 1-mL AGE
- Kyolic Reserve Aged Garlic Extract, 600-mg capsules AGE powder
- Natrol® GarliPure Daily Formula, 600-mg powdered bulb extract (1200 µg allicin yield, 1200 µg thiosulfinates)

- Natrol GarliPure Formula 500, 1000-mg powdered bulb extract (1500 μg allicin yield, 1600 μg thiosulfinates)
- Natrol GarliPure Maximum Allicin Formula, 600-mg powdered bulb extract (3600 μg allicin yield, 3800 μg thiosulfinates)
- Natrol GarliPure Once Daily Potency, 600-mg powdered bulb extract (6000 μg allicin yield, 6060 μg thiosulfinates)
- Natrol GarliPure Organic Formula, 1000-mg organically grown powdered bulb extract (1500 μg allicin yield, 1600 μg thiosulfinates)
- Nature's Plus® Garlite, 500-mg odorless Vegicap
- Nature's Plus Ultra Garlite, 1000-mg deodorized sustained-release tablet
- Nature's Way Garlicin, 300-mg allinaise-rich garlic powder
- Nature's Resource® Garlic Cloves, 400-mg capsules garlic bulb (0.8 mg allicin)
- Nature's Resource Odor-Controlled, 180-mg enteric-coated tablets garlic bulb (1.8 mg allicin)
- One-a-Day® Garlic, 600-mg odor-free softgel, concentrated oil macerate
- Sundown® Herbals Garlic Oil, 3-mg oil (1500-mg garlic clove equivalent)
- Sundown Herbals Garlic Whole Herb, 400-mg tablets garlic clove concentrate
- Sundown Herbals Garlic, 400-mg tablets garlic clove concentrate (1200-mg garlic clove equivalent)
- Sundown Herbals Odorless Garlic, 400-mg tablets garlic clove concentrate (1200-mg garlic clove equivalent)
- Sun Source® Garlique, 400-mg enteric-coated, odor-free tablets, garlic powder (5000 μg allicin yield)
- Wellness GarlicCell, 650-mg enteric-coated tablets, garlic clove (6000 μg allicin yield, 6000 μg thiosulfinates)

4.2. Recommended Daily Doses in Humans

- 4 g of fresh garlic, approx 1 clove (4–12 mg of allicin or 2–5 mg of allicin) *(6)*
- Dehydrated garlic powder, 600–1200 mg in divided doses
- AGE, 1–7.2 g/day
- Fresh air-dried bulb, 2–5 g
- Garlic oil, 2–5 mg
- Dried bulb, 2–4 g times daily
- Tincture (1:5 in 45% alcohol), 2–4 mL three times daily *(6)*

4.3. Garlic Compounds

Raw, intact garlic contains various chemical compounds, all of which are converted to other sulfur-containing compounds when processed. All of these compounds are derived from the compound allicin. Allicin is formed from alliin, by the action of allinase, which is released when garlic is chopped or chewed *(8)*. Allicin is extremely unstable and further breaks down to produce hundreds of organosulfur compounds such as diallyl sulfide (DAS), diallyl disulfide (DADS), diallyl trisulfide, ajoenes, methyl allyl di- and trisulfides, vinyl dithiins, and other sulfur compounds, depending on how the garlic is prepared. By the formation of these compounds, allicin is responsible for most of the biological activity of garlic; however, it is also a major contributor in garlic's characteristic odor *(5)*.

Different methods of processing garlic, resulting in products containing different sulfur-containing thiosulfinate derivatives, have been discussed *(8)*. A bulb of raw garlic, on average will contain up to 1.8% alliin, a small amount of SAC, which is a less-odorous biologically active compound, but no allicin. When garlic is chopped or crushed, 1 mg of alliin is converted to 0.48 mg of allicin *(3)*. Cooking whole or coarsely chopped garlic destroys allinase, the enzyme necessary for production of allicin, ajoene, diallyl sulfide, diallyl disulfide, and vinyl dithiins; only cysteine sulfoxides such as alliin remain.

Crushing or finely chopping garlic followed by boiling in an open container leads to volatilization and loss of many chemically unstable, but potentially medicinal thiosulfinates. Steam distillation produces an oily mass of active compounds including diallyl, methyl allyl, dimethyl, and allyl 1-propenyl oligosulfides that originate from the thiosulfinates. Maceration of garlic in vegetable oil or soybean oil produces vinyl dithiins, ajoenes, and diallyl and methyl allyl trisulfides. These latter two methods are used to prepare some commercially available garlic capsules. When garlic is allowed to ferment (cold aging, AGE products), water-soluble SAC, *S*-allyl-mercaptocysteine, and other biologically active compounds are produced.

Garlic powder is produced by drying and pulverizing sliced or crushed garlic. The drying process is thought to cause powders to lose approximately one-half the amount of alliin found in whole garlic cloves. If dried at low temperatures, the garlic powder will remain odor-free until the product reaches the gastrointestinal (GI) tract after ingestion *(7)*. In contrast to common belief, odorless garlic products still produce the same adverse drug reactions as those nonodorless products. Kwai brand coated odorless garlic powder tablets contain dried garlic powder prepared by freeze-drying fresh garlic *(9)*. After the tablets are ingested, the alliin is converted to allicin in the GI tract by the enzyme allinase, which can come into contact with alliin once the coated tablets disintegrate and mix with intestinal water. Kwai is one of the most common garlic preparations used in studies.

Regardless of the processing procedure used, no garlic preparation available contains allicin, because of its high volatility. Many products report an "allicin yield" potential when consumed. However, studies have shown that allicin is not produced in significant amounts after ingestion of garlic products, which may be owing to inactivation of alliinase in the acidic stomach. Therefore, allicin is likely not an appropriate marker of the potential activity of the product *(5)*. Enteric-coated products may preserve the activity of allinase, by delaying dissolution of the product. AGE products are standardized to SAC, which is found in detectable levels in the body and may therefore be a better standardization marker for garlic products than allicin yield *(5)*.

5. PHARMACOLOGICAL/TOXICOLOGICAL EFFECTS

5.1. Cardiovascular Effects

5.1.1. ANTIOXIDANT AND ANTIATHEROSCLEROTIC EFFECTS

Garlic has been shown to have significant effects on the cardiovascular system. Such areas include improvement in lipids, modest effects on blood pressure, platelet inhibition, antioxidant effects, and a decrease in fibrinolytic activity. In vitro studies have shown garlic possesses specific antiatherosclerotic effects such as reducing inducible nitric oxide synthase (iNOS) mRNA expression *(10)*, inhibition of oxidized low-density lipoprotein (LDL)-induced lactate dehydrogenase (LDH) release and inhibition of oxidized LDL-induced depletion of glutathione *(11)*.

Results from a study in nine subjects found that supplementation with AGE at a dose of 2.4g/day significantly inhibited the oxidation of LDL, but ingesting 6 g/day of crushed raw garlic did not have a significant effect. The authors believe that this difference in response may be owing to the fact that the active ingredient in raw garlic is allicin, whereas SAC is believed to be the active component of AGE in preventing atherosclerosis. However, when compared to α-tocopherol (Vitamin E), which is well documented at preventing lipid oxidation, both AGE and raw garlic were less effective at inhibiting oxidation ($p < 0.05$) *(12)*. In addition, it has been shown that 900 mg/day of garlic powder vs placebo for 4 years caused a significant decrease in arteriosclerotic plaque volume in both men and women with advanced atherosclerotic plaques and at least one cardiovascular risk factor *(13)*.

The effects of garlic as an antioxidant and its ability to alter the atherosclerotic process require additional study. To date, no trials evaluating patient outcomes have been completed.

5.1.2. ANTIHYPERLIPIDEMIC EFFECTS

Garlic as a lipid-lowering agent is perhaps the most studied topic related to its use in cardiovascular health. The mechanism by which garlic lowers lipoprotein levels is not well understood. Animal data shows that garlic significantly decreases hydroxy-methylglutaryl coenzyme A (HMG-CoA) reductase activity, and may have some effects on cholesterol α-hydroxylase, fatty acid synthetase, and pentose-phosphate pathway enzyme activity *(14)*.

A recent meta-analysis using multiple databases, from inception until November 1998, compiled all randomized, double-blind, placebo-controlled trials using monopreparations of garlic, to test the effectiveness of garlic in lowering total cholesterol (TC) *(15)*. Inclusion criteria included trials in which participants had elevated TC, defined as 5.17 mmol/L (200 mg/dL) at baseline, and reported TC levels as an end point. Studies were excluded if they did not contain enough data to compute effect size. Of the 39 garlic-in-hyperlipidemia studies identified, 21 were excluded because they were not placebo-controlled, randomized, double-blinded, did not use a monopreparation of garlic, did not report TC, or have a baseline TC meeting inclusion criteria. An additional five trials did not include enough data to perform statistical pooling. Of the 13 studies cited in the meta-analysis, 10 used Kwai powder tablets in doses of 600, 800, and 900 mg/day. One study used 700 mg of spray-dried powder per day, another used 0.25mg/kg body mass of essential oil, and the other study used 10 mg/day of steam-distilled oil. Study duration ranged from 8 to 24 weeks. Of the 13 trials, 10 required a diagnosis of hypercholesterolemia or hyperlipoproteinemia, whereas the other trials required diagnosis of coronary heart disease, hypertension, or healthy participants. A total of 796 participants were involved, and all trials excluded participants using hypolipidemic drugs. Results showed that TC levels decreased by a modest 5.8% (0.41 mmol/L; 15.7 mg/dL) in participants taking garlic compared to placebo ($p < 0.01$). Of the five methodologically similar trials using Kwai 900 mg/day, no significant difference was seen in reducing total cholesterol with garlic. Additionally, in an analysis of the six trials that controlled for diet, no significant difference was seen in reducing total cholesterol with garlic. The authors also looked at data presented in these studies regarding changes in LDL and high-density lipoprotein (HDL) levels. No significant difference was seen in reducing or increasing these values, respectively *(15)*.

Several more recent studies have confirmed the TC-lowering effect seen in this meta-analysis. A randomized, double-blind, placebo-controlled study in 50 subjects with hypercholesterolemia and LDL levels between 150 and 200 mg/dL, triglycerides less than 300 mg/dL, an average age of 53 years, and who were not using lipid-lowering drugs, evaluated the effect of 300 mg three times daily of garlic powder (Kwai) for 12 weeks. Patients were classified by their LDL pattern A or B. Pattern B LDL has been shown to be more atherogenic than pattern A This study was designed to not only look at the effect of garlic on lipoprotein levels, but also LDL particle size, LDL and HDL subclass distribution, and the effect on lipoprotein(a) [Lp(a)]. The only significant difference found was a significant decrease in LDL peak particle diameter in LDL pattern A. It is unclear what the implications of this decrease in diameter are. Results showed no significant difference in plasma lipid levels, overall LDL peak particle diameter, LDL or HDL subclass distribution, apolipoprotein B, or Lp(a) in the garlic vs placebo groups *(16)*.

Another double-blind, placebo-controlled, randomized study of 34 men, average age of 48 years, with total cholesterol levels between 220 mg/dL and 285 mg/dL evaluated the effects of 7.2 g of AGE daily for 5 months. At 2 and 4 months after beginning the study, no significant difference was seen in TC or LDL cholesterol levels. At 5 months, a significant drop (7% in TC, 10% in LDL) was seen in the garlic group vs placebo. Plasma HDL and triglyceride levels did not change *(17)*.

Overall, there is conflicting data regarding the effects of garlic on serum lipid levels. The diverse nature in the design of these studies makes it difficult to pool data. In addition, the use of various garlic preparations may have differing effects on lipids because of the diverse activity of organosulfide compounds present in each product. However, a larger number of studies have found garlic to provide a significant but small decrease in LDL and total cholesterol when garlic is used for up to 4 months. Further study is needed to determine whether garlic has a prolonged affect on lipids and if the effects are sustainable. Compared to the available lipid-lowering prescription drugs, garlic provides a small-percent decrease in lipid values and has not been shown to have morbidity and mortality benefits in these patients.

5.1.3. PLATELET INHIBITORY AND FIBRINOLYTIC EFFECTS

Platelet inhibition is another widely studied effect of garlic use. Platelet inhibition has been demonstrated in several in vitro and animal studies with fresh garlic cloves *(18)*, ajoene *(8)*, garlic oil *(19)*, and AGE *(20)*. Mechanisms proven by in vitro studies include a dose-dependent, irreversible inhibition of platelet aggregation through almost complete suppression of thromboxane production *(8,19)*, a dose-dependent inhibition of collagen-induced platelet aggregation *(21)*, and inhibition of adenosine diphosphate (ADP) and epinephrine-induced platelet aggregation *(8)*. Multiple mechanisms may be responsible for the platelet inhibitory affects of garlic. It is thought that the inhibition of thromboxane production is caused by inhibition of cyclooxygenase, but not lipoxygenase *(8,19)*; however, some studies question whether garlic inhibits cyclooxygenase. There may be a direct inhibition of thromboxane. β-Thromboglobulin release is decreased, which suggests that the effect may be more on the platelet activation phase *(23)*. The specific components of garlic may also have different effects on the various mechanisms of antiplatelet activity. Some forms of garlic may include adenosine, which increases cyclic adenosine monophosphate (CAMP) levels and thus decreases thromboxane formation *(8)*.

The antiplatelet effects of garlic are thought to be caused by allicin, SAC, adenosine, methyl allyl trisulfide (MATS), diallyl disulfide, and diallyl trisulfide *(8,18,21,24)*. It has been demonstrated that raw garlic extract is more effective than boiled garlic extract in inhibiting platelets ($p < 0.001$) *(21)*. However, a double-blind, randomized, placebo-controlled study found no significant difference in platelet aggregation when subjects took the equivalent of 15 g raw garlic in capsule form. The garlic preparation consisted of garlic cloves homogenized in water and further processed into an oil extract *(25)*. Higher doses of garlic may be needed for inhibition of thromboxane synthesis, whereas lower doses may have other mechanisms. A randomized, placebo-controlled, double-blinded crossover study showed that AGE increased the threshold concentrations needed for ADP-, epinephrine-, and collagen-induced platelet aggregation in human blood. Doses of 7.2 g AGE per day significantly increased the threshold of ADP-induced platelet aggregation ($p < 0.05$), whereas lower doses of 2.4 and 4.6 g AGE per day significantly increased the collagen- and epinephrine-induced threshold. Higher doses of 7.2 g AGE per day did not show a significant difference than the lower doses for the latter two substances. Platelet adhesion to collagen-coated surfaces, fibrinogen, and von Willebrand factor were measured. At a dosage of 4.8–7.2 g AGE per day, adhesion to collagen-coated surfaces was significantly reduced ($p < 0.05$). All doses significantly decreased adhesion to fibrinogen ($p < 0.01$) and only the highest dosage of 7.2 g AGE per day significantly reduced adhesion to von Willebrand factor ($p < 0.05$) *(20)*. Similar results were seen in an earlier study of 15 men with hypercholesterolemia *(26)*.

A double-blind, randomized, matching placebo-controlled parallel group investigation was done to evaluate the effect of 800 mg dried garlic powder for 4 weeks (Kwai/Sapec® [Lichter Pharma, Berlin, Germany]; 300-mg tablets; contains 1.3% alliin, which corresponds to an Allicin release of 0.6%) in patients with an increased risk of juvenile ischemic attack owing to increased circulating platelet aggregates. The ratio of circulating platelet aggregates decreased by 10.3%, and spontaneous platelet aggregation decreased by 56.3% during the treatment period compared to baseline ($p < 0.01$) and placebo ($p < 0.01$). Plasma viscosity also significantly decreased in the garlic group after 4 weeks of treatment compared with baseline and placebo ($p < 0.0001$).These levels returned to pretreatment levels 4 weeks after treatment was stopped *(27)*.

In another study, 800 mg of dried garlic powder (Kwai/Sapec) daily for 15 weeks significantly improved pain-free walking distance in patients with arterial occlusive disease. In this randomized, placebo-controlled, double-blind study, 60 patients underwent 15 weeks of physical therapy, 30 of the subjects received garlic, and 30 baseline-matching patients received an identical placebo. After 6 weeks of treatment, pain-free walking distance was significantly farther in the garlic group ($p < 0.038$). Cholesterol levels ($p < 0.011$), plasma viscosity ($p < 0.0013$), and spontaneous thrombocyte aggregation ($p < 0.013$) were significantly lower in the garlic group *(28)*. This is the only published study addressing a clinically relevant outcome associated with platelet inhibition caused by garlic supplements.

Fibrinolytic effects of garlic have also been evaluated. Garlic oil was shown to increase fibrinolytic activity by 55% ($p < 0.01$) after 3 months of treatment, with 2 g twice daily for 3 months. Fibrinogen was not affected *(24)*. A dried garlic preparation (Sapec) was shown to significantly increase tissue plasminogen activator activity compared to placebo after 1 day and 14 days of treatment *(23)*.

The antiplatelet and antifibrinolytic activity of garlic is of great interest to researchers. Many studies have confirmed these effects as a result of garlic consumption. As with the lipid-lowering effects of garlic, more clinical outcome trials are needed to justify its use in patients with cardiovascular risk factors. In addition, comparative studies with aspirin would be needed to show if there are any benefits to using garlic instead. Because of the demonstrated antiplatelet effect of garlic, its use should be avoided in patients with bleeding disorders and discontinued 1–2 weeks prior to surgery *(4)*.

5.1.4. ANTIHYPERTENSIVE EFFECT

In addition to its effects on lipids and platelet inhibition, garlic has been studied for its effects on lowering blood pressure. A meta-analysis of the effects of garlic on blood pressure was conducted by Silagy and Neil in 1994 *(29)*. Each of the eight randomized studies identified within the analysis used the dried garlic preparation Kwai, 600–900 mg daily (1.8–2.7 g/day fresh garlic), for at least 4 weeks in 415 subjects. Overall, there was an average decrease in systolic blood pressure (SBP) of 7.7 mmHg (95% confidence interval [CI] 4.3–11), and a decrease in diastolic blood pressure (DBP) of 5 mmHg (95% CI 2.9–7.1) in those subjects taking garlic. However, only two of the placebo-controlled trials were limited to hypertensive patients. These studies showed an average decrease in SBP of 11.1 mmHg (95% CI 5–17.2) and a decrease in DBP of 6.5 mmHg (95% CI 3.4–9.6) in those subjects taking the garlic preparation. In a pilot study, 2400 mg dried garlic powder (Kwai) containing 1.3% allicin, was administered to nine patients with persistent severe hypertension (DBP ≥115 mmHg). A statistically significant decrease was seen in the DBP at 5–14 hours ($p < 0.05$), with a maximum decrease at 5 hours after the dose (16 ± 2 mmHg). No significant difference was seen in SBP at any time-point; however, a trend was present *(30)*. Other studies, which had primary outcomes other than blood pressure, have also shown similar findings.

5.2. GI Effects

Garlic was effective against castor oil-induced diarrhea, and relieved abdominal distension/discomfort, belching, and flatulence in 30 patients *(31)*. Small doses of garlic are purported to increase the tone of smooth muscle in the GI tract, whereas large doses decrease such actions *(1)*. An ethanol-chloroform extract of fresh bulb-antagonized acetylcholine and prostaglandin E induced rat fundus smooth muscle contraction at a concentration of 0.002 mg/mL; however, an ethanol extract of fresh garlic bulb caused rat fundus smooth muscle stimulation at a concentration of 0.016 mg/mL *(31)*.

In vitro data shows an antibacterial effect of garlic against *Helicobacter pylori (32–34)*; however, studies in humans with documented *H. pylori* infection showed no in vivo effect on *H. pylori* with dried garlic powder, oil, or freshly sliced cloves *(35–37)* and no effect of garlic oil on symptoms or grade of gastritis *(35)*. This demonstrates the importance of in vivo data with garlic, rather than extrapolating from in vitro studies.

Animal studies in rats show a protective effect of garlic from intestinal damage from methotrexate *(38,39)* and 5-fluorouracil *(39)*, but human data is not available.

5.3. Antimicrobial Activity

5.3.1. ANTIBACTERIAL ACTIVITY

Garlic has in vitro activity against many Gram-negative and Gram-positive bacteria, including species of *Escherichia, Salmonella, Staphylococcus, Streptococcus, Klebsiella, Proteus, Bacillus, Clostridium*, and *Mycobacterium tuberculosis*. Even some bacteria resistant to antibiotics, including methicillin-resistant *Staphylococcus aureus*, multidrug-resistant strains of *Escherichia coli, Enterococcus* spp., and *Shigella* spp. were sensitive to garlic *(40)*. Activity against *H. pylori* is discussed in the GI effects section (*see* Subheading 4.2). A study in 30 subjects was done to determine activity of garlic against oral microorganisms. After using both garlic and chlorhexidine, antimicrobial activity from the subject's saliva was shown against *Streptococcus mutans* and no other oral microorganisms, but adverse effects were significantly higher for garlic *(41)*. Antibacterial activity is thought to be caused by the allicin component of garlic. A characteristic unique to allicin is the low likelihood of most bacteria to develop resistance to it *(40)*. However, more investigation should be done regarding this issue. Data is insufficient for the use of garlic to treat bacterial infections. In vitro data does not always correlate with in vivo clinical data, and such studies are not currently available.

5.3.2. *ANTIFUNGAL ACTIVITY*

Garlic has in vitro antifungal effects against *Cryptococcus neoformans, Candida* spp., *Trichophyton, Epidermophyton, Microsporum, Aspergillus* spp., and *Mucor pusillus (40)*. When five volunteers consumed 10–25 mL of fresh garlic extract, urine samples had antifungal activity, but susceptibility from serum samples dropped significantly *(42)*.

Data is also available suggesting efficacy of topical garlic on fungal infections. For tinea pedis, 1-week topical treatment with ajoene 1% twice daily resulted in mycological cure 60 days later in 100% of patients, compared to 94% for 1% topical terbinafine and 72% for 0.6% topical ajoene *(43)*. Another study showed that 0.6% topical ajoene was as effective as 1% terbinafine cream, both applied twice daily for 1 week, for the treatment of tinea cruris and corposis. After 60 days, effectiveness (clinical plus mycological cure) was 73 vs 71%, respectively *(44)*. In addition, a 0.4% cream was also shown to be effective *(45)*. Although a topical preparation is not available commercially, it could likely be compounded.

5.3.3. *ANTIVIRAL EFFECTS*

Garlic has been shown in in vitro studies to have antiviral activity against several viruses including cytomegalovirus, influenza B, *Herpes simplex* virus types 1 and 2, parainfluenza virus type 3, and human rhinovirus type 2 *(40)*. Antiviral activity is thought to be caused more by the ajoene component than the allicin component of garlic *(46)*.

5.3.4. *ANTIPARASITIC EFFECTS*

Garlic has in vitro activity against *Entamoeba histolytica, Giardia lamblia, Leishmania major, Leptomonas colsoma,* and *Crithidia fasciculate (40,47)*. In vivo and clinical data is needed before garlic can be used for treatment of infections with these organisms.

5.4. Antineoplastic Effects

In vitro and animal studies show that the organosulfur components of garlic suppress tumor incidence in breast, blood, bladder, colon, skin, uterine, esophagus, and lung cancers. Potential mechanisms include decreasing nitrosamine formation, decreased bioactivation of carcinogens, improved DNA repair, immune stimulation, and antiproliferative effects (regulation of cell cycle progression, modification of pathways of signal transduction, and induction of apoptosis) *(47,48)*. Other factors that may play a role in cancer prevention are cytochrome P450 enzyme stimulation, sulfur compound binding, or antioxidant activity *(49)*. The chemical components of garlic that have shown these effects are ajoene, allicin, diallyl sulfide, diallyl disulfide, diallyl trisulfide, SAC, and S-allylmercaptocysteine *(48)*. Heating garlic (microwave or oven) destroys the active allyl sulfur compound formation; however, if crushed garlic is allowed to stand for 10 minutes before heating, the total loss of anticancer activity is prevented *(50)*. Although much of the anticancer data is from in vitro and animal studies, epidemiological studies are available *(51)*.

5.4.1. COLORECTAL CANCER

Several case–control studies and cohort studies were done evaluating dietary raw and cooked garlic consumption and association with colorectal cancer *(52–56)*. The results were mixed but generally positive. One study showed an association of garlic with a reduction in the incidence of colon cancer *(52)*. Another study showed an inverse relationship with rectal cancer in women for garlic consumers but no association in men *(53)*. The third case-control study showed weak evidence of garlic consumption associated with a lower risk of colon cancer for men, although this was not significant, and no effect for women was shown *(54)*. Two cohort studies *(55,56)* showed a non-significant inverse association with colon cancer, although one showed a significant inverse association when limited to just the distal colon (relative risk [RR] 0.52 [95% CI 0.3-0.93]) *(55)*. These studies evaluating dietary garlic consumption cannot be extrapolated to garlic supplements. Only one study evaluated the effects of garlic *supplements* on colon and rectal cancer *(57)*. This cohort study did not show an association between garlic supplement use and colon and rectal cancers. A meta-analysis showed that issues with the studies, including publication bias, heterogeneity of effect estimates, differences in doses, and confounding factors such as total vegetable consumption may not allow for definite conclusions *(58)*. Overall, dietary garlic may have some efficacy in prevention of colorectal cancer, but there is not enough evidence for garlic supplements.

5.4.2. GASTRIC CANCER

Two epidemiological studies show an inverse association between dietary raw and cooked garlic and gastric cancer *(59,60)* and one showed a slight protective effect *(61)*. These results cannot be extrapolated to garlic supplements. One cohort study was done to examine the effects of garlic supplement use on gastric cancer. This study did not show a protective effect of garlic supplements on gastric cancer *(62)*. In fact, there was a small, non-significant increase in risk. A meta-analysis showed that issues with the studies, including publication bias, heterogeneity of effect estimates, differences in doses, and confounding factors such as total vegetable consumption may not allow for definite conclusions *(58)*. Dietary garlic may have some efficacy in the prevention of gastric cancer, but insufficient evidence exists for garlic supplements.

5.4.3. PROSTATE CANCER

High dietary intake of garlic (approximately 1 clove per day) is associated with a 50% reduction in the risk of developing prostate cancer *(63)*. Another study showed that a reduced risk of prostate cancer was associated with both dietary garlic (odds ratio [OR] 0.64; 95% CI 0.38–1.09) and garlic supplements (OR 0.68; 95% CI 0.41–1.1), although both just fell short of statistical significance *(64)*.

5.4.4. OTHER CANCERS

Garlic supplements have been associated with an *increased* risk of lung carcinoma (RR 1.78; 95% CI 1.08–2.92) in a cohort study. However, this was not seen in those using garlic together with any other supplement (RR 0.93; 95% CI 0.46–1.86) *(65)*. A case–control study showed an inverse relationship between dietary garlic consumption and the development of breast cancer *(66)*; however, a cohort study evaluating the effects of garlic supplements did not show a protective effect *(67)*. Insufficient epidemiological evidence exists for the effects of dietary garlic on head and neck cancers *(51)*.

In summary, although the data is encouraging, more studies are needed before definitive conclusions can be made about the effect of garlic on the prevention or the cause (in the case of lung cancer) of cancer, especially with garlic supplements.

5.5. Immunostimulant Effects

Immunostimulant effects of garlic include an increase in proliferation of lymphocyte and macrophage phagocytosis, induction of the infiltration of lymphocytes and macrophages in transplanted tumors, induction of splenic hypertrophy, increased release of interleukin (IL)-2, interferon-γ, and tumor necrosis factor-α, and enhancement of natural killer cell activity. It is thought that these effects may be mechanisms of cancer prevention *(49)*. Lau and colleagues tested an aqueous garlic extract from Japan, the protein fraction isolated from this same extract, and three additional extracts obtained from health food stores in Loma Linda, CA, for ability to stimulate murine T-lymphocyte function and macrophage activity in vitro. Both Japanese extracts were shown to stimulate macrophage activity, and the protein fraction from the Japanese protein extract stimulated lymphocyte activity. Of the three extracts sold in American health food stores, only one stimulated macrophage activity *(68)*. Aged garlic extract was shown in an in vitro study to enhance the proliferation of spleen cells, augment IL-2–induced proliferation, and enhance natural killer cell activity *(69)*.

5.6. Other Effects

AGE, but not fresh garlic, has been shown to have antioxidant effects. The compounds with the highest activity are SAC and *S*-allylmercaptocysteine *(70,71)*. Garlic exerts antioxidant effects by scavenging free radicals, enhancing superoxide dismutase, catalase and glutathione peroxidase, and increasing cellular glutathione. These effects of garlic may play a role in the cardiovascular, antineoplastic, and cognitive effects of garlic *(2)*.

Aged garlic extract has been shown in in vitro and animal studies to protect against liver toxicity from environmental substances, such as bromobenzene *(72)*, protect against cardiotoxicity from doxorubicin *(73)*, and improve age-related spatial memory deficits *(74)*. A placebo-controlled human study showed that garlic may also be useful as a tick repellent *(75)*. In addition, a double-blind, randomized, placebo-controlled human study showed that garlic supplements taken over a 12-week period in the winter significantly reduced the incidence of the common cold ($p < 0.001$), and reduced the duration of symptoms when they occurred ($p < 0.001$) *(76)*.

6. PHARMACOKINETICS

6.1. Absorption

The bioavailability of the garlic component SAC was found to be 64.1, 76.6, and 98.2% in rats after oral administration of 12.5, 25, and 50 mg/kg, respectively. The bioavailability was 103% in mice and 87.2% in dogs. SAC is rapidly absorbed from the GI tract, with a peak plasma concentration occurring at 15 minutes in dogs, 30 minutes at doses of 12.5 mg/kg and 25 mg/kg in rats, and at 1 hour in rats administered 50 mg/kg *(77)*.

6.2. Distribution

Egen-Schwind et al. *(78)* found that 1,2-vinyl dithiin, a component of oily preparations of garlic, accumulates in fatty tissues, whereas 1,3-vinyl dithiin is more hydrophilic and is rapidly eliminated from serum, kidney, and fat tissue. The latter compound was detected in rat liver over the first 24 hours after administration, whereas 1,2-vinyl dithiin was not. Both 1,3-vinyl dithiin and 1,2-vinyl dithiin were detected in the serum, kidney, and fat. In rats, mice, and dogs, SAC is distributed mainly in the liver, kidney, and plasma *(77)*. In rats, SAC levels are highest in the kidney, and plasma and tissue levels peak 15–30 minutes after oral administration.

Garlic apparently distributes into human amniotic fluid and breast milk. Placebo or garlic oil capsules were given to 10 women 45 minutes prior to routine amniotic fluid sampling. Four of the five amniotic fluid samples from the women who had ingested garlic were judged by a blinded panel to have a stronger and more garlic-like odor than a paired amniotic fluid sample from a woman in the placebo group *(79)*. The ingestion of garlic by nursing mothers was shown to significantly change the perceived odor of milk, as well as significantly increase the amount of time the infant spent attached to the nipple while feeding and the number of sucks during feeding. The total amount of milk ingested by the infants was not significantly affected, however *(80)*. In contrast, these authors later found that the ingestion of garlic for 3 days by nursing women decreased the infants' feeding time compared to infants of mothers who had taken placebo *(81)*.

6.3. Metabolism/Elimination

De Rooij et al. *(82)* conducted a study to evaluate the urinary excretion of *N*-acetyl-*S*-allyl-L-cysteine (allylmercapturic acid, ALMA). The importance of this study lies in the use of ALMA as a biomarker for occupational exposure to alkyl halides; if garlic produces detectable urine concentrations of ALMA, garlic consumption could interfere with toxicological studies. Six human volunteers were administered 200 mg of garlic extract in tablet form (Kwai). The volunteers ranged from 20 to 27 years of age, with body weights ranging from 60 to 90 kg. Urine samples were collected prior to administration of the garlic and up to 24 hours postadministration. Gas chromatography-mass spectrometry (GC-MS) was used to evaluate the excretion of ALMA. γ-Glutamyl-*S*-allyl-L-cysteine (GAC) is ALMA's most likely precursor. γ-Glutamine is hydrolyzed from GAC by glutamine-transpeptidase, resulting in *S*-allyl-L-cysteine. This compound then undergoes acetylation via *N*-acetyl transferase to form ALMA. It is difficult to calculate to what extent GAC is excreted as ALMA in the urine because GAC content of garlic varies depending on the product. By assuming that GAC represents 1% of the dry weight of garlic bulbs, and that the tablets represented 100% dry garlic, the researchers approximated that 10% of GAC is excreted as ALMA within the first 24 hours of garlic ingestion. The average elimination half-life of ALMA was 6.0 ± 1.3 hours *(82)*.

N-acetyl-*S*-(2-carboxypropyl) cysteine, *N*-acetyl-*S*-allyl-L-cysteine (ALMA), and hexahydrohippuric acid were identified in the urine of humans ingesting garlic or onions *(83)*. It is important to note that the study participants' urine contained *N*-acetyl-*S*-(2-carboxypropyl)-cysteine at baseline in minute amounts, even before garlic ingestion, but increased after ingestion of garlic or onions. As with the study by De Rooij, the importance of these findings lies in the use of urinary excretion of mercapturic acids as a marker for industrial exposure to halogenated alkanes, such as vinyl chloride. Elimination of other garlic components has also been studied. Allicin is metabolized in rat liver homogenate more rapidly than the vinyl dithiins, the main constituents of oily preparations of garlic.

As discussed in Subheading 6.2. Distribution, 1,2-vinyldithiin is lipophilic and tends to accumulate in fat, whereas 1,3-vinyldithiin is less lipophilic and more quickly eliminated from the serum, fat, and kidney. Both vinyldithiins can be detected in the serum, fat, and kidney using GC-MS for at least 24 hours after oral administration *(78)*.

SAC is thought to undergo first-pass metabolism in rats based on nonlinear increases in AUC (area under the plasma concentration vs time curve) after oral administration. SAC is likely metabolized to ALMA by acetyltransferase in the liver and kidney. The high concentration of SAC in rat kidney has been attributed to conversion of ALMA back into SAC by kidney acylase. Of the full SAC dose, 30–50% is excreted in the urine of rats as ALMA, and less than 1% of the dose is excreted as unchanged SAC in the urine and bile. In mice, both SAC (16.5%) and the *N*-acetylated metabolite (7.2%) are excreted in the urine, whereas in dogs, less than 1% of the dose was found in the urine as either SAC or ALMA. The half-life of SAC in rats ranges from 1.49 hours with an intravenous dose of 12.5 mg/kg, to 2.33 hours with an oral dose of 50 mg/kg. In mice, the half-life of SAC is 0.77 hour when given orally, and 0.43 hour for intravenous administration, and in dogs approx 10 hours after either oral or intravenous administration *(77)*.

ALMA is also detectable in human urine and concentrations in blood increase in response to ingesting garlic *(20)*. Therefore, because SAC is found in many garlic preparations, it may be the best standardization compound and compliance marker for garlic preparations.

7. ADVERSE EFFECTS AND TOXICITY

Garlic is most commonly consumed as a food, rather than as a supplement. According to the Food and Drug Administration (FDA), chopped garlic and oil mixes left at room temperature have the ability to result in fatal botulism food poisoning *(84)*. Such products need to be kept refrigerated, especially those that do not contain acidifying agents such as phosphoric or citric acid. *Clostridium botulinum* bacteria are dispersed throughout the environment, but are not dangerous in the presence of oxygen. The spores produce a deadly toxin in anaerobic, low-acid conditions. The garlic-in-oil mixture provides the environment for the spores to produce their toxin, leading to botulism. At least 40 cases of this poisoning were reported in the late 1980s.

Since the effects of garlic as a medicinal agent have been studied, reports of common side effects have been reported. The most common of which is malodorous breath and body odor. This effect generally can last many hours after garlic consumption and is not removed by brushing teeth or bathing. One study attempted to reveal the mechanism behind this unpopular effect. Air from the mouth and lungs, as well as urine samples were analyzed for sulfur-containing gases (hydrogen sulfide, methanethiol, allyl mercaptan, allyl methyl sulfide, allyl methyl disulfide, and allyl disulfide) after garlic ingestion. Most of the gas levels were present in higher levels in mouth air than lung air or urine up to 3 hours after ingestion and decreased thereafter. However, allyl methyl sulfide concentrations remained high in mouth and lung air, and urine. This indicates that this gas was absorbed and released from the lungs and in the urine. The authors concluded that systemic absorption of allyl methyl sulfide was responsible for the prolonged odor caused by garlic consumption and therefore explained why oral hygiene could not abolish the smell *(85)*.

Many of the cardiovascular trials reported side effects of garlic use, with the most frequently reported being GI symptoms and garlic breath. In addition, rash and prolonged oozing from a razor cut were reported in one of these studies *(86)*. Other commonly described side effects associated with garlic use include GI effects such as abdominal pain, fullness, anorexia, and flatulence.

Coagulation dysfunctions have also been reported, such as postoperative bleeding and prolonged clotting time *(87)*. One case of spinal epidural hematoma associated with excessive garlic ingestion has been reported. An 87-year-old man who reported to consume an average of four gloves of garlic per day to prevent heart disease, presented to the emergency room with acute onset abdominal discomfort and bilateral sensory and motor paralysis in the lower extremities. Prothrombin time was 12.7 seconds and partial thromboplastin time was 22.3 seconds. The patient had no additional risk factors for bleeding and was taking no other medications that would affect bleeding tendency. The platelet inhibition caused by garlic was determined to be the cause *(88)*.

In 2001, Hoshino et al. *(89)* investigated whether different garlic preparations have undesirable effects on the GI mucosa in dogs. When administered directly to the stomach, AGE did not produce any changes compared to control to the mucosa, whereas boiled garlic powder caused redness, and raw garlic powder caused redness and erosion of the mucosa. Pulverized enteric-coated tablets caused redness and a loss of epithelial cells. Although these findings were significant, further study in humans should be done to confirm the relevance of these findings.

In addition, garlic has been shown to change the odor of breast milk in lactating women, as well as alter the sucking patterns of nursing infants (ref. *80*; *see* Subheading 7.2).

Reports of adverse effects in garlic studies are inconsistent. Studies using AGE have reported fewer side effects and toxicities than those using other garlic preparations *(5)*. Therefore, the frequency and severity of effects seen with garlic may vary with the type of preparation used.

7.1. Garlic Allergy

Allergic reactions to garlic have also been reported in the literature. Garlic allergy can manifest as occupational asthma, contact dermatitis, urticaria, angioedema, rhinitis, and diarrhea. A 35-year-old woman experienced several episodes of urticaria and angioedema associated with ingestion of raw or cooked garlic, as well as urticaria from touching garlic. Two garlic extracts as well as fresh garlic produced a 4+ reaction on skin prick tests (SPTs) in this patient, but no other food allergens produced positive results. The patient's symptoms were immunoglobulin E (IgE)-mediated, but she also produced specific IgG, which confounded the results of IgE testing *(90)*. A group of 12 garlic workers with respiratory symptoms associated with garlic exposure underwent SPTs using garlic powder in saline, commercial garlic extract, and various other possible allergens; bronchial provocation tests with garlic powder; oral challenge with garlic dust; and specific IgE testing using the CAP (CAP System; Pharmacia, Uppsala, Sweden) methodology. Patients were classified into two groups depending on the results of the bronchial provocation tests. Seven patients had positive responses (rhinitis or asthma) to the inhalation challenge test, and were designated as Group 1. Six of these patients reacted to the garlic SPT, and five had specific garlic IgE. In addition, six patients had specific IgE to onion, three to leeks, and four to asparagus. In Group 2 (patients who did not respond to the inhalation challenge), one patient had a positive response to the garlic SPT, one to the onion SPT, and two to the leek SPT. None had garlic or onion IgE. Three patients in Group 1 reported that in the past, they had experienced urticaria, asthma, angioedema, and anaphylaxis after garlic ingestion. Two of these patients were administered garlic orally in increasing doses up to 1600 mg. The patient who had reported anaphylaxis tolerated the full dose, whereas the patient who reported urticaria developed a 35% decrease in forced expiratory volume in 1 second (FEV1) and angioedema of the eyelids at a dose of 500 mg. Using immunoblot and IgE immunoblot inhibition analysis, the investigators also attempted to elucidate the specific garlic component to which the patients reacted. Using pooled sera from Group 1, the investigators found that several garlic allergens cross-react with grass and Chenopodiaceaepollens *(91)*.

A group of 50 catering workers with eczema or dermatitis of the hand or arm were studied for suspected occupational dermatitis. All workers were prick tested with foods that commonly irritated their hands at work, as well as patch tested with garlic 50% in arachis oil, onion 50% in arachis oil, and pieces of the same prick test foods. Seven workers reacted to 50% garlic in oil and one reacted to whole garlic *(92)*.

Housewives were found to be more likely to experience contact dermatitis of the hand than those exposed to garlic in other job settings such as chef, agricultural, and industrial positions. A group of 93 patients were patch tested with diallyl disulfide. Of these, 22.6% tested positive for allergy, 79.5% of whom were women. Dermatological eruptions were primarily located on the hand; however, lesions were also seen on the feet, head, legs, and in widespread distribution *(93)*.

Other cases of occupational allergy and asthma associated with garlic extract include an 11-year-old boy who helped with garlic harvesting on his parents' farm and a 15-year-old who helped collect and store garlic *(94)*; a 49-year-old proprietor of a spice marketing and packing firm *(95)*; a 30-year-old electrician working in a spice processing plant *(96)*; and a 16-year-old who had helped his father load stored garlic into a van for several years *(97)*. Symptoms described included wheezing *(95)*; cough, dyspnea, and chest tightness *(96)*; rhinitis *(94,95,97)*; and conjunctivitis *(94)*. Garlic allergy was confirmed using a wide variety of tests including scratch testing *(94)*; SPT *(95–97)*; IgE to garlic using radioallegosorbent test (RAST) *(96)*, polystyrene tube solid phase radioimmunoassay technique *(95)*, CAP system *(97)*; oral challenge *(96)*; bronchial provocation *(94 97)*; and basophil degranulation *(94)*. Test methodologies are detailed in the references cited. Patients with occupational garlic allergy are often allergic to other foods as well as to airborne allergens, including peanuts, onion, ragweed pollen (95), asparagus, and chives *(96)*.

7.2. Topical Reactions

Topically applied garlic can cause "garlic burns" as well as allergic garlic dermatitis. A 17-month-old infant suffered partial thickness burns when a plaster made of garlic in petroleum jelly was applied to the skin for 8 hours *(98)*. Another infant, age 6 months, suffered garlic burns when his father, disappointed that no antibiotics had been prescribed for a treatment of suspected aseptic meningitis, applied crushed garlic cloves by adhesive band to the wrists for 6 hours (99). After 1 week, a round ulceration 1 cm in diameter surrounded by a slightly raised, erythematous border was noted on the left wrist. A similar, more superficial lesion was also seen on the right wrist. When questioned, the parents explained that these ulcerations were the residual blisters that had formed after garlic application. The author of this case report described this reaction as a second-degree chemical burn. An allergic mechanism was ruled out because the infant had not previously been exposed to garlic or onions. A patch test was not done for ethical reasons. Although Garty hypothesized that the infants' delicate skin predisposed them to garlic burns, such reactions have also been reported in older children and adults. For example, a 6-year-old child developed a necrotic ulcer on her foot after her grandmother applied crushed garlic under a bandage as a remedy for a minor sore (100).

A 38-year-old woman developed a garlic burn after applying a poultice made from fresh, uncooked garlic to her breast for treatment of a self-diagnosed *Candida* infection secondary to breastfeeding her 6-month-old son (101). Despite a burning sensation upon application, she left the poultice in place for 2 days. The infant continued to feed with no apparent adverse effects. She presented to the emergency room 2 days after removal of the poultice. Physical exam revealed that the area where the poultice had been applied appeared as a burn with skin loss, ulceration, crusting, hyperpigmentation, granulation tissue, serous discharge, minor bleeding, and erythema on the periphery. The area was tender. The patient was treated with 1% silver sulfadiazine cream.

Another adult suffered garlic burns after applying a compress of crushed garlic wrapped in cotton to her chest and abdomen for 18 hours (102). The erythematous, blistering rash was in a dermatomal distribution on the right side of the patient's chest and upper abdomen, approximating the dermatomal distribution of thoracic segments 8 and 9. She reported that the pain had been present for 1 week and had a stabbing quality. She was initially diagnosed with *Herpes zoster* and was prescribed acyclovir before admitting to use of topical garlic after further questioning. Biopsy revealed full thickness necrosis, many pyknotic nuclei, and focal separation of the necrotic epidermis from the dermis. The burns healed with scarring. The patient refused patch testing, and specific IgE RAST testing to garlic was negative. The nonspecific appearance of garlic burns has been exploited. Three soldiers applied fresh ground garlic to their lower legs and antecubital fossa to produce an erythematous, vesicular rash in an effort to avoid military duty (103).

Eight patients who developed contact dermatitis after rubbing cut fresh garlic cloves on fungal skin infections responded to a topical fluorinated steroid but had negative garlic patch tests, suggesting irritation rather than allergy *(104)*. Patch testing with 1% diallyl disulfide in petrolatum has also been recommended when allergy is suspected *(105)*.

8. INTERACTIONS

Garlic has antiplatelet properties, and can increase the risk of bleeding when used together with drugs with antiplatelet and anticoagulant effects, such as aspirin, clopidogrel, ticlopidine, dipyridamole, heparins, and warfarin *(3)*. Increased international normalized ratio (INR), has been reported when garlic was added to warfarin *(106)*. Garlic supplements that contain allicin can induce the cytochrome P450 3A4 (CYP 3A4) isoenzyme and can result in clinically important decreases in concentrations of drugs metabolized by this enzyme. This interaction was proven with saquinavir *(107)*. However, a garlic preparation containing alliin and alliinase (which formed one-half the amount of allicin stated on the label) did not significantly inhibit CYP 3A4, which was proven by a lack of interaction with the drug alprazolam *(108)*. It is not known whether the differences in the preparations and dose of allicin or some other factor in the metabolism of saquinavir, such as P-glycoprotein, are responsible for this effect. Until more data is available, it would be prudent to avoid or use caution when allicin is used together with some drugs metabolized by CYP 3A4, including protease inhibitors, cyclosporine, ketoconazole, itraconazole, glucocorticoids, oral contraceptives, verapamil, diltiazem, lovastatin, simvastatin, and atorvastatin.

9. REPRODUCTION

Insufficient data is available regarding the effects and safety of garlic use in pregnant and lactating women. (*See* Subheading 6.2. for information about garlic distribution in breast milk.)

10. REGULATORY STATUS

The oil, extract, and oleo resin have been deemed generally recognized as safe as food substances by the FDA, and garlic is also regulated as a dietary supplement in the United States. Garlic is approved in Germany as a nonprescription drug. In Canada, garlic is approved as a food supplement; garlic is on the general sale list in the United Kingdom; in France it is accepted for the treatment of minor circulatory disorders; and in Sweden it is classified as a natural product *(109)*.

REFERENCES

1. Tyler VE, ed. Garlic. In: *The Honest Herbal. A Sensible Guide to the Use of Herbs and Related Remedies, 3rd edition.* New York: Pharmaceutical Products Press, 1993.
2. Borek C. Antioxidant health effects of aged garlic extract. J Nutr 2001;131:1010S–1015S.
3. Natural Medicines Comprehensive Database. http://www.naturaldatabase.com/. Date accessed: June 11, 2006.
4. Rottblatt M, Ziment I, eds. Garlic *(Allium Sativum).* In: *Evidence Based Herbal Medicine.* Philadelphia: Hanley & Belfus, Inc., 2002, pp. 193–200.
5. Amagase H, Petesch BL, Matsuura H. Intake of garlic and its bioactive components. J Nutr 2001;131:955S–962S.
6. Blumenthal M, ed. Garlic. In: *The complete German Commission E Monographs. Therapeutic Guide to Herbal Medicines.* Austin: American Botanical Council, 1998.
7. Agency for Healthcare Research and Quality. Garlic: effects on cardiovascular risks and disease, protective effects against cancer, and clinical adverse effects. Summary, evidence report/technology assessment. www.ahrq.gov/clinic/epcsums/garlicsum.htm. Date accessed: June 11, 2006.
8. Srivastava KC, Tyagi OD. Effects of a garlic-derived principle (ajoene) on aggregation and arachidonic acid metabolism in human blood platelets. Prostagland Leukotr Essent Fatty Acids 1993;49:587–595.
9. Isaacsohn JL, Moser M, Stein EA, et al. Garlic powder and plasma lipids and lipoproteins: a multicenter, randomized, placebo controlled trial. Arch Intern Med 1998;158:1189–1194.
10. Dirsch V, Kiemer A, Wagner H, Vollmar A. Effect of allicin and ajoene, two compounds of garlic, on inducible nitric oxide synthase. Atherosclerosis 1998;139:333–339.
11. Ide N, Lau BHS. Aged garlic extract attenuates intracellular oxidative stress. Phytomedicine 1999;6(2):125–131.
12. Munday JS, James KA, Fray LM, Kirkwood SW, Thompson KG. Daily supplementation with aged garlic extract, but not raw garlic, protects low density lipoprotein against in vitro oxidation. Atherosclerosis 1999;143:399–404.
13. Koscielny J, KluBendorf D, Latza R, et al. The antiatherosclerotic effect of Allium sativum. Atherosclerosis 1999;144:237–249.
14. Qureshi AA, Din ZZ, Abuirmeileh N, et al. Suppression of avian hepatic lipid metabolism by solvent extracts of garlic: impact on serum lipids. J Nutr 1983;113:1746–1755.
15. Stevinson C, Pittler MH, Ernst E. Garlic for treating hypercholesterolemia: a meta-analysis of randomized clinical trials. Ann Intern Med 2000;133:420–429.
16. Superko HR, Krauss RM. garlic powder, effect on plasma lipids, postprandial lipemia, low-density lipoprotein particle size, high-density lipoprotein subclass distribution and lipoprotein(a). J Am Coll Cardiol 2000;35(2):321-326.
17. Yeh Y, Liu L. Cholesterol-lowering effect of garlic extracts and organosulfur compounds: human and animal studies. J Nutr 2001;131:989S–993S.

18. Boullin DJ. Garlic as a platelet inhibitor. Lancet 1981;1:776–777.
19. Makheja AN, Vanderhoek JY, Bailey JM. Inhibition of platelet aggregation and thromboxane synthesis by onion and garlic. Lancet 1979;1:781.
20. Steiner M, Li W. Aged garlic extract, a modulator of cardiovascular risk factors: a dose-finding study on the effects of AGE on platelet functions. J Nutr 2001;131:980S–984S.
21. Ali M, Bordia T, Mustafa T.Effect of raw versus boiled aqueous extract of garlic and onion on platelet aggregation. Prostagland Leukot Essen Fatty Acids 1999;60(1):43–47.
22. Srivastava KC. Effects of aqueous extracts of onion, garlic, and ginger on platelet aggregation and metabolism of arachidonic acid in the blood vascular system: in vitro study. Prostagland Leukotr Med 1984;13:227–235.
23. Legnani C, Frascaro M, Guazzaloca G, et al. Effects of a dried garlic preparation on fibrinolysis and platelet aggregation in healthy subjects. Drug Res 1993;43(I2):119–122.
24. Bordia A, Verma SK, Srivastava KC. Effect of garlic (Allium sativum) on blood lipids, blood sugar, fibrinogen and fibrinolytic activity in patients with coronary artery disease. Prostagland Leukotr Essen Fatty Acids 1998;58:257–263.
25. Morris J, Bure V, Mori T, et al. Effects of garlic extract on platelet aggregation: a randomized placebo-controlled double-blind study. Clin Exper Pharm Phys 1995;22:414–417.
26. Steiner M, Lin RS. changes in platelet function and susceptibility of lipoproteins oxidation associated with administration of aged garlic extract. J Cardio Pharm 1998;31:904–908.
27. Kiesewetter H, Jung E, Jung EM, et al. Effect of garlic coated tablets in peripheral arterial occlusive disease. Clin Investig 1993;71:383–386.
28. Kiesewetter H, Jung E, Jung EM, et al. Effect of garlic on platelet aggregation in patients with increased risk of juvenile ischaemic attack. Eur J Clin Pharmacol 1993;43:333–336.
29. Silagy CA, Neil HA. A meta-analysis of the effect of garlic on blood pressure. J Hypertens 1994;12(40):463–468.
30. McMahon FG, Vargas R. Can garlic lower blood pressure? A pilot study. Pharmacotherapy 1993;13(4):406–407.
31. Ross IA, ed. *Allium sativum*. In: *Medicinal Plants of the World. Chemical Constituents, Traditional and Modern Medicinal Uses*. Totowa: Humana Press, 1998.
32. O'Gara EA, Hill DJ, Maslin DJ. Activities of garlic oil, garlic powder, and their diallyl constituents against *Helicbacter pylori*. Appl Environ Microbiol 2000;66:2269–2273.
33. Jonkers D, van den Broek E, van Dooren I, et al. Antibacterial effect of garlic and omeprazole on *Helicobacter pylori*. J Antimicrob Chemother 1999;43:837–839.
34. Sivam GP. Protection against *Helicobacter pylori* and other bacterial infections by garlic. J Nutr 2001;131:1106S–1108S.
35. Aydin A, Ersoz G, Tekesin O, et al. Garlic oil and helicobacter pylori infection. Am J Gastro 2000;95:563–564.

36. Graham DY, Anderson SY, Lang T. Garlic or jalapeno peppers for treatment of *Helicobacter pylori* infection. Am J Gastroenterol 1999;94:1200–1202.
37. Ernst E. Is garlic an effective treatment for *Helicobacter pylori* infection? Arch Intern Med 1999;159:2484–2485.
38. Horie T, Matsumoto H, Kasagi M, et al. Protective effect of aged garlic extract on the small intestinal damage of rats induced by methotrexate administration. Planta Med 1999;65:545–548.
39. Horie T, Awazu S, Itakura Y, et al. Alleviation by garlic of antitumor drug-induced damage to the intestine. J Nutr 2001;131:1071S–1074S.
40. Ankri S, Mirelman D. Antimicrobial properties of allicin from garlic. Microbes Infect 1999;2:125–129.
41. Groppo FC, Famacciato JC, Simoes RP, et al. Antimicrobial activity of garlic, tea tree oil and chlorhexidine against oral microorganisms. Int Dent J 2002;52:433–437.
42. Caporaso N, Smith SM, Eng RHK. Antifungal activity in human urine and serum after ingestion of garlic (Allium sativum). Antimicrob Agents Chemother 1983;23:700–702.
43. Ledezma E, Marcano K, Jorquera A, et al. Efficacy of ajoene in the treatment of tinea pedis: a double-blind and comparative study with terbinafine. J Am Acad Dermatol 2000;43:829–832.
44. Ledezma E, Lopez JC, Marin P, et al. Ajoene in the topical short-term treatment of tinea cruris and tinea corporis in humans: randomized comparative study with terbinafine. Arzneim-Forsch/Drug Res 1999;49:544–547.
45. Ledezma E, DeSousa L, Jorquera A, et al. Efficacy of ajoene an organosulphur derived from garlic, in the short-term therapy of tinia pedis. Mycoses 1996;39:393–395.
46. Weber ND, Andersen DO, North JA, et al. In-vitro virucidal effects of allium sativum (garlic) extract and compounds. Planta Med 1992;58:417–423.
47. Pinto JT, Rivlin RS. Antiproliferative effects of allium derivatives from garlic. J Nutr 2001;131:1058S–1060S.
48. Milner JA. A historical perspective on garlic and cancer. J Nutr 2001;131:1027S–1031S.
49. Lamm DL, Riggs DR. Enhanced immunocompetence by garlic: role in bladder cancer and other malignancies. J Nutr 2001;131:1067S–1070S.
50. Song K, Milner JA. The influence of heating on the anticancer properties of garlic. J Nutr 2001;131:1054S–1057S.
51. Fleischauer AT, Arab L. Garlic and cancer: a critical review of the epidemiologic literature. J Nutr 2001;131:1032S–1040S.
52. Iscovich JM, L'Abbe KA, Castelleto R, et al. Colon cancer in Argentina I: risk from intake of dietary items. Int J Cancer 1992;51:851–857.
53. Hu J, Liu Y, Zhao T, et al. Diet and cancer of the colon and rectum: a case-control study in China. Int J Epidemiol 1991;20:362–367.
54. Le Marchand L, Hankin JH, Wilkens LR, et al. Dietary fiber and colorectal cancer risk. Epidemiology 1997;8:658–665.
55. Steinmetz KA, Kushi LH, Bostick RM, et al. Vegetables, fruit and colon cancer in the Iowa Women's Health Study. Am J Epidemiol 1994;139:1–15.

56. Giovannucci E, Rimm EB, Stampfer MJ, et al. Intake of fat, meat, and fiber in relation to risk of colon cancer in men. Cancer Res 1994;54:2390–2397.
57. Dorant E, van den Brandt PA, Goldbohm RA. A prospective cohort study on the relationship between onion and leek consumption, garlic supplement use and the risk of colorectal carcinoma in The Netherlands. Carcinogenesis 1996;17:477–484.
58. Fleischauer AT, Poole C, Arab L. Garlic consumption and cancer prevention: meta-analyses of colorectal and stomach cancers. Am J Clin Nutr 2000;72:1047–1052.
59. You WC, Blot WJ, Chang YS, et al. Allium vegetables and reduced risk of stomach cancer. J Natl Cancer Inst 1989;81:162–164.
60. Buaitti E, Palli D, Decarli A, et al. A case-control study of diet and gastric cancer in Italy. Int J Cancer 1989;44:611–616.
61. Hansson LE, Nyren O, Bergstrom R, et al. Diet and risk of gastric cancer: a population-based case-control study in Sweden. Int J Cancer 1993;55:181–189.
62. Dorant E, van den Brandt PA, Goldbohm RA, et al. Consumption of onions and a reduced risk of stomach carcinoma. Gastroenterology 1996;110:12–20.
63. Hsing AW, Chokkalingam AP, Gao YT, et al. Allium vegetables and risk of prostate cancer: a population based study. J Natl Cancer Inst 2002;94:1648–1651
64. Key TJ, Silcocks PB, Davey GK, et al. A case-control study of diet and prostate cancer. Br J Cancer 1997;76:678–687.
65. Dorant E, van den Brandt PA, Goldbohm RA. A prospective cohort study on Allium vegetable consumption, garlic supplement use, and the risk of lung carcinoma in The Netherlands. Cancer Res 1994;54:6148–6153.
66. Levi F, La Vecchia C, Gulie C, et al. Dietary factors and breast cancer risk in Vaud, Switzerland. Nutr Cancer 1993;19:327–335.
67. Dorant E, van den Brandt PA, Goldbohm RA. Allium vegetable consumption, garlic supplement intake, and female breast carcinoma incidence. Breast Cancer Res Treat 1995;33:163–170.
68. Lau BHS, Yamasaki T, Gridley DS. Garlic compounds modulate macrophage and T-lymphocyte functions. Mol Biother 1991;3:103–107.
69. Kyo E, Uda N, Kasuga S, et al. Immunomodulatory effects of aged garlic extract. J Nutr 2001;131:1075S–1079S.
70. Thomson M, Ali M. Garlic (Allium sativum): a review of its potential use as an anti-cancer agent. Curr Cancer Drug Targets 2003;3:67–81.
71. Banerjee SK, Mukherjee PK, Maulik SK. Garlic as an antioxidant: the good, the bad and the ugly. Phytother Res 2003;17:97–106.
72. Wang BH, Zuzel KA, Rahman K, et al. Treatment with aged garlic extract protects against bromobenzene toxicity to precision cut rat liver slices. Toxicology 1999;132:215–25.
73. Mostafa MG, Mima T, Ohnishi ST, et al. S-allylcysteine ameliorates doxorubicin toxicity in the heart and liver in mice. Planta Med 2000;66:148–151.
74. Moriguchi T, Saito H, Nishiyama N. Aged garlic extract prolongs longevity and improves spatial memory deficit in senescence-accelerated mouse. Biol Pharm Bull 1996;19:305–307.
75. Stjernberg L, Berglund J. Garlic as an insect repellent. JAMA 2000;284:831.
76. Josling P. Preventing the common cold with a garlic supplement: a double-blind, placebo-controlled survey. Adv Ther 2001;18:189–193.

77. Nagae S, Ushijima M, Hatono S, et al. Pharmacokinetics of the garlic compound S-allylcysteine. Planta Med 1994;60:214–217.
78. Egen-Schwind C, Eckard R, Jekat FW, et al. Pharmacokinetics of vinyldithiins, transformation products of allicin. Planta Med 1992;58:8–13.
79. Mennella JA, Johnson A, Beauchamp GK. Garlic ingestion by pregnant women alters the odor of amniotic fluid. Chem Senses 1995;20:207–209.
80. Mennella JA, Beauchamp GK. Maternal diet alters the sensory qualities of human milk and the nursling's behavior. Pediatrics 1991;88(4):737–744.
81. Mennella JA, Beauchamp GK. The effects of repeated exposure to garlic-flavored milk on the nursling's behavior. Pediatr Res 1993;34:805–808.
82. De Rooij BM, Boogaard PJ, Rijksen DA, et al. Urinary excretion of N-acetyl-S-allyl-L-cysteine upon garlic consumption by human volunteers. Arch Toxicol 1996;70:635–639.
83. Jandke J, Apiteller G. Unusual conjugates in biological profiles originating from consumption of onions and garlic. J Chromatogr 1987;421:1–8.
84. Lecos C. Chopped garlic in oil mixes. www.fda.gov. March 1989. Last Accessed August 2006/bbs/topics/NEWS/NEW00130 html.
85. Suarez F, Springfield J, Furne J, et al. Differentiation of mouth versus gut as site of origin of odiferous breathe gases after garlic ingestion. Am J Physiol 1999;276:G425–G430.
86. Jain AK, Vargas R, Gotzkowsky S, et al. Can garlic reduce levels of serum lipids? A controlled clinical study. Am J Med 1993;94:632–635.
87. Ackermann RT, Mulrow CD, Ramirez G, et al. Garlic shows promise for improving some cardiovascular risk factors. Arch Intern Med 2001;161:813–824.
88. Rose KD, Croissant PD, Parliament CF, et al. Spontaneous spinal epidural hematoma with associated platelet dysfunction from excessive garlic ingestion: a case report. Neurosurgery 1990;26(5):880–882.
89. Hoshino T, Kashimoto N, Kasuga S. Effects of garlic preparations on the gastrointestinal mucosa. J Nutr 2001;131:1109S–1113S.
90. Asero R, Mistrello G, Roncarolo D, et al. A case of garlic allergy. J Allergy Clin Immunol 1998;101:427–428.
91. Anibarro B, Fontela JL, De La Hoz F. Occupational asthma induced by garlic dust. J Allergy Clin Immunol 1997;100:734–738.
92. Cronin E. Dermatitis of the hands in caterers. Contact Derm 1987;17:265–269.
93. Fernandez-Volmediano JM, Armario-Hita JC, Manrique-Plaza A. Allergic contact dermatitis from diallyl disulfide. Contact Derm 2000;42:108–109.
94. Couturier P, Bousquet J. Occupational allergy secondary to inhalation of garlic dust. J Allergy Clin Immunol 1982;70:145.
95. Falleroni AE, Zeiss R, Levitz D. Occupational asthma secondary to inhalation of garlic dust. J Allergy Clin Immunol 1981;68:156–160.
96. Lybarger JA, Gallagher JS, Pulver DW, et al. Occupational asthma induced by inhalation and ingestion of garlic. J Allergy Clin Immunol 1982;69:448–454.
97. Armentia A. Can inhalation of garlic dust cause asthma? Allergy 1996;51:137–138.
98. Parish RA, McIntyre S, Heimbach DM. Garlic burns: a naturopathic remedy gone awry. Pediatr Emerg Care 1987;3:258–260.

99. Garty B. Garlic burns. Pediatrics 1993;91:658–659.
100. Canduela V, Mongil I, Carrascosa M, et al. Garlic: always good for the health [letter]? Br J Dermatol 1995;132:161–162.
101. Roberge RJ, Leckey R, Spence R, et al. Garlic burns of the breast. Am J Emerg Med 1997;15:548–549.
102. Farrell AM, Staughton RCD. Garlic burns mimicking herpes zoster. Lancet 1981;317:1195.
103. Kaplan B, Schewach-Millet M, Yorav S. Facial dermatitis induced by application of garlic. Int J Dermatol 1990;29:15:75–76.
104. Lee TY, Lam TH. Contact dermatitis due to topical treatment with garlic in Hong Kong. Contact Derm 1991;24:193–196.
105. Delaney TA, Donnelly AM. Garlic dermatitis. Austr J Dermatol 1996;37:109–110.
106. Sunter W. Warfarin and garlic. Pharm J 1991;15:722.
107. Piscitelli SC, Burstein AH, Welden N, et al. The effect of garlic supplements on the pharmacokinetics of saquinavir. Clin Infect Dis 2002;34:234–238.
108. Markowitz JS, DeVane L, Chavin KD, et al. Effects of garlic (allium sativum) supplementation on cytochrome P450 2D66 and 3A4 activity in healthy volunteers. Clin Pharmacol Ther 2003;74:170–177.
109. Blumenthal M, ed. Garlic cloves. In: *Popular Herbs in the U.S. Market*. Austin: American Botanical Council, 1997.

Chapter 9

Ginger

Douglas D. Glover

SUMMARY

Ginger has been promoted for a variety of medical conditions and is purported to have carminative, diaphoretic, spasmolytic, expectorant, peripheral circulatory stimulant, astringent, appetite stimulant, anti-inflammatory, diuretic, and digestive effects. It is most commonly used in the United States for its antinauseant effects to relieve and prevent motion sickness, and relieve morning sickness in pregnancy. Ginger has compared favorably to a variety of other antinauseant therapeutic agents including metoclopramide, dimenhydrinate, promethazine, and scopolamine. Studies assessing therapeutic benefit for other uses such as an anti-inflammatory or antimutagenic agent are less impressive. Ginger has enjoyed a long history of safe use and concerns over a theoretical interaction with antiplatelet drugs has not been confirmed in clinical practice or adverse event reports.

Key Words: *Zingiber officinale*; nausea; motion sickness.

1. HISTORY

Ginger is a perennial plant with thick tuberous rhizomes from which an above-ground stem rises approx 3 feet *(1)*. The plant produces an orchidlike flower *(2)* with petals that are greenish-yellow streaked with purple *(3)*. Ginger is cultivated in areas of abundant rainfall (at least 80 inches/year) *(3)*. Native to southern Asia, ginger is cultivated in tropical areas such as Jamaica, China, Nigeria, and Haiti *(1)*. Ginger was introduced to Jamaica and the West Indies by Spaniards in the 16th century, and exports from Jamaica to the rest of the world amount to more than two million pounds per year *(4)*.

From Forensic Science and Medicine:
Herbal Products: Toxicology and Clinical Pharmacology, Second Edition
Edited by: T. S. Tracy and R. L. Kingston © Humana Press Inc., Totowa, NJ

Ginger is an ingredient in more than one-half of all traditional Chinese medicines *(5)*, and has been used since the 4th century BCE *(4)*. Marco Polo documented its use in India in the late 13th century *(4)*. African and West Indies cultures have also used ginger medicinally *(5)*, and the Greeks and Romans used it as a spice *(4)*. The Chinese used ginger for stomach aches, diarrhea, nausea, cholera, bleeding *(1)*, asthma, heart conditions, respiratory disorders *(3)*, toothache, and rheumatic complaints *(5)*. In China, the root and stem are used to combat aphids and fungal spores *(2)*. Ginger is purported to have use as a carminative, diaphoretic, spasmolytic, expectorant, peripheral circulatory stimulant, astringent, appetite stimulant, antiinflammatory agent, diuretic, and digestive aid *(3)*. It has also been used to treat migraines, fever, flu, amenorrhea *(3)*, snake bites, and baldness *(1)*.

2. CURRENT PROMOTED USES

In the United States, ginger is promoted to relieve and prevent nausea caused by motion sickness, morning sickness, and other etiologies. Additionally, in Germany it is promoted for use against nervousness, coughing, urinary tract conditions, and sore throat *(6)*.

3. SOURCES AND CHEMICAL COMPOSITION

Zingiber officinale Roscoe, *Zingerberis rhizoma*, ingwerwurzelstok *(7)*, Jamaican ginger, African ginger, cochin ginger *(8)*, *Zingiber capitatum*, *Zingiber zerumbet* Smith *(2)*, calicut, gengibre, gingembre, jenjibre, zenzero *(3)*.

4. PRODUCTS AVAILABLE

The best quality ginger comes from Jamaica and consists of whole ginger with the epidermis completely peeled from the rhizomes and dried in the sun for 5 or 6 days, although high-quality, partially scraped ginger used pharmaceutically also comes from Bengal and Australia *(3)*. Extracts are prepared from the unpeeled root, as essential oil can be lost from peeled ginger *(1)*.

Ginger is commercially available in the United States as the dried powdered root, syrup, tincture, capsules, tablets, tea, oral solution, powder for oral solution, as a spice, and in candy, ice cream, and beer *(3)*.

Ginger root is available from several manufacturers as a tea, liquid extract, and as 50-, 250-, 400-, 470-, 500-, 535-, and 550-mg capsules.

Examples include:

- Alvita® Teas Ginger Root tea bag
- Breezy Morning Teas® Jamaican Ginger tea bag

- Celestial Seasonings® Ginger Ease™ Herb tea bag
- Aura Cacia Essential Oil Ginger
- Abunda Life Chinese Ginger powder
- Frontier Ginger — Hawaiian Root capsule
- Nature's Herbs® Ginger Root, 535-mg capsule
- Nature's Way® Ginger Root, 550-mg capsule
- Nature's Plus® Liquid Ginger Extract, 4% volatile oils
- Health Plus Ginger Root extract, 50-mg capsule
- Nature's Answer® Ginger Root Low Alcohol (Liquid)
- Nature's Answer Ginger Root Alcohol Free (Liquid)
- Nature's Herbs Ginger Root Extract (Liquid)
- Nature's Way Ginger Extract (Liquid)
- Quanterra™ Stomach Relief, 250-mg dried ginger root powder (Zintona®) capsule

5. PHARMACOLOGICAL/TOXICOLOGICAL EFFECTS

5.1. Gastrointestinal Effects

A study was conducted to evaluate the effect of ginger on the nystagmus response to vestibular or optokinetic stimuli, as measured by electro-nystagmographic (ENG) techniques *(9)*. Study subjects were screened prior to study enrollment and were excluded if they responded abnormally to vestibular or optokinetic tests. A total of 38 subjects, 20 women and 18 men between the ages of 22 and 34, were given 1 g of ginger (Zintona™), 100 mg dimenhydrinate, or placebo in a double-blind, crossover fashion 90 minutes prior to each test. Ginger had no effect on the ENG, in contrast to dimenhydrinate, which decreased nystagmus response to caloric, rotary, and optokinetic stimulations. Therefore, the authors considered a central nervous system (CNS) effect had been ruled out as ginger's antiemetic mechanism of action, and a direct gastrointestinal effect was proposed.

The antimotion sickness effect of ginger was also compared to that of dimenhydrinate (Dramamine®) in 18 male and 18 female college students who were self-rated as having extreme or very high susceptibility to motion sickness *(10)*. The subjects were given either two ginger capsules (940 mg), one dimenhydrinate capsule (100 mg), or two placebo capsules (powdered chickweed herb [*Stellaria media*]). Subjects were led blindfolded to a previously concealed rotating chair 20 to 25 minutes after consuming the capsule(s). None of the dimenhydrinate or placebo subjects were able to remain in the chair a full 6 minutes, and three patients in the placebo group vomited. One-half of the ginger subjects stayed the full 6 minutes. It was concluded that 940 mg of ginger was superior to 100 mg of dimenhydrinate in preventing motion sickness. It is important to note that none of the subjects in the dimenhydrinate group specifically asked to have the test terminated; the test was stopped by the investigator because of the magnitude of the subjects' self-reported "intensity of stomach feeling." Although the study subjects were blinded not only to the treatments used, but also to the purpose of the study, it is unclear if the investigator was also blinded.

Anesthesiologists appreciate the fact that individuals who experience motion sickness are also at risk of having postoperative nausea and vomiting that may persist for days after surgery. Application of a scopolamine transdermal patch behind the ear for 3 days beyond surgery may serve as a useful adjunct to antiemetic therapy and eradicate this problem *(11)*.

The efficacy of Ginger as a single agent was compared to various drugs alone or in combination to prevent motion sickness in a double-blind, placebo-controlled study *(12)*. Three doses of ginger were investigated and, in the opinion of the authors, neither dose of ginger alone was more effective than placebo. Dimenhydrinate, promethazine, scopolamine, and *d*-amphetamine were effective as single agents. The efficacy of the first three was enhanced by addition of *d*-amphetamine to the regimen. Most effective in preventing motion sickness with limited side effects in this study was a combination of scopolamine 0.6 mg and 10 mg *d*-amphetamine.

The efficacy of ginger as an antiemetic has been studied *(13)* and compared to metoclopramide after major gynecologic surgery in a double-blind, placebo-controlled, randomized study. Premedication with either powdered ginger or a placebo capsule and 10 mg intravenous metoclopramide or placebo was given 60 to 90 minutes prior to the operation. Surgical time lasted between 50 and 60 minutes and the anesthesia time exceeded 1 hour in all cases. Postoperative pain was managed with papaverine or acetaminophen, and postoperative nausea or vomiting was managed with metoclopramide. The incidence of postoperative nausea or vomiting was similar (28 and 30%) in the groups that had received ginger or metoclopramide and considerably greater in those who had received the placebo (51%).

Another placebo-controlled study tested the effectiveness of ginger in preventing postoperative nausea and vomiting *(14)*. This randomized, double-blind study included 108 subjects slated for elective gynecologic laparoscopy, a procedure generally shorter than that of the previous study. The number of subjects provided 80% power to detect a reduction in the incidence of nausea from 30 to 20%. All patients received 10 mg of diazepam orally and were randomized to receive two 500-mg ginger capsules, one 500-mg ginger capsule and one placebo capsule, or two placebo capsules 1 hour prior to surgery. Nausea, when present, was rated on a scale of 1 to 3 (mild, moderate, severe).

Although there was a trend favoring ginger, the difference was not statistically significant ($p = 0.36$). The investigators concluded that neither dose of ginger was effective in preventing postoperative nausea and vomiting. Blinding may have been problematic in this study because of the characteristic taste and smell of ginger, which was noted by one of the patients. Adverse effects were reported by five of the ginger patients and consisted of flatulence and a bloated feeling, heartburn (two patients), nausea, and burping. One patient in the placebo group complained of "feeling windy and having the urge to burp."

A third placebo-controlled study tested the efficacy of ginger in prevention of postoperative nausea and vomiting *(15)*. This randomized, double-blind study consisted of 120 subjects slated for elective gynecologic diagnostic-laparoscopy. The subjects were given either 1 g of ginger, 100 mg of metoclopramide, or a placebo (1 g of lactose) 1 hour prior to surgery. The incidence of nausea and vomiting with metoclopramide was 27%, 21% with ginger, and 41% with placebo. Ginger was similar in effectiveness to metoclopramide in preventing postoperative nausea and vomiting ($p = 0.34$) and significantly more effective than lactose ($p = 0.006$), the placebo.

Data from the previous three randomized controlled trials on postoperative nausea were appropriate for meta-analysis. The pooled absolute risk reduction for the incidence of postoperative nausea proved the difference between the groups treated with ginger and placebo to lack significance. These values indicate a point of the number-needed-to-treat of 19 and a 95% confidence interval that also includes the possibility of no benefit *(16)*.

More recently, Visalyaputra and associates examined the efficacy of a 2-g dose of ginger root, compared to placebo and intravenous droperidol, and a combination of both oral ginger and intravenous droperidol to reduce postoperative nausea and vomiting. The authors concluded that neither ginger root capsules nor administration of a combination of intravenous droperidol and oral ginger lowered the incidence of postoperative nausea and vomiting in women having gynecologic diagnostic laparoscopy *(17)*.

A randomized, double-blind crossover study was conducted to determine the efficacy of ginger in treating hyperemesis gravidarum *(18)*. A total of 30 pregnant women at less than 20 weeks gestation previously admitted to the hospital for hyperemesis gravidarum participated in the study. The treatment included a 250-mg ginger capsule or a placebo (lactose) capsule three times a day for the first 4 days. After a 2-day washout period, the subjects received the alternate treatment for 4 days. Ginger was significantly more efficacious in reducing symptoms of hyperemesis gravidarum than placebo ($p = 0.035$).

Lastly, Vutyavanich and colleagues *(19)*, conducted a randomized, double-blind, placebo-controlled trial to study 70 women with a lesser degree of nausea and vomiting who did not require hospital admission for hyperemesis gravidarum. All had registered prior to 17 weeks gestation and met the author's criteria for exclusion of other medical causes of nausea and vomiting. Subjects received capsules containing 250 mg powdered ginger 4 times daily or an identical-appearing placebo capsule. Prior to the day of entry, each subject graded the degree of nausea and vomiting she experienced on a scale of 0 to 10. Subjects were dispensed 18 capsules of powdered ginger or placebo, advised to record the number of vomiting episodes twice daily (at noon and bedtime), and to return the 5-item Likert scale with packaging and unused capsules (if any) in a week. After a 2-day washout period, they started the second 4-day course of study drug. Outcomes: of the 32 women in the ginger group, all had one or more episodes of vomiting in the 24 hours before treatment. Only two of the placebo group had no vomiting during this time frame. Of those who received powdered ginger, vomiting was significantly less than in the placebo group. By calculating the exact number of vomiting episodes in the treatment group vs the placebo group, those receiving powdered ginger had a greater reduction in vomiting than those receiving placebo. Of the ginger-treated women, 87% were symptomatically improved as compared to 29% of the placebo group. All patients in the ginger group were compliant with the treatment regimen, as compared with 85% of the placebo group. Adverse affects in this study were minimal. Of those receiving ginger, one experienced heartburn, another abdominal discomfort, and a third had diarrhea for 1 day. The incidence of cephalagia in both groups was equal *(19)*.

Ginger root has been studied as prophylaxis against seasickness *(20)* in a randomized, placebo-controlled trial. A group of 80 naval cadets who were inexperienced in sailing in heavy seas received either 1 g of powdered ginger root or placebo as the ship encountered heavy seas for the first time. Scorecards were kept for the next 4 hours regarding four symptoms of seasickness: nausea, vomiting, vertigo, or cold sweats. The cadets continued their assigned tasks throughout the study. All but one scorecard was valuable. Outcomes: 48 of the 79 cadets reported symptoms of seasickness (61%) and 31 (16 in the ginger group and 15 in the placebo group) reported no symptoms at all. Five subjects in the placebo group vomited more than once but none of the ginger group was so afflicted. Although all seasickness symptoms were less severe in the ginger group, the different was not statistically significant for nausea and vertigo.

A comparative study of motion sickness has been conducted *(21)* in which powdered ginger was compared with scopolamine or placebo. A group of 28 subjects sat in a rotating chair to an end point of motion sickness short of vomiting. Antimotion sickness was defined as activity allowing a greater number of head motions than the placebo. Electrical activity of the stomach was monitored by positioning electrodes over the epigastric area. Outcomes: Powdered ginger provided no protection against motion sickness; however, subjects were able to perform an average of 147.5 more head movements after receiving 0.6 mg scopolamine orally than placebo. The rate of gastric emptying was significantly delayed when tested immediately, but quickly recovered. The authors concluded ginger does not posess antimotion sickness activity nor does it significantly alter gastric function during motion sickness.

The effect of powdered ginger root on gastric emptying rate has been studied in a double-blind, random, controlled, crossover trial of 16 healthy volunteers. The subjects received either 1000 mg powdered ginger root or placebo, and gastric emptying was monitored using the oral acetaminophen absorption model. Powdered ginger did not alter gastric emptying. The authors concluded the antiemetic effect of ginger was not related to its effect on gastric emptying *(22)*.

However, there is lack of consensus regarding the mechanism of the antiemetic effect of ginger. Is this effect a result of vestibular input to the vomiting center of the brain via muscarinic acetylcholine receptors, or is it a direct effect on the stomach? Studies in rats have shown 6-gingerol enhances gastrointestinal transport of a charcoal meal. It has been suggested Phillips' study failed to demonstrate this effect because of inadequate dose of 6-gingerol. Additionally, both 6-gingerol, shogaol, and galanolactone *(23)* have anti-5-hydroxytryptamine (5HT) activity in isolated guinea pig ileum. Lastly, available data is inadequate to clarify the significance of CNS activity *(24)*.

Abrupt discontinuation or noncompliance with serotonin reuptake inhibitor (SRI) treatment regimens may result in a recently described "SRI Discontinuation Syndrome," characterized by disequilibrium, dizziness, vertigo, and ataxia. Although no randomized, placebo-controlled studies have been published regarding this entity, case reports of successful alleviation of its symptoms by ginger root have been emerging *(25)*. The dose of ginger root most frequently utilized to treat this syndrome is 500 to 1000 mg three times daily.

5.2. Anti-Inflammatory Activity

Ginger components 6-gingerol, 6-dehydrogingerdione, 10-dehydrogingerdione, 6-gingerdione, and 10-gingerdione inhibit prostaglandin synthetics in vitro *(26)*. The latter four components were found to have greater potency as prostaglandin inhibitors than indomethacin. In an additional study of ginger's ability to affect arachidonic acid metabolism in human platelets and rat aorta, an aqueous extract of ginger was able to inhibit production of thromboxane and prostaglandins in a dose-dependent manner *(27)*. Ginger appears to act as a dual inhibitor of both cyclooxygenase and lipooxygenase to inhibit leukotriene synthesis *(6)*.

Altman and Marcussen evaluated 247 patients with osteoarthritis and moderate to severe knee pain in a randomized, double-blind, placebo-controlled, multicenter, parallel-group, 6-week study. The results of their investigation revealed small but statistically insignificant benefits of ginger over placebo *(6)*.

Another placebo-controlled, 3-week treatment crossover study with a 1-week washout period between treatments studied the same ginger extract compound compared to ibuprofen and placebo. Pain relief by ibuprofen was significantly greater than placebo, but a difference between ginger extract and placebo was lacking *(28)*.

The antiinflammatory effects of ginger oil on arthritic rats were studied *(29)*. A 0.05-mL suspension of heat-killed *Mycobacterium tuberculosis bacilli* in liquid paraffin (5 mg/mL) was injected into the knees and paws to induce arthritis in treatment rats. Rats were randomized to receive 33 mg/kg of ingwerol (ginger oil obtained by steam distillation of dried ginger root), 33 mg/kg of eugenol (a component of clove oil purported to have antiinflammatory activity), or normal saline orally for 26 days, beginning just prior to the induction of arthritis. Compared to normal saline, both treatments were effective in decreasing both knee and paw swelling.

5.3. Migraine Prevention

A case reported the use of ginger for the prevention of migraines *(30)*. A 42-year-old woman suffered migraine with aura once or twice every 2 or 3 months for 10 years. Because the frequency and duration of migraine increased, the patient was prescribed 500–600 mg of powdered ginger to be taken at the onset of aura, then every 4 hours for the next 3–4 days. The patient reported some relief within 30 minutes of the first dose. Then she added uncooked fresh ginger to her diet. In a 13-month period, she reported only six migraines. These results should be confirmed in a double-blind, controlled trial.

5.4. Cardiovascular Effects

In vitro studies of gingerol using canine cardiac tissue and rabbit skeletal muscle demonstrated Ca^{2+}-adenosine triphosphatase (ATPase) activation in the cardiac and skeletal sarcoplasmic reticulum (SR) *(31)*. Gingerol (3–30

μM) increased Ca2+-ATPase pumping rate in a dose-dependent manner. A 100-fold dilution with fresh saline solution of 30 μM gingerol completely reversed Ca^{2+}-ATPase activation. The investigators concluded that gingerol may be a useful pharmacological tool in the study of regulatory mechanisms of the SR Ca^{2+} pumping systems, and their effect on muscle contractility.

Another in vitro study examined the effect of 6-, 8-, and 10-gingerol on isolated left atria of guinea pigs *(32)*. The study found the gingerols had a dose-dependent positive inotropic effect that was evident at doses as low as 105, 10-6, and 3 × 10-5 g/mL for 6-, 8-, and 10-gingerol, respectively. Thus, 8-gingerol was the most potent gingerol in regard to cardiotonic activity.

In vitro, aqueous ginger extract has dose-dependent antithromboxane synthetase activity that correlates with its ability to inhibit aggregation of human platelets in response to adenosine diphosphate, collagen, and epinephrine *(27)*. However, this may not be clinically significant; inhibition of platelet aggregation has been demonstrated in humans only after consumption of 5 g of raw ginger daily for 1 week *(33)*. A single 2-g dose of dried ginger did not affect platelet function *(34)*.

5.5. Mutagenicity

A study showed that 6-gingerol and 6-shogaol isolated from *Z. officinale* using column chromatography were mutagenic at 700 μM in the Hs30 strain of *Escherichia coli (35)*. 6-Gingerol was noted to be a potent mutagen whereas 6-shogaol was less mutagenic. Another study documented the antimutagenicity of zingerone, another ginger component, in addition to the mutagenicity of gingerol and shogaol in *Samonella typhimurium* strains TA 100, TA 1535, TA 1538, and TA 98 *(36)*. Gingerol and shogaol activated by rat liver enzymes at doses of 5-200 μg/plate mutated strains TA 100 and TA 1535, whereas zingerone was nonmutagenic in all four strains. Zingerone also suppressed the mutagenicity of gingerol and shogaol in a dose-dependent manner. Although all three compounds are similar in chemical structure, zingerone has a shorter side chain than the mutagenic compounds; thus the side chains may be responsible for the mutagenic activity of gingerol and shogaol.

Pyrolysates of cigarettes, fish, and meats have been found to have potent carcinogenic capability. Research in Japan *(37)* found evidence that vegetables, such as cabbage and ginger, contain antimutagenic factors that suppress mutagenesis. Another study suggests that ginger juice contains more antimutagenic than mutagenic substance(s), and thus has the capability to suppress mutagenesis by the contained pyrosylates *(38)*.

6. PHARMACOKINETICS

No human studies of the pharmacokinetics of any ginger components have been conducted. Only one study in rats has been conducted to examine the pharmacokinetics of a ginger component, 6-gingerol. This study observed that after IV administration, the plasma concentration-time curve was best described by a two-compartment model with a rapid terminal elimination half-life of 7.2 minutes and a total body clearance of 16.8 mL/minute / kg. The protein binding of 6-gingerol was approx 92%.

7. ADVERSE EFFECTS

7.1. Case Reports of Toxicity Caused By Commercially Available Products

Consumption of Jamaica ginger, an alcoholic ginger extract that was popular as a beverage in the rural southern United States during prohibition, resulted in a peripheral polyneuritis *(39)*. In reported cases, the first symptom to appear was sore calves for 1 or 2 days. After the soreness disappeared, walking became notably difficult for the case subjects. Subjects could not walk without the aid of a cane or crutches within 1 week. Bilateral weakness of the upper and lower extremities and foot drop, without sensory disturbance or pain, was a common physical finding. The skin on the feet was noted to be red and glossy, but not swollen. Deep tendon reflexes were inconsistent among patients; ankle jerks were not present in any subject, but some had normal knee reflexes. There were no cranial nerve deficits. Although the beverage contained 60–90% alcohol, alcoholic neuropathy was ruled out as an etiology of the syndrome because of the sporadic nature of Jamaica ginger consumption. Jamaica ginger was eventually exonerated as cause of the neuropathy and an adulterating agent, triorthocresyl phosphate, was identified as the putative toxin *(40)*. This chemical had been added to the beverage presumably as a tasteless substitute for the oleo resin of ginger so that the product would be more palatable. Additional research with alcohol and true United States Pharmacopeia (USP) ginger fluid extract failed to produce paralysis. Subsequent reports of cases of neuropathy associated with the use of ginger in any form have not been forthcoming.

8. DRUG INTERACTIONS

Although no drug interactions with ginger have been reported, caution should be exercised with patients taking anticoagulants and antiplatelet drugs because of its potential antiplatelet effect *(3)*.

9. REPRODUCTION

In addition to its mutagenic activity, concern has been raised that the receptor binding of testosterone may be affected in the fetus because of ginger's inhibition of thromboxane synthetase *(41)*. Commission E contraindicates ginger's use during pregnancy for morning sickness, although this contraindication has been disputed by some owing to the lack of reported adverse effects despite its long history of use in pregnancy in traditional Chinese medicine *(7)*.

10. REGULATORY STATUS

In Austria and Switzerland, ginger is registered as an over-the-counter drug indicated for the prevention of motion sickness, nausea, and in Austria, for vomiting in febrile pediatric patients. Australia's Therapeutic Goods Administration's Listed Products category includes ginger as an acceptable active ingredient. Likewise, in the United Kingdom, ginger is on the General Sale List of the Medicines Control Agency. In Belgium, ginger rhizome is permitted as a traditional digestive aid, and the German Commission E approves ginger for dyspeptic complaints and the prevention of motion sickness *(3)*. Ginger is listed as an official monograph in the USP-National Formulary *(42)*. Ginger is regulated as a dietary supplement in the United States. It is also considered "generally recognized as safe as a food substance" by the FDA *(8)*.

REFERENCES

1. Leung AY. Ginger. In: *Encyclopedia of Common Natural Ingredients Used in Food, Drugs, and Cosmetics*. New York: John Wiley and Sons, 1980, p. 1845.
2. Anonymous, ed. *The Lawrence review of natural products*. St. Louis: Facts and Comparisons, 1991.
3. United States Pharmacopeial Convention (USP). Ginger. In: *Botanical monograph series*. Rockville: United States Pharmacopeial Convention, 1998.
4. Tyler VE, Brady LR, Robbers JE, eds. *Pharmacognosy, 8th edition*. Philadelphia: Lea and Febiger, 1981, p. 156.
5. Awang DVC. Ginger. CPJ 1992;125:309–311.
6. Altman RD, Marcussen KC. Effects of ginger extract on knee pain in patients with osteoarthritis. Arthritis Rheum 2001;44(11):2531–2538.
7. Blumenthal M, ed. *The Complete German Commission E Monographs: Therapeutic Guide to Herbal Medicines*. Austin: American Botanical Council, 1998.
8. Tyler VE, ed. Digestive system problems. In: *Herbs of Choice: The Therapeutic Use of Phytomedicinals*. Binghamton: Pharmaceuticals Products Press, 1994.

9. Holtmann S, Clarke AH, Scherer H, Hohn M. The anti-motion sickness mechanism of ginger. A comparative study with placebo and dimenhydrinate. Acta Otolaryngol 1989;108:168–174.

10. Mowrey DB, Clayson DE. Motion sickness, ginger, and psychophysics. Lancet 1982;1:6557.

11. Price NM, Schmitt LG, McGuire J, et al. Transdermal scopolamine in prevention of motion sickness at sea. Clin Pharmacol Ther 1981;29:414–419.

12. Wood CD, Manno JE, Wood MJ, Manno BR, Mims ME. Comparison of efficacy of ginger with various antimotion sickness drugs. Clin Res Pr Drug Regul Aff 1988;6(2):129–136.

13. Bone ME, Wilkinson DJ, Young JR, McNeil J, Charlton S. Ginger root—a new antiemetic. Anesthesia 1990;45:669–671.

14. Arfeen Z, Owen H, Plummer JL, Ilsley AH, Sorby-Adams RAC, Doecke CJ. A double-blind randomized controlled trial of ginger for the prevention of postoperative nausea and vomiting. Anaesth Intens Care 1995;23:449–452.

15. Phillips S, Ruggier R, Hutchinson SE. *Zingiber officinale* (ginger)—an antiemetic for day case surgery. Anaesthesia 1993;48:715–717.

16. Ernst E, Pittler MH. Efficacy of ginger for nausea and vomiting: a systemic review of randomized clinical trials. Br J Anaesth 2000;84(3):367–371.

17. Visalyaputra S, Petchpaisit N, Somcharoen K, Choavaratana R. The efficacy of ginger root in the prevention of postoperative nausea and vomiting after outpatient gynaecological laproscopy. Anaesthesia 1998; 53:486–510,

18. Fisher-Rasmussen W, Kjaer SK, Dahl C, Asping U. Ginger treatment of hyperemesis gravidarum. Eur J Obstet Gynecol Reprod Biol 1990;38:19–24.

19. Vutyavanich T, Kraisarin T, Ruangsri RA. Ginger for nausea and vomiting in pregnancy: randomized, double-masked, placebo-controlled trial. Obstet Gynecol 2001;97(4):577–582.

20. Grontved A, Brask T, Kambskard J, Hentzer E. Ginger root against seasickness. Acta Otolaryngol 1988;105:45–49.

21. Stewart JJ, Wood MJ, Wood CD, Mims ME. Effects of ginger on motion sickness susceptibility and gastric function. Pharmacology 1991;42:111–120.

22. Phillips S, Hutchinson S, Ruggier R. *Zingiber officinale* does not affect gastric emptying rate. Anaesthesia 1993;48:393–395.

23. Huang Q, Iwamoto M, Aoki S, et al Anti-5-hydroxytryptamine, effect of galanolactone, diterpenoid isolated from ginger. Chem Pharm Bull 1991;39(2):397–399.

24. Lumb AB. Mechanism of antiemetic effect of ginger. Anaesthesia 1993;48:1118.

25. Schechter JO. Treatment of disequilibrium and nausea in the SRI Discontinuation Syndrome. J Clin Psychiatry 1998;59:431–432.

26. Kiuchi F, Shibuya M, Sankawa U. Inhibitors of prostaglandin biosynthesis from ginger. Chem Pharmacol Bull 1982;30:754–757.

27. Srivastava KC. Effects of aqueous extracts of onion, garlic and ginger on platelet aggregation and metabolism of arachidonic acid in the blood vascular system: in vitro study. Prostagland Leukotr Med 1984;13:227–235.

28. Bliddal H, Rosetzsky A, Schlichting P, et al. A randomized, placebo-controlled, cross-over study of ginger extracts and ibuprofen in osteoarthritis. Osteoarthritis Cartilage 2000;8:9–12.
29. Sharma JN, Srivastava KC, Gan EK. Suppressive effects of eugenol and ginger oil on arthritic rats. Pharmacology 1994;49:314–318.
30. Mustafa T, Srivastava KC. Ginger (*Zingiber officinale*) in migraine headache. J Ethnopharmacol 1990;29:267–273.
31. Kobayashi M, Shoji N, Ohizumi Y. Gingerol, a novel cardiotonic agent, activates Ca^{2+} pumping ATPase in skeletal and cardiac sarcoplasmic reticulum. Biochim Biophys Acta 1987;903:96–102.
32. Shoji N, Iwasa A, Takemoto T, Ishida Y, Ohizumi Y. Cardiotonic principles of ginger (*Zingiber officinale Roscoe*). J Pharm Sci 1982;71:1174–1175.
33. Srivastava KC. Effect of onion and ginger consumption on platelet thromboxane production in humans. Prostagland Leukotr Essent Fatty Acids 1989;35:183–185.
34. Lumb AB. Effect of dried ginger on human platelet function. Thromb Haemost 1994;71:110–111.
35. Nakamura II, Yamamoto T. The active part of the [6] gingerol molecule in mutagenesis. Mutat Res 1983;122:87–94.
36. Nagabhushan M, Amonkar AJ, Bhide SV. Mutagenicity of gingerol and shogaol and antimutagenicity of zingerone in salmonella/microsome assay. Cancer Lett 1987;36:221–233.
37. Kada T, Kazuyoshi M, Inuoue T. Anti-mutagenic action of vagetable factor(s) on the mutagenic principle of tryptophan pyrolysate. Mutat Res 1987;53:351–353.
38. Nakamura H, Yamamoto T. Mutagen and anti-mutagen in ginger, *Zingerber officinale*. Mutat Res 1982;103:119–126.
39. Harris S. Jamaica ginger paralysis (a peripheral polyneuritis). South Med J 1930;23:375–380.
40. Valaer P. The examination of cresyl-bearing extracts of ginger. Am J Pharm 1930;102:571–574.
41. Backon J. Ginger in preventing nausea and vomiting of pregnancy: a caveat due to its thromboxane synthetase activity and effect on testosterone binding [letter]. Eur J Obstet Gynecol Reprod Biol 1991;42:163–164.
42. United States Pharmacopoeia (USP). USP, *28th edition, 2091*.

Chapter 10

Saw Palmetto

Timothy S. Tracy

SUMMARY

Administration of saw palmetto can be effective in treating symptoms of benign prostatic hyperplasia, and its use may have benefit in stimulating hair growth. Adverse effects of saw palmetto therapy are usually mild and its use does not appear to result in significant drug interactions.

Key Words: *Serenoa repens*; BPH; prostate health; antiandrogens.

1. HISTORY

Saw palmetto is a dwarf palm tree that grows in Texas, Florida, Georgia, and southern South Carolina *(1)*. The tree grows up to 6 feet tall and has wide leaves divided into fan-shaped lobes that are gray to blue-green in color. The plant produces purple-black berries from September to January *(2)*.

The earliest known use of saw palmetto was in the 15th century BC in Egypt to treat urethral obstruction *(2)*. The Native Americans also used saw palmetto to treat genitourinary conditions *(1)*. In the early 20th century, it was used in conventional medicine as a mild diuretic and as a treatment for benign prostatic hypertrophy (BPH) and chronic cystitis *(3)*. Historically, saw palmetto has also been used to increase sperm production, increase breast size, and increase sexual vigor *(4)*. Early settlers in the United States observed that animals that ate the berries grew fat and healthy, and by the 1870s saw palmetto was purported to improve general health, reproductive health, disposition, and body weight, and to stimulate appetite *(2)*.

From Forensic Science and Medicine:
Herbal Products: Toxicology and Clinical Pharmacology, Second Edition
Edited by: T. S. Tracy and R. L. Kingston © Humana Press Inc., Totowa, NJ

2. CURRENT PROMOTED USES

Saw palmetto is promoted as a treatment for BPH, to improve prostate health and urinary flow, and to improve reproductive and sexual functioning, as well as stimulate hair growth.

3. SOURCES AND CHEMICAL COMPOSITION

Serenoa repens (Bartram) Small, *Sabal serrulata* (Michaux) Nichols, *Serenoa serrulatum* Schultes *(5)*

4. PRODUCTS AVAILABLE

Saw palmetto is commercially available alone and in combination products including capsules, gelcaps, and tablets. There are more than 100 commercial products containing saw palmetto as the sole ingredient or as a combination product.

5. PHARMACOLOGICAL/TOXICOLOGICAL EFFECTS

5.1. In Vitro/Animal Studies

Saw palmetto's benefits in treatment of BPH are hypothesized to be caused in part by antiandrogen effects *(6)*. Saw palmetto is a multisite inhibitor of androgen action. In an in vitro study *(7)*, a liposterolic saw palmetto extract called Permixon® was shown to compete with a radiolabeled synthetic androgen for the cytosolic androgenic receptor of rat prostate tissue. Another in vitro study found that saw palmetto lipid extract inhibits 5α-reductase, the enzyme responsible for the conversion of testosterone to its active metabolite dihydrotestosterone (DHT); inhibits 3-ketosteroid reductase, the enzyme responsible for DHT metabolism to other active androgens; and blocks androgen receptors *(8)*. Saw palmetto may also improve BPH signs and symptoms by inhibiting estrogen receptors in the prostate *(6)*. A study of the effects of saw palmetto on cancer cell lines *(9)* has demonstrated that saw palmetto can inhibit 5α-reductase activity without affecting prostate-specific antigen (PSA) expression, confirming that saw palmetto can be administered without interfering with this biomarker (PSA) of tumor progression. Finally, it has recently been demonstrated in vitro that saw palmetto extracts do not affect α1-adrenoceptor subtypes, suggesting that its primary mechanism of action is on androgen metabolism *(10)*.

An in vivo study in rats evaluated the effects of saw palmetto and cernitin (another natural product) and finasteride on prostate growth *(11)*. In castrated rats who were given testosterone, all three treatments significantly reduced prostate size as compared to rats (castrated + testosterone) who were not given any treatment. Though finasteride produced the greatest effect on prostate size, no statistical difference was noted among any of the three treatments.

Anti-inflammatory effects of saw palmetto also have been hypothesized to improve BPH symptoms *(6)*. An acidic, lipophilic saw palmetto extract (Talso®) was shown in vitro to inhibit both the cyclooxygenase and 5-lipoxygenase pathways, preventing the formation of inflammatory-producing prostaglandins and leukotrienes *(12)*. Finally, saw palmetto has been purported to stimulate immune function *(13)*.

5.2. Human Studies

Saw palmetto extract in a dose of 160 mg or placebo three times daily was administered to 35 elderly men, and prostatic tissue was collected *(14)*. The investigators found that some component of the saw palmetto extract inhibits nuclear estrogen receptors in the prostates of patients with BPH patients.

Clinically, 160 mg of Permixon® twice daily was superior to placebo in a double-blind trial in 110 men with BPH *(15)*. A statistically significant ($p <$ 0.001) benefit compared to placebo was seen in nocturia, flow rate, postvoid residual, self-rating, physician rating, and dysuria. Compared with baseline, both placebo and saw palmetto were beneficial in improving nocturia ($p <$ 0.001), but only saw palmetto improved flow rate and postvoid residual compared to baseline ($p < 0.001$). Headache was the only adverse effect. A double-blind study *(16)* compared Proscar® (finasteride, a prescription 5α-reductase inhibitor), 5 mg daily, with Permixon, 160 mg twice daily for 6 months. Both finasteride and saw palmetto improved International Prostate Symptom Score (I-PSS) and quality of life compared to baseline, with no statistical difference between the two treatments. Finasteride improved peak urinary flow rate more than saw palmetto ($p = 0.035$), and residual volume was decreased more with finasteride than with saw palmetto ($p = 0.017$). Finasteride decreased prostate volume more than saw palmetto ($p < 0.001$), and only finasteride decreased PSA compared with baseline ($p < 0.001$). Although only one patient in each treatment group withdrew because of sexual problems, the finasteride patients experienced a statistically significant deterioration in the sexual function score compared with baseline ($p < 0.01$). Twice as many patients withdrew from the saw palmetto group because of side effects (28 vs 14), but there were no statistically significant differences noted between the two groups in regard to any adverse effect. Hypertension was the most common adverse effect, occurring in 3.1% of the saw palmetto patients and 2.2% of the finasteride patients. Other adverse effects included decreased libido, abdominal pain, impotence, back pain, diarrhea, flulike illness, urinary retention, headache, nausea, constipation, and dysuria. A drawback of this study is that no placebo group was included; more data on the efficacy of these two drugs compared to placebo are needed.

The findings of Carraro and colleagues discussed previously suggest that Permixon does not affect PSA. These results were confirmed by an in vitro study in which Permixon 10 µg/mL (calculated plasma concentration achieved with therapeutic doses), did not interfere with secretion of PSA *(17)*. These findings imply that PSA can continue to be used for prostate cancer screening in men taking saw palmetto.

Another randomized, double-blind, placebo-controlled trial of saw palmetto for the treatment of lower urinary tract symptoms also demonstrated its usefulness in these types of conditions *(18)*. These investigators studied 85 men, randomized to receive either saw palmetto or placebo for 6 months. Effectiveness was monitored using the I-PSS, a sexual function questionnaire and urinary flow rate. Results of these studies demonstrated that the I-PSS symptom score decreased (i.e., improved) from 16.7 to 12.3 in those subjects receiving saw palmetto, whereas the symptom score decreased from 15.8 to 13.6 in the placebo group ($p = 0.038$). No significant difference was noted in the quality of life component of the I-PSS. Also, no differences were noted in either the sexual function questionnaire score or peak urinary flow rate between the saw palmetto and placebo groups. This study demonstrated that saw palmetto administration for 6 months resulted in an improvement in symptoms associated with BPH but not in sexual function or peak flow rate.

Several well-conducted studies of saw palmetto effect on BPH symptoms have been conducted using combination products that may have additional active ingredients. Marks and colleagues *(19)* evaluated the effectiveness of a saw palmetto herbal blend (saw palmetto, nettle root extract, pumpkin seed oil extract, lemon bioflavonoid extract, Vitamin A, and other minor ingredients) in subjects with symptomatic BPH. Using a double-blind, placebo-controlled trial design, 44 subjects were investigated ($n = 21$ in the saw palmetto herbal blend group and $n = 23$ in the placebo group) following treatment for 6 months. Prostate epithelial contraction was noted where percent epithelium decreased from 17.8% at baseline to 10.7% at 6 months in the saw palmetto herbal blend treatment group ($p < 0.01$). Saw palmetto treatment increased the percent of atrophic glands from 25% to 41% ($p < 0.01$). Neither treatment (saw palmetto or placebo) altered PSA or prostate volume. Another group of investigators studied the effect of saw palmetto herbal blend (same ingredients as previously mentioned) on nuclear measurements of DNA content in men with symptomatic BPH (20). Using nuclear morphometric descriptors (NMDs) (size, shape, DNA content, and textural features) of the nucleus of prostatic tissue, 6-month treatment of saw palmetto herbal blend was compared to placebo control. After 6 months, 25 of the 60 NMDs were significantly different in the saw palmetto treatment group, whereas none were changed in the placebo group. These investigators then used four of these 25 altered NMDs to develop a multivariate model that was proposed to be predictive of treatment effect. These investigators proposed that saw palmetto herbal blend treatment alters the DNA chromatin structure and organization of prostate epithelial cells. Using a different combination formula containing saw palmetto (saw palmetto, cernitin, β-sitosterol, and Vitamin E), Preuss and colleagues conducted a 3-month randomized, placebo-controlled trial in 127 subjects of this formulation in the treatment of BPH symptoms (21). These investigators found that treatment with this saw palmetto-containing product results in a statistically significant decrease in nocturia severity ($p < 0.001$, daytime frequency ($p < 0.04$) and the American Urological Association symptom index was significantly improved ($p < 0.001$). No change in PSA measurements, maximal and average urinary flow rates, or residual volumes was noted. Furthermore, no adverse effects were noted in either group.

A number of studies evaluating the effects of saw palmetto on BPH and urinary symptoms have been conducted that lack placebo controls, comparing saw palmetto to another agent or looking at longitudinal effect. Kaplan and colleagues (22) conducted a 1-year prospective trial of saw palmetto vs finasteride for the treatment of category III prostatitis/chronic pelvic pain syndrome. Finasteride significantly decreased (~25%) the National Institutes of Health Chronic Prostatitis Symptom Index score, whereas saw palmetto had no effect on this measure. Finasteride also improved quality of life and pain measures but not urination. These authors concluded that saw palmetto resulted in no appreciable long-term improvement in category III prostatitis/chronic pelvic pain syndrome. However, this study suffers from lack of placebo control. Al-Shukri et al. (23) studied the effects of Permixon on lower urinary tract symptoms caused by benign prostatic hyperplasia. These investigators administered Permixon 160 mg twice daily for 9 weeks and compared the results to a control group who received no treatment at all. Permixon treatment increased maximum flow rate by 6% ($p < 0.001$), decreased maximum detrusor pressure by 12.8% ($p < 0.001$) and reduced residual urine volume by 12.6% ($p < 0.001$). Furthermore, the I-PSS and the quality of life score improved (26.8 and 18.2%, respectively) in treated patients. Control subjects exhibited no change in any of these parameters. Again, this study lacks a placebo treatment group to assess any possible placebo effects. Treatment with Permixon of symptoms related to BPH has also been evaluated in a 2-year study of effect on symptoms, quality of life, and sexual function (24). These investigators studied 150 men receiving Permixon 160 mg twice daily for 2 years. The I-PSS score improved by 41% at the end of 2 years. Approximately one-half of the subjects had improvements in obstructive and irritative symptoms, and a 40% improvement in quality of life score was noted. Again, no control group and no placebo group were used and thus any placebo effect could not be discerned.

A meta-analysis (6) of randomized trials comparing saw palmetto to placebo or other therapy was recently published. The authors concluded that despite methodology problems, saw palmetto appears to improve urologic symptoms and urinary flow to an extent similar to that of finasteride, but with fewer adverse effects.

Interestingly, saw palmetto has also been studied in a randomized, double-blind, placebo-controlled trial for the treatment of androgenetic alopecia (25). Subjects received either active formulation or placebo for an average of 5 months. In the 10 subjects studied, six of the subjects (60%) were determined to have significant improvement in hair growth as assessed by both the investigators and the subject.

6. Adverse Effects and Toxicity

Clinical studies have reported very few adverse effects that are of a mild nature (usually gastric distress or headache) following saw palmetto administration at normal doses. One randomized, double-blind study of finasteride, tamsulosin, and saw palmetto for 3 months observed no differences among the three treatments in terms of the effectiveness measures and no change in sexual function in those individuals receiving saw palmetto, though ejaculation disorders were noted as the most common side effect in those individuals receiving either tamsulosin or finasteride (26).

6.1. Case Reports of Toxicity Caused By Saw Palmetto Products

A case of toxicity associated with the use of Prostata®, a preparation containing saw palmetto, zinc picolinate, pyridoxine, L-alanine, glutamic acid,

Apis mellifica pollen, silica, hydrangea extract, *Panax ginseng*, and *Pygeum africanum*, was reported in the Annals of Internal Medicine *(27)*.

A 65-year-old man developed acute and protracted cholestatic hepatitis after taking Prostata. The man stopped taking the product after 2 weeks of use because he developed jaundice and severe pruritus. On physical exam, the patient's abdomen was not tender and his liver and spleen were not palpable. Lab results were as follows: bilirubin 8.2 mg/dL, aspartate aminotransferase 1238 IU/L, alanine aminotransferase 1364 IU/L, alkaline phosphatase 179 IU/L, γ-glutamyl transferase 391 IU/L, hematocrit 41%, leukocyte count 3.3 × 103/mm3, platelet count 153,000 cells/mm3, serum protein 6.3 g/dL, albumin 3.6 g/dL, carcinoembryonic antigen less than 2 mg/μL. Serological testing was negative for hepatitis A virus immunoglobulin M (IgM), hepatitis B surface antigen, cytomegalovirus IgM, and hepatitis C virus antibodies. The patient was negative for antinuclear antibodies and antismooth muscle antibodies, but positive for antimitochondrial antibodies. Liver enzyme levels remained abnormal for more than 3 months. Liver biopsy was done after 2 months and showed parenchymal infiltrate of neutrophils and lymphocytes that involved the portal tracts, early bridging, and mild periportal fibrosis. There was no evidence of bile duct damage, cirrhosis, or granulomas. The authors postulated that the patient's cholestasis was an extension of saw palmetto's estrogenic or antiandrogen effect.

In another case report, a 53-year-old white male with meningioma developed intraoperative hemorrhage during surgery for resection of the tumor *(28)*. During the surgery, the patient began experiencing substantial bleeding that was difficult to control. The patient was given 4 L of crystalloid fluids, 4 U of packed red blood cells, 3 U of pooled platelets, and 3 U of fresh frozen plasma. The estimated blood loss was approx 2000 mL. The patient had not received any preoperative thromboprophylaxis and all clotting tests were normal prior to the procedure. However, after surgery, the bleeding time was several times longer than normal and eventually became normal after 5 days. No medications that could have resulted in excessive bleeding were discovered as being taken; however, upon further questioning the patient disclosed that he had been taking saw palmetto for BPH, but had not mentioned it to the physician. A conclusive cause-effect relationship was not established, however.

7. PHARMACOKINETICS/TOXICOKINETICS

Limited pharmacokinetic data are available because saw palmetto is a mixture of various compounds *(29)*. With respect to absorption of saw palmetto components, a mean peak plasma drug concentration of 2.6 mg/L of the "second component" with a high-performance liquid chromatography retention time of 26.4 minutes was measured in 12 healthy young men after a single oral dose of 320 mg of saw palmetto. The time to peak concentration occurred 1.5 hours after administration *(30)*. A 640-mg rectal dose of saw palmetto extract produced a peak of 2.6 µg/mL occurring 3 hours after the dose *(31)*. Rat studies indicate that prostate concentrations are higher than those achieved in other genitourinary tissues or in the liver *(29)* suggesting selective distribution to tissues of interest. The elimination half-life of the "second component" discussed previously was 1.9 hours and the mean area under the concentration vs time curve (AUC) was 8.2 mg/(L·hour) after a single oral dose of 320 mg *(30)*. The AUC of the "second component" produced by a 640-mg rectal dose of saw palmetto extract was 10 mg/(L·hour), and plasma levels were detectable up to 8 hours postdose *(31)*.

8. Drug Interactions

A study of the in vitro ability of saw palmetto to inhibit the metabolic activity of cytochromes P450 3A4, 2D6, and 2C9 was recently studied *(32)*. These investigators found that saw palmetto had no effect on the metabolism of model substrates of cytochrome P450 3A4 and 2D6. However, saw palmetto was noted to be a potent inhibitor of cytochrome P450 2C9 activity in vitro.

Two recent in vivo studies have attempted to evaluate the effect of saw palmetto administration on cytochrome P450 metabolic activity in humans. Markowitz and colleagues *(33)* studied the ability of saw palmetto administration for fourteen days to inhibit the metabolism of dextromethorphan and alprazolam, probe substrates for cytochrome P450 2D6 and 3A4, respectively. This study in six male and six female healthy volunteers found no effect of chronic saw palmetto administration on the metabolism or elimination of either probe substrate and, thus, no effect on either cytochrome P450 2D6 or 3A4 activity. In a similar study, Gurley and colleagues *(34)* evaluated the effect of saw palmetto on cytochrome P450 1A2, 2D6, 2E1, and 3A4 activity in vivo. A group of 12 healthy volunteers were administered saw palmetto for 28 days and phenotyped for each of the previously listed enzyme activities before and after saw palmetto administration. No significant effect of saw palmetto on any of the phenotypic ratios was noted, suggesting that saw palmetto has no effect on in vivo cytochrome P450 1A2, 2D6, 2E1, or 3A4 activity. These results of Markowitz et al. *(33)* and Gurley et al. *(34)* confirm the in vitro results of Yale and Glurich *(32)* regarding these enzymes. However, no in vivo studies have been conducted to date to evaluate whether saw palmetto affects cytochrome P450 2C9 activity, as suggested by Yale and Glurich. This has particular clinical significance because warfarin and phenytoin (both agents with narrow therapeutic indices) are metabolized by P450 2C9. Thus, clinical studies to evaluate this potential interaction are needed.

9. REPRODUCTION

Because of its potential effects on 5α-reductase enzymes, analogous to finasteride, saw palmetto should not be used during pregnancy.

10. REGULATORY STATUS

The German Commission E lists saw palmetto as an approved herb. The berry is the only part of the plant approved for use. The approved uses include urination problems associated with BPH stages I and II and urination problems associated with prostate adenoma. This evaluation is based on reasonable proof of safety and efficacy *(35)*.

Saw palmetto is considered a dietary supplement by the Food and Drug Administration *(3)*. Saw palmetto was previously included in the National Formulary (NF) and the United States Pharmacopeia, but was deleted in 1950 and 1916, respectively. Saw palmetto was deleted because no active ingredient could be found to account for its use *(2)*. Saw palmetto was again included in the NF as an official monograph in 1998 *(4)*.

REFERENCES

1. Chavez ML. Saw palmetto. Hosp Pharm 1998;33:1335–1361.
2. Nemecz G. Saw palmetto. US Pharm 1998;23:97–98, 100–102.
3. Tyler VE, ed. *The Honest Herbal, 3rd edition*, Binghamton: Pharmaceutical Products Press, 1993.
4. Anonymous. Saw palmetto. In: *The Lawrence Review of Natural Products*. St. Louis: Facts and Comparisons, 1994.
5. United States Pharmacopoeia. *National Formulary, 18th edition, Supplement 9*, Rockville: United States Pharmacopeial Convention, 1998.
6. Wilt TJ, Ishani A, Stark G, MacDonald R, Lau J, Mulrow C. Saw palmetto extracts for treatment of benign prostatic hyperplasia. JAMA 1998;280:1604–1609.
7. Briley M, Carilla E. Fauran F. Permixon, a new treatment for benign prostatic hyperplasia, acts directly at the cytosolic androgen receptor in rat prostate [abstract]. Br J Pharmacol 1983;79:327P.
8. Sultan C, Terraza A, Devillier C, et al. Inhibition of androgen metabolism and binding by a liposterolic extract of "Serenoa repens B" in human foreskin fibroblasts. J Steroid Biochem 1984;20:515–519.

9. Habib FK, Ross M, Ho CKH, Lyons V, Chapman K. Serenoa repens (Permixon®) inhibits the 5α-reductase activity of human prostate cancer cell lines without interfering with PSA expression. Int J Cancer 2005;114:190–194.

10. Goepel M, Dinh L, Mitchell A, Schafers RF, Rubben H, Michel MC. Do saw palmetto extracts block human α1-adrenoceptor subtypes in vivo? Prostrate 2001;46:226–232.

11. Talpur N, Echard B, Bagchi D, Bagchi M, Preuss HG. Comparison of saw palmetto (extract and whole berry) and cernitin on prostate growth in rats. Mol Cell Biochem 2003;250:21–26.

12. Breu W, Hagenlocher M, Redl K, Tittel G, Stadler F, Wagner H. Antiphlogistic activity of an extract from Sabal serrulata fruits prepared with supercritical carbon dioxide. In-vitro inhibition of cyclooxygenase and 5-lipoxygenase metabolism. Arzneim Forsch 1992;42:547–551.

13. Blumenthal M, Riggins CW. Saw palmetto berry. In: *Popular Herbs in the US Market. Therapeutic Monographs*. Austin: American Botanical Council, 1997.

14. Di Silverio F, D'Eramo G, Lubrano C, et al. Evidence that Serenoa repens extract displays an antiestrogenic activity in prostatic tissue of benign prostatic hypertrophy patients. Eur Urol 1992;21:309–314.

15. Champault G, Patel JC, Bonnard AM. A double-blind trial of an extract of the plant Serenoa repens in benign prostatic hyperplasia. Br J Clin Pharmacol 1984;18:461–462.

16. Carraro J, Raynaud J, Koch G, et al. Comparison of phytotherapy (Permixon®) with finasteride in the treatment of benign prostatic hyperplasia; a randomized international study of 1,098 patients. Prostate 1996;29:231–240.

17. Bayne CW, Donnelly F, Ross M, Habib FK. Serenoa repens (Permixon): a 5alpha-reductase types I and II inhibitor—new evidence in a coculture model of BPH. Prostate 1999;40:232–241.

18. Gerber GS, Kusnetsov D, Johnson BC, Burstein JD. Randomized, double-blind, placebo-controlled trial of saw pollmetto in men with lower urinary tract symptoms. Urology 2001;58:960–965.

19. Marks LS, Partin AW, Epstein JI, et al. Effects of saw palmetto herbal blend in men with symptomatic benign prostatic hyperplasia. J Urol 2000;163:1451–1456.

20. Veltri RW, Marks LS, Miller MC, et al. Saw palmetto alters nuclear measurements reflecting DNA content in men with symptomatic BPH: evidence for a possible molecular mechanism. Urology 2002;60:617–622.

21. Preuss HG, Marcusen C, Regan J, Klimberg IW, Welebir TA, Jones WA. Randomized trial of a combination of natural products (cernitin, saw palmetto, β-sitosterol, vitamin E) on symptoms of benign prostatic hyperplasia (BPH). Int Urol Nephrol 2001;33:217–225.

22. Kaplan SA, Volpe MA, Te AE. A prospective, 1-year trial using saw palmetto versus finasteride in the treatment of category III prostatitis/chronic pelvic pain syndrome. J Urol 2004;171:284–288.

23. Al-Shukri SH, Deschaseaux P, Kuzmin IV, Amdiy RR. Early urodynamic effects of the lipido-sterolic extract of *Serenoa repens* (Permixon®) in patients with lower urinary tract symptoms due to benign prostatic hyperplasia. Prostate Cancer Prostatic Dis 2000;3:195–199.

24. Pytel YA, Vinarov A, Lopatkin N, Sivkov A, Gorilovsky L, Raynaud JP. Long-term clinical and biologic effects of the lipidosterolic extract of *Serenoa repens* in patients with symptomatic benign prostatic hyperplasia. Adv Natural Ther 2002;19:297–306.
25. Prager N, Bickett K, French N, Marcovici G. J Altern Complement Med 2002;8: 143–152.
26. Zlotta AR, Teillac P, Raynaud JP, Schulman CC. Evaluation of male sexual function in patients with lower urinary tract symptoms (LUTS) associated with benign prostatic hypertrophy (BPH) treated with a phytotherapeutic agent (Permixon®), tamsulosin or finasteride. Eur Urol 2005;48:269–276.
27. Hamid S, Rojter S, Vierling J. Protracted cholestatic hepatitis after the use of Prostata. Ann Intern Med 1997;127:169–179.
28. Cheema P, El-Mefty O, Jazieh AR. Intraoperative haemorrhage associated with the use of extract of saw palmetto herb: a case report and review of literature. J Intern Med 2001;250:167–169.
29. Plosker GL, Brogden RN. Serenoa repens (Permixon®). A review of its pharmacology and therapeutic efficacy in benign prostatic hyperplasia. Drug Aging 1996;9:379–395.
30. De Bernardi di Valserra M, Tripodi AS. Rectal bioavailability and pharmacokinetics of Serenoa repens new formulation in healthy volunteers. Arch Med Intern 1994;46:77–86.
31. De Bernardi Di Valerra M, Tripodi AS, Contos S, Germogli R. Serenoa repens capsules: a bioequivalence study. Acta Toxicol Ther 1994;15:21–39.
32. Yale SH, Glurich I. Analysis of the inhibitory potential of *Ginkgo biloba*, *Echinacea purpurea*, and *Serenoa repens* of the metabolic activity of cytochrome P450 3A4, 2D6, and 2C9. J Altern Comp Med 2005;11:433–439.
33. Markowitz JS, Donovan JL, DeVane CL, et al. Multiple doses of saw palmetto (*Serenoa repens*) did not alter cytochrome P450 2D6 and 3A4 activity in normal volunteers. Clin Pharmacol Ther 2003;74:536–542.
34. Gurley BJ, Gardner SF, Hubbard MA, et al. In vivo assessment of botanical supplementation on human cytochrome P450 phenotypes: *Citrus aurantum*, *Echinacea purpurea*, milk thistle, and saw palmetto. Clin Pharmacol Ther 2004;76:428–440.
35. Blumenthal M. Therapeutic guide to herbal medicine. In: *The Complete German Commission E Monographs*. Austin: American Botanical Council, 1998.

Chapter 11

Panax ginseng

Timothy S. Tracy

SUMMARY

Ginseng is commonly used for a variety of conditions where it is purported to have positive effects on mental, physical, and sexual performance. There are some data to suggest it may have some small positive effects on mental and sexual activities, but the data remain conflicting. Ginseng also can reduce glycemic concentrations after glucose challenge. It should be used with caution in patients receiving anticoagulants as reports have suggested it may reduce the effect of warfarin.

Key Words: Hyperglycemia; adaptogen; cognition; coagulation.

1. HISTORY

Panax ginseng is a perennial herb that starts flowering in its fourth year *(1)*. It grows in the United States, Canada, and the mountainous forests of eastern Asia *(2)*. The translucent, yellowish-brown roots are harvested when plants reach between 3 and 6 years of age *(2)*. This herb has been used in the Orient for 5000 years as a tonic *(3)*. According to traditional Chinese medicine's "philosphy of opposites," American ginseng (*Panax quinquefolius* L.) is a "cool" or "yin" tonic used to treat "hot" symptoms such as stress, insomnia, palpitations, and headache, whereas Asian ginseng (*P. ginseng* L.) is "hot" or "yang" and is used to treat "cold" diseases *(4)*. In the Orient, ginseng is considered a cure-all. This stems from the "Doctrine of Signatures," because the root is said to resemble a man's appearance and is therefore useful to treat all of man's ailments *(5)*. Throughout history, the root has been

From Forensic Science and Medicine:
Herbal Products: Toxicology and Clinical Pharmacology, Second Edition
Edited by: T. S. Tracy and R. L. Kingston © Humana Press Inc., Totowa, NJ

used as a treatment for asthenia, atherosclerosis, blood and bleeding disorders, colitis, and relief of symptoms associated with aging, cancer, and senility *(5)*. Ginseng is also widely believed to be an aphrodisiac *(6)*.

2. CURRENT PROMOTED USES

Ginseng is promoted as a tonic capable of invigorating the user physically, mentally, and sexually. It is also said to possess antistress activity, or to serve as an "adaptogen," improve glycemic control and stimulate immune function. Claims that ginseng can improve athletic performance, enhance longevity, or treat toxic hepatitis are not supported by human trials.

3. SOURCES AND CHEMICAL COMPOSITION

Korean ginseng, Asian ginseng, Oriental ginseng, Chinese ginseng *(7)*, Japanese ginseng, American ginseng *(8)*. Note that the term "ginseng" can refer to the species of the genus *Panax*, as well as to *Eleutherococcus senticosus* (Siberian or Russian ginseng) *(8)*. Unless otherwise noted, the information in this monograph refers specifically to species of the genus *Panax*. Depending on the particular botanical reference, there are three to six different species of *Panax* ginseng, and three with purported medicinal benefits: *P. ginseng* (Chinese or Korean ginseng), *Panax pseudoginseng* (Japanese ginseng), and *Panax quinquefolium* (American ginseng) *(8)*. In this chapter, the term "*Panax ginseng*" will be used to refer to these species, and "Siberian ginseng" will be used to refer to *E. senticosus*. The chemical composition of Siberian ginseng differs from that of *P. ginseng (8)*; thus, the distinction between the two is important in a discussion of therapeutic and adverse effects.

4. PRODUCTS AVAILABLE

Two commercial forms of the herb are available. "White" ginseng consists of the dried root and "red" ginseng is prepared by steaming the fresh, unpeeled root before drying *(9)*. Many different formulations of the herb are available including capsules, gelcaps, powders, tinctures, teas, slices to eat in salads, and whole root to chew. There are also a wide variety of products that claim to contain ginseng such as ginseng cigarettes, toothpaste, cosmetics, soaps, beverages (including beer), candy, baby food, gum, candy bars, and coffee. Prices vary widely based on the quantity and quality of the ginseng root used *(10)*. Tinctures are more expensive but last for years. Powder capsules are cheaper but have a shelf-life of only 1 year *(11)*.

One of the problems in the manufacture of ginseng is the lack of quality control and standardization *(7)*. Although the amount of ginsenosides, the purported active ingredients, ranges widely among brands and often differs from the content stated on the label, testing by Consumer Reports revealed that the amount of ginsenosides in Ginsana®, the ginseng market leader in the United States, is well standardized *(12)* (*see* Section 9 for discussion of factors affecting ginsenoside content). The manufacturer (Pharmaton, Ridgefield, CT) claims that each Ginsana capsule contains 100 mg of standardized, concentrated ginseng *(13)*. A study *(14)* of the Swedish Ginsana product revealed consistency in ginsenoside content between batches. Ginsana is available in the United States in softgel capsules and chewy squares. The capsules are green because chlorophyll is added. Other brands of ginseng are most commonly available in capsule or tablet form and are usually brown. Dosage strengths normally range between 50 mg and 300 mg of *P. ginseng* extract per capsule or tablet. Also, several combination products are available. For example, Ginkogin® is a combination of Panax ginseng, Ginkgo biloba, and garlic. There are other types of ginseng on the market including Siberian, Brazilian, and Indian ginseng. These are not of the genus *Panax* and do not contain ginsenosides *(15)*.

5. PHARMACOLOGICAL/TOXICOLOGICAL EFFECTS

5.1. Endocrine Effects

P. ginseng may exert hypoglycemic effects possibly by accelerating hepatic lipogenesis and increasing glycogen storage *(16–18)*. In a study of 36 newly diagnosed patients with type II diabetes, ginseng at a dose of 200 mg daily exerted a statistically significant benefit on glycosylated hemoglobin (HbA1c) compared to 100 mg of ginseng daily or placebo after 8 weeks of therapy, and patients receiving 100 mg of ginseng had smaller mean fasting blood glucose levels than patients taking 200 mg of ginseng or placebo *(18)*. The actual difference among the mean HbA1c in the three groups was small; the 200-mg ginseng group had a mean glycosylated hemoglobin of 6 vs 6.5% for the 100-mg ginseng and placebo groups. Likewise, the actual difference among mean fasting blood glucose in the three groups was small; the mean fasting blood glucose was 7.7 mmol/L for the 100-mg ginseng group, 7.4 mmol/L for the 200-mg ginseng group, and 8.3 mmol/L for the placebo group at the end of the study. The observed differences might be attributed to differences in body weight among the three groups. The small study sample limits the generalizability of these results. Vuksan and colleagues observed that whether given concurrently or prior to glucose challenge in patients with type 2 diabetes, ginseng blunted the glycemic response by approx 20% *(19)*. In nondiabetic individuals, reduction in glycemic response was only noted when ginseng was administered 40 minutes prior to the glucose challenge. In a related study, investigators demonstrated that dose of ginseng but not timing of administration resulted in a statistically significant reduction in postprandial glycemia in patients with type 2 diabetes following a glucose challenge *(20)*. At 120 minutes postchallenge, reductions in incremental glycemica as much as 60% were noted. Again, these same investigators studied 10 nondiabetic individuals who received different doses of ginseng at different times prior to glucose challenge *(21)*. Compared with placebo, all doses of ginseng reduced the glycemic response up to 90 minutes in some cases. However, time of administration had no effect. Ironically, these same investigators later reported that the effect of ginseng on postprandial glycemia in healthy individuals was time of administration-dependent but not dose-dependent *(22)*, conflicting with their previous reports. Vuksan and colleagues reported that a batch of ginseng that was lower in ginsenosides than previous batches had no effect on postprandial glycemia *(23)*. Finally, it has been reported that different types of ginseng can have differing effects on postprandial glycemia (decreasing, null or increasing) and that these divergent effects may be related to the ginsenoside composition in the preparations *(24)*. Thus, at the present time, it is difficult to predict the effects of ginseng administration on glycemia because varying effects may be noted depending on the composition and preparation of the ginseng.

All the ginsenosids (saponins) so tested have shown antifatigue actions in mice *(25)*. This may reflect the purported "adaptogenic" action of ginseng, which can be defined as an increase in resistance to stresses and is thought to be secondary to normalization of body processes through regulation of the production of various hormones *(4)*. In evaluating the administration of Siberian ginseng for treatment of chronic fatigue syndrome, Hartz and colleagues found no measurable positive effect in those individuals receiving ginseng as compared to subjects receiving placebo *(26)*.

With respect to increasing exercise performance, Hsu and colleagues reported that ginseng attenuated the formation of creatine kinase induced by submaximal exercise in subjects undergoing a treadmill test *(27)*. However, no increase in aerobic work capacity was noted. In a related study of exercise performance effects, Siberian ginseng administration had no effect on steadystate substrate utilization or any physiological measure in individuals undergoing prolonged cycling exercise *(28)*. The study was conducted in a randomized, double-blind, placebo-controlled fashion and followed 7 days of treatment with either ginseng or placebo.

Ginseng appears to have a modulating effect on the hypothalamic-pituitary-adrenal axis by inducing secretion of adrenocorticotropic hormone from the anterior pituitary to increase plasma cortisol (29,30), perhaps accounting for improvement in 11 quality of life measurements in a large double-blind study using ginseng extract G115 (31).

Although many products containing ginseng are marketed specifically for postmenopausal women, a recent review concluded that there is insufficient evidence that ginseng is effective for treatment of menopausal symptoms (11). In vitro, Siberian ginseng extract, but not P. ginseng extract, binds to estrogen receptors. Both extracts have affinity for progestin, glucocorticoid, and mineralocorticoid receptors (32). A recent study reported that a morning/evening formulation containing ginseng and other constituents relieved menopausal symptoms, but no placebo control was included so it is difficult to tell whether the effect was caused by the formulation or a placebo effect (33).

5.2. Neurological Effects

Commercially available P. ginseng products have been reported to have stimulant effects on the central nervous system (CNS) in humans (34) (see Section 5). In animal models, ginseng extracts have been shown to have CNS-stimulant effects (35). Ginsenoside Rg1 inhibits neuronal apoptosis in vitro (35), and ginsenoside Rb1 reverses short-term memory loss in rats (4).

It has been suggested that ginseng may hold promise for the treatment of dementia in humans (4,36). To this end, a number of studies have been performed to evaluate the effects of ginseng on cognition. Wesnes and colleagues studied the memory-enhancing effects of either P. ginseng or Ginkgo biloba in healthy middle-aged volunteers (37). These investigators found that administration of either agent resulted in a small but statistically significant improvement in the Index of Memory Quality (~7.5%) as compared to placebo. In a similar study, another group studied the effects of either ginseng, G. biloba, or the combination on the modulation of cognition and mood in healthy young adults (38). These investigators found that all three treatments improved secondary memory performance and that ginseng administration elicited some improvement in the speed of performing memory tasks and the accuracy of attentional tasks. Only ginkgo elicited a self-rated improvement in mood. Scholey and Kennedy (39) again studied the effects of P. ginseng and G. biloba on several tests of cognitive demand. Increasing doses of ginseng improved accuracy but slowed responses on the Serial Sevens test and the combination product caused a sustained improvement in the number of Serial Sevens responses. This was accompanied by improved accuracy on this same test, again in a dose-dependent fashion. These same investigators later conducted a study of the effects of *P. ginseng* as compared to guarana in several cognitive performance tests *(37)*. Again, ginseng administration led to an improvement the speed of attention task performance, but little evidence of increased accuracy was noted.

However, two studies have also suggested that administration of ginseng (or a combination of ginseng and *G. biloba*) has no effect on cognition (and mood). Hartley and colleagues evaluated the effects of a 6- or 12-week course of a ginkgo/ginseng combination product (Gincosan®) on the mood and cognition of postmenopausal women *(41)*. Subjects were administered a battery of mood, somatic anxiety, sleepiness, and menopausal symptom tests. The Gincosan treatment had no measurable effect on any parameter. In a similar study of ginseng administration, investigators found no effect of ginseng on positive affect, negative affect, or total mood disturbance in a randomized, placebo-controlled, double-blind trial *(42)*. Persson and colleagues studied the memory-enhancing effects of either ginseng or *G. biloba* taken over a sustained period of time (mean intake time of 5.3 months) in healthy community-dwelling volunteers *(43)*. No improvement in memory performance evaluated by eight separate tests was noted in either the group receiving ginseng or the group receiving *G. biloba*. Thus, it appears that conflicting results still exist as to the ability of ginseng to improve memory and cognition; however, even in those studies demonstrating a positive effect, the enhancement was generally small in magnitude.

The administration of ginseng has also been studied in the treatment of attention-deficit hyperactivity disorder (ADHD). Lyon et al., conducted a pilot study ($n = 36$) evaluating the effects of a combination product containing ginseng and ginkgo for the treatment of ADHD *(44)*. The investigators reported improvement in 31–67% of the subjects depending on the outcome measure; however, no placebo control was included, so it is difficult to ascertain if the effect was caused by the treatment or a placebo effect.

5.3. Cardiovascular Effects

In animal studies, ginsenoside Rb1 decreases blood pressure, perhaps owing to relaxation of smooth muscle *(25)*. In humans, small studies suggest ginseng may decrease systolic blood pressure at a dose of 4.5 g/day *(45)*, and enhance the efficacy of digoxin in class IV heart failure *(46)*. In contrast, ginsenoside Rg1 has been purported to have hypertensive effects *(4)*. Finally, it has been reported that ginseng has no effect on blood pressure in individuals with hypertension *(47)*.

An in vitro study using a crude extract of ginseng saponins and rabbit corpus-cavernosal smooth muscle suggests that some component of ginseng may be a nitric oxide donor, capable of causing relaxation of smooth muscle in the corpus carvernosum *(48)*. This finding might provide a scientific basis for claims that ginseng enhances sexual potency, and for the results of a study that showed increased penile rigidity and girth compared to placebo or trazodone in patients with erectile dysfunction *(49)*.

Red ginseng powder may be useful in hyperlipidemia; it was shown to decrease triglycerides as well as increase high-density lipoprotein (HDL) in a pilot study *(50)*. A previous rat study lends validity to ginseng's ability to decrease triglyceride levels *(16)*, but a study in patients with diabetes showed no effect on total cholesterol, low-density lipoprotein (LDL), HDL, or triglyceride levels *(18)*.

5.4. Hematological Effects

P. ginseng may inhibit platelet aggregation by regulating the levels of cGMP and thromboxane A2 *(51)*.

5.5. Immunological Effects

Red ginseng stimulates accumulation of neutrophils in a dose-dependent manner following intraperitoneal injections in mice *(52)*. Data show *P. ginseng* extracts are also able to stimulate an immune response in humans. Chemotaxis of polymorphonuclear cells was increased compared to placebo. Both the phagocytosis index and fraction were enhanced in the ginseng groups and intracellular killing was increased compared to the placebo group. Total lymphocytes and helper T-cells were increased as well *(53)*. There have been other reports of increases in cell-mediated immunity as well as natural-killer cell activity *(54)*.

Predy and colleagues evaluated the ability of ginseng to prevent upper respiratory infections in a randomized, placebo-controlled trial *(55)*. Administration of ginseng for 4 months resulted in a reduction in both the mean number of colds experienced, the number of individuals experiencing two or more colds, and the total number of days of cold symptoms. Similarly, McElhaney et al. studied the ability of ginseng to prevent acute respiratory illness in institutionalized older adults *(56)*. The incidence of confirmed influenza cases was lower in the ginseng-treated group as compared to placebo treatment.

5.6. Antineoplastic Effects

Data from in vitro studies, animal models, case-control studies, and cohort studies suggest ginseng may prevent or ameliorate various cancers. These studies have been reviewed in detail elsewhere *(57–59)*. Suh and colleagues studied the effects of red ginseng on the recurrence of cancer after curative resection in patients with previous gastric cancer during postoperative chemotherapy *(60)*. Survival rate was approximately twice that of control, but placebo treatment was not used. Prospective, placebo-controlled studies of ginseng's ability to prevent or treat cancer are lacking.

5.7. Case Reports of Toxicity Caused By Commercially Available Products

In 1979 the term "ginseng abuse syndrome" (GAS) was coined as the result of a study *(34)* of 133 people who had been using a variety of ginseng preparations for at least 1 month. Most study subjects experienced CNS excitation and arousal. A total of 14 patients experienced GAS, defined as hypertension, nervousness, sleeplessness, skin eruptions, and morning diarrhea. Five of these subjects also exhibited edema. The effects of ginseng on mood appeared to be dose-dependent; four patients experienced depersonalization and confusion at doses of 15 g, and depression was reported following doses greater than 15 g. A total of 22 subjects experienced hypertension. All of the patients experiencing GAS or hypertension were also using caffeinated beverages. Six other subjects also experienced GAS but were considered "atypical" because they were either using Siberian ginseng instead of *P. ginseng*, or were injecting ginseng, and thus were not included in the study results. One subject experienced anaphylaxis followed by confusion and hallucinations after injection of 2 mL of ginseng extract. The average daily dose of the 14 patients experiencing GAS was 3 g of ginseng root, and most users reported titrating the dose to minimize nervousness and tremor. One subject experienced hypotension, weakness, and tremor when ginseng use was abruptly discontinued. The author compared ginseng's effects to those of high doses of corticosteroids. GAS seemed to be found predominantly during the first year of use, possibly because by the 18-month follow-up visit, ginseng use had declined to an average of 1.7 g daily, and by the 24-month visit, one-half of the patients with GAS had discontinued ginseng use, and 21% of the remaining subjects had stopped using it. Eight subjects were still experiencing diarrhea and nervousness at the 2-year follow-up. Because this study was not controlled, the existence of GAS has been questioned *(6)*.

Hypertension, shortness of breath, dizziness, inability to concentrate, a loud palpable fourth heart sound, "thrusting" apical pulse, and hypertensive changes on fundal examination were reported in a 39-year-old man who had taken various ginseng products for 3 years *(61)*. His blood pressure measured 140/100 mmHg on three occasions over 6 weeks, and when referred for management of his hypertension it was 154/106 mmHg. He was advised to discontinue the ginseng products, and 5 days later was normotensive at 140/85 mm Hg. At 3-month follow-up, he remained normotensive and his other symptoms had resolved. No attempt was made to confirm the identity or composition of the ginseng products.

An episode of Stevens-Johnson syndrome was reported in a 27-year-old man following ginseng administration (two pills a day for 3 days). Infiltration of the dermis by mononuclear cells was noted. The patient recovered completely within 30 days *(62)*.

An association between ginseng and mastalgia has been reported. A 70-year-old woman developed swollen, tender breasts with diffuse nodularity after using a *P. ginseng* powder (Gin Seng) for 3 weeks. Symptoms ceased following discontinuation of the herb and reappeared with two additional rechallenges. Prolactin levels were within normal limits *(63)*.

A 72-year-old woman experienced vaginal bleeding after taking 200 mg daily of a Swiss-Austrian geriatric formulation of ginseng (Geriatric Pharmaton, Bernardgrass, Austria) for an unspecified time *(64)*. In a similar case, a 62-year-old woman had undergone a total hysterectomy 14 years previously and had been taking Rumanian ginseng alternating with Gerovital® every 2 weeks for 1 year *(65)*. The patient derived a marked estrogenic effect from the product based on microscopy of vaginal smears as well as the gross appearance of the vaginal and cervical epithelium. The patient was dechallenged from the products for 5 weeks, rechallenged with Gerovital for 2 weeks, then rechallenged with ginseng for 2 weeks. Estrone, estradiol, and estriol levels were essentially unchanged over this time period, but the estrogenic effects on the vaginal smear coincided with ginseng use. Using gas chromatography, the investigators found no estrogen in the tablets the patient had been taking. They did discover that a crude methanolic extract of the ginseng product competed with estradiol for the estrogen and progesterone binding sites in human myometrial cytosol.

A 44-year-old woman who had experienced menopause at age 42 experienced three episodes of spotting associated with use of Fang Fang ginseng face cream (Shanghai, China). Interestingly, these episodes of bleeding were associated with a decrease in follicle-stimulating hormone levels and a disordered proliferative pattern on endometrial biopsy. The woman discontinued use of the cream and experienced no further bleeding *(66)*. Whether the products used in these reports of vaginal bleeding and mastalgia contained *P. ginseng* or Siberian ginseng (*E. senticosus*) was not investigated. Whether *Panax* or Siberian ginseng causes estrogenic effects requires further study.

Maternal ingestion of 650 mg of Siberian ginseng (Jamieson Natural Sources, Toronto) twice daily was associated with androgenization in a neonate *(67)*. The product had been taken for the previous 18 months, including the pregnancy. During pregnancy, the mother noted increased and thicker hair growth on her head, face, and pubic area, and had experienced repeated premature uterine contractions during late pregnancy. At birth, the Caucasian child weighed 3.3 kg, had thick black pubic hair, hair over the entire forehead, and swollen red nipples. The woman continued to take the ginseng product for 2 weeks after the baby's birth, during which time she breast-fed the baby. She was advised to discontinue the product when the baby was 2 weeks old, and his pubic and forehead hair began to fall out. By 7.5 weeks of age, hair was scant, but his testes were enlarged. Weight gain was 1.1 kg during the first 3.5 weeks of life, and 1.4 kg during the next 3.5 weeks. At age 7.5 weeks, his weight (5.8 kg), length (60.6 cm), and head circumference (41.5 cm) were at or above the 97th percentile. At that time, testosterone, 17-hydroxyprogesterone, and cortisol levels were normal. Subsequent information did not confirm the product's androgenic effects. A sample of the raw material used in manufacturing the preparation used by this patient was identified as *Periploca sepium* (Chinese silk vine), not Siberian ginseng. No androgenic effects were noted in rats administered the manufacturer's sample *(68)*. *P. sepium* ("jia-pi") was reported previously to be mislabeled as Siberian ginseng ("wu-jia-pi"), perhaps owing to similarities in the Chinese terms for these herbs *(69)*.

6. DRUG INTERACTIONS

A probable interaction between warfarin and apanax ginseng product has been reported *(13)*. A 47-year-old man with a St. Jude-type mechanical aortic valve had been controlled on warfarin with an international normalized ratio (INR) of 3.1 (goal 2.5–3.5). He experienced a subtherapeutic INR of 1.5 following 2 weeks of ginseng administration (Ginsana three times daily). Other medications included 30 mg of diltiazem three times daily, nitroglycerin as needed, and 500 mg of salsalate three times daily as needed. He had been on all of these medications for at least 3 years before the abrupt change in his INR. Discontinuation of ginseng resulted in an increase in INR to 3.3 within

2 weeks. In this regard, a randomized, double-blind, placebo-controlled trial was undertaken to study the effects of ginseng on warfarin and INR (70). Coadministration of ginseng statistically significantly reduced the INR by –0.19 (95% confidence interval, –0.36 to –0.07) as well as reduced the INR area under the curve (AUC) and the AUC of warfarin. It should be noted that this study involved healthy volunteers and though they received ginseng for 2 weeks, they only received warfarin for three days prior to administration of ginseng and thus, steady-state warfarin concentrations were not likely achieved. In contrast, an open-label, randomized, three-way crossover study evaluated the effects of 1 week of either ginseng or St. John's wort on the INR and pharmacokinetics of warfarin following a single dose of warfarin 25 mg (71). These investigators found no effect of ginseng on either the INR or the pharmacokinetics of (S)-warfarin (the more active enantiomer) or its (S)-7-hydroxywarfarin metabolite.

In a phenotypic trait measure study of effects of various herbal preparations on cytochrome P450 enzyme activity, Gurley and colleagues evaluated the effects of ginseng administration on CYP1A2, CYP2D6, CYP2E1, and CYP3A4 activity in healthy human volunteers (72). Metabolism of probe drugs for each of these enzymes was studied in the absence and presence of ginseng administered for 28 days. Ginseng administration had no effect on the metabolism of any of the probe drugs, suggesting that ginseng administration will not result in drug interactions with drugs metabolized by CYP1A2, CYP2D6, CYP2E1, or CYP3A4. However, the enzyme that is responsible for the metabolism of (S)-warfarin is CYP2C9 (see previous section) and was not evaluated in this study. These findings of lack of effect on CYP2D6 and CYP3A4 were corroborated by a similar study that found no effect of ginseng administration on the activity of either of these two enzymes (73).

Manic-like symptoms were reported in a patient treated with phenelzine and ginseng. The symptoms disappeared with cessation of the herbal therapy (74). Users should also exercise caution if ginseng is taken in combination with caffeinated beverages; as discussed in Section 5, hypertension and nervousness have been reported when the two are combined (34).

Although Siberian ginseng is not of the same genus as *P. ginseng*, it may be confused with and substituted for *P. ginseng*, and thus a discussion of drug interactions with Siberian ginseng is warranted. Siberian ginseng has been reported to inhibit the metabolism of hexobarbital in mice by 66% (75). Siberian ginseng ingestion was associated with elevated digoxin levels in a 74-year-old man whose digoxin levels had been maintained between 0.9 and 2.2 ng/L (normal, 0.6–2.6 ng/L) for more than 10 years. He was asymptomatic for digoxin toxicity despite a level of 5.2 ng/L. Electrocardiogram, potassium level, and serum creatinine level were normal. The level decreased on dechallenge and increased on rechallenge. The product was analyzed for digoxin or digitoxin contamination, but none was found. The product was not analyzed to determine if it did in fact contain Siberian ginseng. It was hypothesized that some component of Siberian ginseng might impair digoxin elimination or interfere with the digoxin assay. The type of digoxin assay used in this case was not specified *(76)*. To this end, the effect of different types of ginseng on assays of digoxin concentration has now been studied extensively *(77)*. These investigators observed that apparent digoxin-like immunoreactivity was observed when ginseng was studied with a fluorescence polarization immunoassay (FPIA) technique and modest immunoreactivity with microparticle enzyme immunoassay (MEIA) methods using serum spiked with ginseng. Interestingly, when serum from patients receiving digoxin was studied and ginseng was then spiked into the samples, falsely high digoxin concentrations were measured with FPIA but falsely lower concentrations were measured using MEIA. Using the Tina-quant assay, no interference was noted with any of the ginseng preparations.

7. PHARMACOKINETICS

The structures and nomenclature of the chemical constituents of Panax ginseng have been discussed elsewhere *(78)* (*see also* Section 9).

7.1. Absorption

β-Sitosterol is a steroid sapogenin that has been isolated from ginseng. Approximately 50-60% of a dose of β -sitosterol is absorbed from the gastrointestinal tract in rats *(79)*. After oral administration of radiolabeled ginsenoside Rg1, blood radioactivity peaked at 2.1 hours. Bioavailability was 49% *(80)*.

7.2. Distribution

Studies of the distribution of [3H]ginsenoside Rg1 following intravenous injection have been performed in mice *(80)*. Tissue radioactivity was greatest in the kidney, followed by the adrenal gland, liver, lungs, spleen, pancreas, heart, testes, and brain. Plasma protein binding was 24%, and tissue protein binding was 48% in the liver, 22% in testes, and 8% in the brain.

7.3. Metabolism/Elimination

The blood radioactivity decreased in a triphasic manner after intravenous injection of [3H]ginsenoside Rg1 to mice *(80)*. Other Chinese studies have characterized the biotransformation of ginsenoside 20(S)-Rg2, one of the main constituents of ginseng roots and leaves. Its metabolism is complex and involves multiple hydrolysis reactions in the gastrointestinal tract. Metabolites of 20(S)-Rg2 include 20(S)-Rh1 and 20(S)-protopanaxatriol. Details of the biotransformation of 20(S)-Rg2 and chemical structures of the ginsenosides are available in the cited reference *(80)*.

Corroborating two rat studies *(81,82)* suggesting that only trace amounts of ginsenosides are excreted in the urine, low levels of ginsenoside aglycones were identified using gas chromatography-mass spectroscopy to analyze urine samples of 65 athletes claiming to have ingested ginseng within the 10 days prior to urine collection *(14)*. An aglycone (molecule from which the sugar moiety has been removed) of ginsenosides, 20(S)-protopanaxatriol, was found at concentrations between 2 and 35 ng/mL in approx 90% of the urine samples studied. Another aglycone, 20(S)-protopanaxadiol, was barely detectable despite the fact that the ginsenosides from which it is derived were the major ginsenosides found in the commercially available Swedish ginseng products analyzed by the investigators.

This indicates that these two ginsenosides have different pharmacokinetics. Because the actual amount of ginseng ingested and the time since ingestion were unknown, little else can be inferred from these data.

8. REGULATORY STATUS

The German Commission E approves *P. ginseng* as a nonprescription drug for use as a "tonic for invigoration and fortification in times of fatigue and debility, for declining capacity for work and concentration, and also for use during convalescence *(83)*. In the United States, ginseng is regulated as a dietary supplement.

REFERENCES

1. Leung A, ed. *Encyclopedia of Common Natural Ingredients Used in Food, Drugs, and Cosmetics*. New York: John Wiley and Sons, 1980.
2. Tyler VE, Brady LR, Robbers JE, eds. *Pharmacognosy, 8th edition*. Philadelphia: Lea and Febiger, 1981.
3. Chong S, Oberholzer V. Ginseng—is there a use in clinical medicine? Postgrad Med J 1988;64:841–846.
4. Awang DVC. The anti-stress potential of North American ginseng. J Herbs Spices Med Plants 1998;6:87–91.
5. Anonymous. Ginseng. In: *Lawrence Review of Natural Products*. St. Louis: Facts and Comparisons, 1990.

6. Tyler V, ed. *The Honest Herbal, 3rd edition*. Binghamton: Pharmaceutical Products Press, 1993.
7. Muller J, Clauson K. Top herbal products encountered in drug information requests (part 1). Drug Benefit Trends 1998;10:43–50.
8. Awang DVS. Maternal use of ginseng and neonatal androgenization [letter]. JAMA 1991;265:1828.
9. Blumenthal M, ed. *Popular Herbs in the U.S. Market*. Austin: American Botanical Council, 1997.
10. Kennedy B. Herb of the month: ginseng. Total Health 1995;17:48.
11. Schiedermayer D. Little evidence for ginseng as treatment for menopausal symptoms. Altern Med Alert 1998;1:77–78.
12. Anonymous. Ginsana: Tonic or dud? Consum Rep Health 1998;10(7):2.
13. Janetzky K, Morreale A. Probable interaction between warfarin and ginseng. Am J Health Syst Pharm 1997;54:692–693.
14. Cui J, Garle M, Bjorkhem I, Eneroth P. Determination of aglycones of ginsenosides in ginseng preparations sold in Sweden and in urine samples from Swedish Athletes. Scand J Clin Lab Invest 1996;56:151–160.
15. Tyler V. Ginseng: king of zing? Prevention 1997;49:69.
16. Yokozawa T, Seno H, Oura H. Effect of ginseng extract on lipid and sugar metabolism. Chem Pharmacol Bull 1975;23:3095–3100.
17. Oshima Y, Konno C, Hikino H. Isolation and hypoglycemic activity of panaxans I, J, K, and L, glycans of panax ginseng roots. J Ethnopharmacol 1985;14:255–259.
18. Sotaniemi E, Haapakoski E, Rautio A. Ginseng therapy in non-insulin-dependent diabetic patients. Diabetes Care 1995;18:1373–1375.
19. Vuksan V, Sievenpiper JL, Koo VYY, et al. American ginseng (Panax *quinquefolius* L) reduces postprandial glycemia in nondiabetic subjects and subjects with Type 2 diabetes mellitus. Arch Intern Med 2000;160:1009–1013.
20. Vuksan V, Stavro MP, Sievenpiper JL, et al. Similar postprandial glycemic reduction with escalation of dose and administration time of American ginseng in Type 2 diabetes. Diabetes Care 2000;23:1221–1226.
21. Vuksan V, Stavro P, Sievenpiper JL, et al. American ginseng improves glycemia in individuals with normal glucose tolerance: effect of dose and time escalation. J Am Coll Nutr 2000;6:738–744.
22. Vuksan V, Sievenpiper JL, Wong J, et al. American ginseng (*Panax quinquefolius* L.) attenuates postprandial glycemia in a time-dependent but not dose-dependent manner in healthy individuals. Am J Clin Nutr 2001;73:753–758.
23. Sievenpiper JL, Arnason JT, Leiter LA, Vuksan V. Variable effects of American ginseng: a batch of American ginseng (*Panax quinquefolius* L.) with a depressed ginsenoside profile does not affect postprandial glycemia. Eur J Clin Nutr 2003;57:243–248.
24. Sievenpiper JL, Arnason JT, Leiter LA, Vuksan V. Decreasing, null and increasing effects of eight popular types of ginseng on acute postprandial glycemic indices in healthy humans: the role of ginsenosides. J Am Coll Nutr 2004;23(3):248–258.

25. Kaku T, Miyata T, Uruno T, Sako I, Kinoshita A. Chemico-pharmacological studies on saponins of panax ginseng. Arzneim Forsch 1975;25:539–547.
26. Hartz AJ, Bentler S, Noyes R, et al. Randomized controlled trial of Siberian ginseng for chronic fatigue. Psychol Med 2004;34:51–61.
27. Hsu CC, Ho MC, Lin LC, Su B, Hsu MC. American ginseng supplementation attenuates creatine kinase level induced by submaximal exercise in human beings. World J Gastroenterol. 2005;11(34):5327–5331.
28. Eschbach LC, Webster MJ, Boyd JC, McArthur PD, Evetovich TK. The effect of Siberian ginseng (Eleutherococcus Senticosus) on substrate utilization and performance during prolonged cycling. J Sports Nutr Exer Metab 2000;10:444–451.
29. Hiai S, Yokoyama H, Oura H. Features of ginseng saponin-induced corticosterone secretion. Endocrinol Jpn 1979;26:737–740.
30. Fulder S. Ginseng and the hypothalamic-pituitary control of stress. Am J Chin Med 1981;9:112–118.
31. Caso Maraso A, Vargas Ruiz R, Salas Villagomez A, Begona Infante C. Double-blind study of a multivitamin complex supplemented with ginseng extract. Drugs Exp Clin Res 1996;22:323–329.
32. Pearce PT, Zois I, Wynne KN, Funder JW. Panax ginseng and Eleutherococcus senticosus extracts—in vitro studies on binding to steroid receptors. Endocrinol Jpn 1982;29:567–573.
33. Sun J. Morning/evening menopausal formula relieves menopausal symptoms. A pilot study. J Altern Complement Med. 2003;9:403–409.
34. Siegel R. Ginseng abuse syndrome. Problems with the panacea. JAMA 1979;241:1614–1615.
35. Takagi K, Saito H, Nabata H. Pharmacological studies of panax ginseng root: estimation of pharmacological actions of panax ginseng root. Jpn J Pharmacol 1972;22:245–259.
36. Li J, Zhang X, Zhang J. Study of the anti-apoptotic mechanism of ginsenoside Rg1 in cultured neurons. Acta Pharmacol Sin 1997;32:406 410.
37. Wesnes KA, Ward T, McGinty A, Petrini O. The memory enhancing effects of a Ginko biloba/Panax ginseng combination in healthy middle-aged volunteers. Psychopharmacology 2000;152:353–361.
38. Kennedy DO, Scholey AB, Wesnes KA. Modulation of cognition and mood following administration of single doses of Ginkgo biloba, ginseng, and a ginkgo/ginseng combination to healthy young adults. Physiol Behav 2002;75(5):739–751.
39. Scholey AB, Kennedy DO. Acute, dose-dependent cognitive effects of *Ginkgo biloba, Panax ginseng* and their combination in healthy young volunteers: differential interactions with cognitive demand. Hum Psychopharmacol Clin Exp 2002;17:35–44.
40. Kennedy DO, Haskell CF, Wesnes KA, Scholey AB. Improved cognitive performance in human volunteers following administration of guardna (*paullinia cupana*) extract: comparsion and interaction with *panax ginseng*. Parmacol Biochem Behav 2004;79:401–411.

41. Hartley DE, Elsabagh S, File SE. Gincosan (a combination of ginkgo biloba and panax ginseng): the effects on mood and cognition of 6 and 12 weeks' treatment in post-menopausal women. Nutr Neurosci 2004;7(5–6):325–333.

42. Cardinal BJ, Engels HJ. Ginseng does not enhance psychological well-being in healthy, young adults: results of a double-blind, placebo-controlled, randomized clinical trial. J Am Diet Assoc 2001;101:655–660.

43. Persson J, Bringlov E, Nilsson LG, Nyberg L. The memory-enhancing effects of ginseng and ginkgo biloba in healthy volunteers. Psychopharmacology 2004;172:430–434.

44. Lyon MR, Cline JC, Totosy de Zepetnek J, Shan JJ, Pang P, Benishin C. Effect of the herbal extract combination Panax quinquefolium and ginko biloba on attention-deficit hyperactivity disorder: a pilot study. J Psychiatry Neurosci 2001:26(3):221–228.

45. Han KH, Choe SC, Kim HS, et al. Effect of red ginseng on blood pressure in patients with essential hypertension and white coat hypertension. Am J Chin Med 1998;26:199–209.

46. Ding DZ, Shen TK, Cui YZ. [Effects of red ginseng on the congestive heart failure and its mechanism]. Chung Kuo Chung Hsi I Chieh Ho Tsa Chih 1995;15:325–327.

47. Stavro 2005

48. Kim HJ, Woo DS, Lee G, Kim JJ. The relaxation effects of ginseng saponin in rabbit corporal smooth muscle: is it a nitric oxide donor? Br J Urol 1998;82:744–748.

49. Choi HK, Seong DH, Rha KH. Clinical efficacy of Korean red ginseng for erectile dysfunction. Int J Impot Res 1995;7:181–186.

50. Yamamoto M, Uemura T, Nakama S, Uemiya M, Kumagai A. Serum HDL-cholesterol-increasing and fatty liver-improving actions of panax ginseng in high cholesterol diet-fed rats with clinical effect on hyperlipidemia in man. Am J Chin Med 1983;11:96–101.

51. Park H, Rheem M, Park K, Nam K, Park KH. Effect of nonsaponin fraction from panax ginseng on cGMP and thromboxane A2 in human platelet aggregation. J Ethnopharmacol 1995;49:157–162.

52. Toda S, Kimura M, Ohnishi M. Induction of neutrophil accumulation by red ginseng. J Ethnopharmacol 1990;30:315–318.

53. Scaglione F, Ferra F, Dugnan S, Falchi M, Santoro G, Fraschi F. Immunomodulatory effects of two extracts of panax ginseng. Drugs Exp Clin Res 1990;16:537–542.

54. Singh V, Agarwal S, Gupta B. Immunomodulatory activity of panax ginseng extract. Planta Med 1984;50:462–465.

55. Predy GN, Goel V, Lovlin R, Donner A, Stitt L, Tapan KB. Efficacy of an extract of North American ginseng containing poly-furanosyl-pyranosyl-saccharides for preventing upper respiratory tract infections: a randomized controlled trial. CMAJ 2005;173(9):1043–1048.

56. McElhaney JE, Gravenstein S, Cole SK, et al. A placebo-controlled trial of a proprietary extract of North American ginseng (CVT-E002) to prevent acute respiratory illness in institutionalized older adults. J Am Geriatr Soc 2004;52:13–19.

57. Xiaoguang C, Hongyan L, Xiaohong L, et al. Cancer chemopreventive and therapeutic activities of red ginseng. J Ethnopharmacol 1998;60:71–78.

58. Ahn YO. Diet and stomach cancer in Korea. Int J Cancer 1997;Suppl 10:7–9.
59. Yun TK. Experimental and epidemiological evidence of the cancer-preventive effects of Panax ginseng C.A. Meyer. Nutr Rev 1996;54(11 Pt 2):S71–S81.
60. Suh SO, Kroh M, Kim NR, Joh YG, Cho MY. Effects of red ginseng upon postoperative immunity and survival in patients with stage III gastric cancer. Am J Chinese Med 2002;30(4):483–494.
61. Hammond TG, Whitworth JA. Adverse reactions to ginseng [letter]. Med J Aust 1981;1:492.
62. Dega H, Laporte J, Frances C, Herson S, Chosidow O. Ginseng as a cause for Stevens-Johnson syndrome? Lancet 1996;347:1344.
63. Palmer BV, Montgomery ACV, Monteiro JCMP. Gin Seng and mastalgia. Br Med J 1978;1:1284.
64. Greenspan EM. Ginseng and vaginal bleeding. JAMA 1983;249:2018.
65. Punnonen R, Lukola A. Oestrogen-like effect of gingseng. Br Med J 1980;281:1110.
66. Hopkins MP, Androff L, Benninghoff AS. Ginseng face cream and unexplained vaginal bleeding. Am J Obstet Gynecol 1988;159:1121–1122.
67. Koren G, Randor S, Martin S, Danneman D. Maternal use of ginseng and neonatal androgenization [letter]. JAMA 1990;264:2866.
68. Waller DP, Martin AM, Farnsworth NR, Awang DVC. Lack of androgenicity of Siberian ginseng. JAMA 1992;267:692–693.
69. Awang DVS. Maternal use of ginseng and neonatal androgenization [letter]. JAMA 1991;266:363.
70. Yuan CS, Wei G, Dey L, et al. Brief communication: American ginseng reduces warfarious effect in healthy patients: a randomized, controlled trial. Ann Intern Med 2004;141:23–27.
71. Jiang X, Williams KM, Liauw WS, et al. Effect of St. John's wort and ginseng on the pharmacokinetics and pharmacodynamics of warfarin in healthy subjects. Br J Clin Pharmacol 2004;57(5):592–599.
72. Gurley BJ, Gardner SF, Hubbard MA, et al. Cytochrome P450 phenotypic ratios for predicting herb-drug interactions in humans. Clin Pharmacol Ther 2002;72:276–287.
73. Donovan JL, DeVane CL, Chavin KD, Taylor RM, Markowitz JS. Siberian ginseng (Eleutheroccus senticosus) effects on CYP2D6 and CYP3A4 activity in normal volunteers. Drug Metab Dispos 2003;31(5):519–522.
74. Jones B, Runikis A. Interaction of ginseng with phenelzine. J Clin Psychopharmacol 1987;7:201–202.
75. Mendon PJ, Ferguson PW, Watson CF. Effects of Eleutherococcus senticus extracts on hexobarbital metabolism in vivo and in vitro. J Ethnopharmacol 1984;10:235–241.
76. McRae S. Elevated serum digoxin levels in a pateint taking digoxin and siberian ginseng. Can Med Assoc J 1996;155:293–295.
77. Dasgupta A, Reyes MA. Effect of Brazilian, Indian, Siberian, Asian, and North American ginseng on serum digoxin measurement by immunoassays and binding of digoxin-like immunoreactive components of ginseng with Fab fragment of antidigoxin antibody (Digibind). Am J Clin Pathol 2005;124:229–236.

78. Phillipson JD, Anderson LA. Ginseng-quality, safety and efficacy? Pharm J 1984;232:161–165.
79. Schon N, Engelhardt P. Tierexperimentelle untersuchungen zur frage der resorption von b-sitosterin. Arzneim Forsch 1960;10:491–496.
80. Liu C, Xiao P. Recent advances on ginseng research in China. J Ethnopharmacol 1992;36:27–38.
81. Odani T, Tanizawa H, Takino Y. Studies on the absorption, distribution, excretion, and metabolism of ginseng saponins. II. The absorption, distribution, excretion of ginsenoside-Rg1 in the rat. Chem Pharamcol Bull 1983;31:292–298.
82. Odani T, Tanizawa H, Takino Y. Studies on the absorption, distribution, excretion, and metabolism of ginseng saponins. III. The absorption, distribution, excretion of ginsenoside-Rb1 in the rat. Chem Pharamcol Bull 1983;31:1059–1066.
83. Blumenthal M, ed. Ginseng root. In: The complete German Commission E monographs. Austin: American Botanical Council, 1998.

Chapter 12

Cranberry

Timothy S. Tracy

SUMMARY

It appears that cranberry juice may be effective in preventing the recurrence of urinary tract infections, but not in treating urinary tract infections. It is generally well tolerated and relatively free of adverse effects. There have been case reports of coadministration of cranberry juice and warfarin resulting in bleeding events, but this potential interaction remains to be conclusively established.

Key Words: *Vaccinium macrocarpon*; prophylaxis; urinary tract infection; kidney stones.

1. HISTORY

Cranberry (*Vaccinium macrocarpon*) is a small evergreen shrub that grows in mountains, forests and damp bogs from Alaska to Tennessee. Native Americans introduced the Europeans to cranberry as a food, dye, and medicine *(1)*. In the 1920s, canned cranberry sauce was introduced, and in the 1940s, cranberry juice became commercially available. Cranberry has been used to prevent and treat urinary tract infections since the 19th century *(2)*.

2. CURRENT PROMOTED USES

Cranberry juice has been widely used for the prevention, treatment, and symptomatic relief of urinary tract infections *(3)*. Also, cranberry juice has been given to patients to help reduce urinary odors in incontinence *(4–6)*. Another potential benefit of the use of cranberry is a decrease in the rates of kidney stone formation *(7–9)*.

From Forensic Science and Medicine:
Herbal Products: Toxicology and Clinical Pharmacology, Second Edition
Edited by: T. S. Tracy and R. L. Kingston © Humana Press Inc., Totowa, NJ

3. PRODUCTS AVAILABLE

Cranberry is available in a variety of forms such as fresh or frozen cranberries, cranberry juice cocktail, other cranberry drinks, cranberry sauce, and powder in hard or soft gelatin capsules *(2,10)*. Cranberries are approx 88% water and contain flavonoids, anthrocyanins (odain), cetechin, triterpinoids, γ-hydroxybutyric acid, citric acid, malic acid, glucuronic acid, quinic acid, benzoic acid, ellagic acid, and vitamin C *(2)*. Fresh or frozen cranberries are a good source of cranberry because they contain pure fruit; however, because of their high acidity and extremely sour taste, they are less readily used in clinical practice *(1)*. Pure cranberry juice is tart like lemon juice because of the high citric and quinic acid content *(2)*. Cranberry juice cocktail is more palatable, but is only 25–33% juice and contains corn syrup as a sweetener *(2,10)*, whereas other cranberry juice drinks contain as little as 10% juice *(2)*. These sweetened beverages are relatively high in calories (approx 140 kcal per 8-oz serving) *(2)* and could cause weight gain in a patient consuming the juice for medicinal purposes *(10)*. Another drawback to sweetened beverages is that, theoretically, the sugar could act as a source of food for uropathogens *(10)*. Cranberry sauce consisting of sweetened or gelled berries at a concentration one-half that of cranberry juice cocktail is also readily available to consumers *(2)*. Cranberry capsules are a sugar-free source of cranberry. Hard gelatin capsules contain more crude fiber and organic acids than cranberry juice cocktail, whereas the soft gelatin capsules contain soybean oil and have only 8% of the total organic acids found in fresh cranberries *(10)*. It takes 12 capsules of cranberry powder to equal 6 fluid oz of cranberry juice cocktail *(10)*.

In the various studies and consumer references, many dosages and dosing regimens have been reported for the use of cranberry in prevention of renal calculi, prevention of urinary odor, and prevention and treatment of urinary tract infections.

3.1. Dosages Used or Recommended in Clinical Studies and Case Reports

Prevention of urinary tract infection: 8 oz of cranberry juice four times a day for several days, then twice daily *(7)*; 300 mL/day as cranberry juice cocktail *(11)*.

Treatment of urinary tract infection: 6 oz cranberry juice of daily for 21 days *(12)*; cranberry juice 6 oz twice daily *(13)*.

Reduction of urinary odors: 16 oz of cranberry juice daily *(4)*; 3 oz of cranberry juice daily, then increased by 1 oz each week to a maximum of 6 oz daily *(5)*.

Prevention of urinary stones: 1 qt of cranberry juice cocktail daily *(8)*; 8 oz of cranberry juice four times a day for several days, then 8 oz twice daily *(7)*.

3.2. Dosages in Lay References

Prevention of urinary tract infection: 3 oz daily as a cocktail *(1)*.
Treatment of urinary tract infection: 12–32 oz daily as a cocktail *(1)*.

3.3. Various Brand Name Products of Cranberry Available

Nature's Resource® contains 405 mg of standardized cranberry juice concentrate per capsule. The recommended dose is two to four capsules three times a day with water, at meals. The label also recommends drinking a full glass of water when taking the capsules and drinking 6–8 oz of liquids per day.

Spring Valley® contains 475 mg of cranberry fruit per capsule. The recommended dose is two to four capsules three times a day, preferably with meals.

Cranberry Fruit Sundown® Herbals contain 425 mg of cranberry fruit per capsule. The recommended dose is two to four capsules up to three times a day as needed.

Celestial Seasonings® Cranberry contains 400 mg of cranberry extract standardized to more than 35% organic acids. The recommended dose is one capsule every day as needed with a full glass of water.

Ocean Spray® Cranberry Juice Cocktail is 27% cranberry juice. It contains filtered water, high fructose corn syrup, cranberry juice concentrate, and ascorbic acid.

4. PHARMACOLOGICAL/TOXICOLOGICAL EFFECTS

4.1. Antimicrobial Activity

Controversy exists on the pharmacological mechanism of cranberry. In the mid-19th century, German researchers discovered hippuric acid in the urine of people who ate cranberries *(2)*. From the 1920s through the 1970s, many researchers thought that hippuric acid produced a bacteriostatic effect by acidifying the urine *(11,14,15)*. The ability of cranberry to prevent renal calculi has also been attributed to its ability to decrease urine pH and inhibit bacterial growth *(7,8,16)*. Not all studies documented a change in urinary pH with cranberry administration, so a parallel line of thinking suggested that hippuric acid, which was structurally similar to mandelic acid, inhibited bacterial multiplication *(15)*. It was found that the concentration of hippuric acid in the urine rarely reached a concentration necessary for bacteriostatic effects *(15)*. Because hippuric acid is a weak acid, it exists in equilibrium with its conjugate base, and requires a urine pH of at least 5.0 to produce the minimum bacteriostatic hippuric acid concentration. Thus, these researchers felt that both urine pH and hippuric acid concentration were important for the bacteriostatic effect of cranberry. More recently, however, studies have shown that the mechanism of action of cranberry is the inhibition of bacterial adherence to mucosal surfaces *(3,11,17–19)*. One study proposed that there are two substances in cranberry juice cocktail, fructose and a glycoprotein, responsible for inhibiting adherence of *Escherichia coli* to mucosal cells *(18)*.

E. coli is responsible for 85% of urinary tract infections *(20)*. Virtually all *E. coli* express type 1 fimbrae, and most uropathogenic *E. coli* express P fimbriae, which are responsible for mediating the adherence of the bacteria to uroepithelial cells *(18)*. Fructose is responsible for inhibiting the adherence of type-1-fimbriated *E. coli*, whereas a polymeric compound inhibits P-fimbriated *E. coli (18)*. Recently, a study *(21)* identified this polymeric compound as condensed tannins (proanthocyanidins) based on the ability of proanthocyanidins purified from cranberries to inhibit the ability of P-fimbriated *E. coli* to attach to isolated uroepithelial cells at concentrations of 10–50 μg/mL. Blueberries, another member of the *Vaccinium* genus, may be a more palatable source of proanthicyanidins.

Epidemiological data *(22)* and data from a double-blind, placebo-controlled trial *(11)* support the use of cranberry juice to prevent urinary tract infections, although in the latter study differences in baseline characteristics between study groups may have influenced the results. Cranberry extract in capsule form was more effective than placebo in preventing recurrent urinary tract infections in a small study *(23)*.

Another potential benefit to the use of cranberry is its antiviral effect. One study *(24)* evaluated the ability of various commercial juices and beverages to inactivate poliovirus type I (Sabin) in vitro. Cranberry juice had some antiviral activity that was noted to be enhanced at pH 7.0 *(24)*. The antiviral effect of commercial juices is thought to be caused by polyphenols, including tannins, which form complexes with viruses *(24)*.

4.2. Gastrointestinal Effects

The ingestion of large amounts of cranberry (>3–4 L/day) may result in diarrhea and other gastrointestinal symptoms *(6)*.

4.3. Renal Effects

Ammoniacal fermentation, or alkalinization and decomposition of urine, is responsible for the foul odor of urine *(4)*. The results of one study *(4)* found that a single dose of 16 oz of cranberry juice lowered the urine pH of six men with chronic urinary tract disorders, and decreased ammoniacal odor and turbidity. The urine pH of five of six men free of urinary tract infections was also lowered with this dose. In another study *(5)*, hospital personnel noted a decrease in urine odor in the geriatric wards of a nursing home, but a change in urine pH or change in ammonia levels in the air could not be detected. Other subjective comments by nursing home personnel included a decrease in complaints among patients who had experienced burning upon urination, and more frequent voiding.

Another potential effect of the use of cranberry is in the management of calculus formation because of the association between alkalinization of the urine and stone formation *(7–9,16)*. A specially prepared sweetened cranberry juice consisting of 80% juice was administered to 41 people who were randomly assigned to ingest 150, 180, 210, or 240 mL of the juice with each meal for 1 week *(25)*. Each subject served as his or her own control. Urine pH was measured by the subjects at each voiding, and a urine sample was collected daily after the evening meal. Mean urine pH was decreased to a statistically significant extent with cranberry juice ingestion compared to baseline. The decrease was not dose-related. Cranberry juice had some effect in lowering daily fluctuations in urine pH, but this effect again was not dose-related. The effect of cranberry juice on urine pH persisted throughout the experimental period (i.e., the kidney did not compensate for changes in pH). Side effects included weight gain and increased frequency of bowel movements.

In another study of cranberry's effect on urinary pH *(20)*, two 6-oz servings of cranberry juice daily for 20 days were able to lower urinary pH more than orange juice in eight patients with multiple sclerosis, but were unable to lower pH consistently to below 5.5.

5. REGULATORY STATUS

In the United States, cranberry is considered a dietary supplement and food *(26)*.

REFERENCES

1. Tyler VE. *The Honest Herbal, 3rd edition*. Binghamton: Pharmaceutical Products Press, 1993.

2. Siciliano A. Cranberry. Herbal Gram 1996;38:51–54.
3. Sobota AE. Inhibition of bacterial adherence by cranberry juice: potential use for the treatment of urinary tract infections. J Urol 1984;131:1013–1016.
4. Kraemer RJ. Cranberry juice and the reduction of ammoniacal odor of urine. Southwest Med 1964;45:211–212.
5. DuGan CR, Cardaciotto PS. Reduction of ammoniacal urinary odors by the sustained feeding of cranberry juice. J Psychiatr Nurs 1966;8:467–470.
6. Anonymous. Cranberry. In: *Lawrence Review of Natural Products*. St. Louis: Facts and Comparisons, 1994.
7. Sternlieb P. Cranberry juice in renal disease. N Engl J Med 1963;268:57.
8. Zinsser HH, Seneca H, Light I, et al. Management of infected stones with acidifying agents. NY State J Med 1968;68:3001–3009.
9. Light I, Gursel E, Zinnser HH. Urinary ionized calcium in urolithiasis. Effect of cranberry juice. Urology 1973;1:67–70.
10. Hughes BG, Lawson LD. Nutritional content of cranberry products [letter]. Am J Hosp Pharm 1989;46:1129.
11. Avorn J, Monane M, Gurwitz JH, Glynn RJ, Choodnovskiy I, Lipstiz L. Reduction of bacteriuria and pyruia after ingestion of cranberry juice. JAMA 1994;271:751–754.
12. Papas PN, Brusch CA, Ceresia GC. Cranberry juice in the treatment of urinary tract infections. Southwest Med J 1966;47:17–20.
13. Moen DV. Observations on the effectiveness of cranberry juice in urinary infections. Wisc Med J 1962;61:282–283.
14. Blatherwick NR, Long ML. Studies of urinary acidity. II. The increased acidity produced by eating prunes and cranberries. J Biol Chem 1923;57:815–818.
15. Bodel PT, Cotran R, Kass EH. Cranberry juice and the antibacterial action of hippuric acid. J Lab Clin Med 1959;54:881–888.
16. Walsh B. Urostomy and urinary pH. J ET Nurs 1992;9:110–113.
17. Schmidt DR, Sobota AE. An examination of the anti-adherence activity of cranberry juice on urinary and nonurinary bacterial isolates. Microbios 1988;55:173–181.
18. Zafriri D, Ofek I, Adar R, Pocino M, Sharon N. Inhibitory activity of cranberry juice on adherence of type I and type P fimbriated Escherichia coli to eucaryotic cells. Antimicrob Agents Chemother 1989;33:92–98.
19. Ofek I, Goldhar J, Zafriri D, Lis H, Adar R, Sharon N. Anti-Escherichia coli adhesion activity of cranberry and blueberry juices [letter]. N Engl J Med 1991;324:1599.
20. Schultz A. Efficacy of cranberry juice and ascorbic acid in acidifying the urine in multiple sclerosis subjects. J Comm Health Nurs 1984;1:159–169.
21. Howell AB, Vorsa N. Inhibition of the adherence of P-fimbriated Escherichia coli to uroepithelial-cell surfaces by proanthocyanidin extracts from cranberries. N Engl J Med 1998;339:1085–1086.
22. Foxman B, Geiger AM, Palin K, Gillespie B, Koopman JS. First-time urinary tract infection and sexual behavior. Epidemiology 1995;6:162–168.
23. Walker EB, Barney DP, Mickelsen JN, Walton RJ, Mickelsen RA Jr. Cranberry concentrate: UTI prophylaxis [letter]. J Fam Pract 1997;45:167–168.

24. Konowalchuk J, Speirs JI. Antiviral effect of commercial juices and beverages. Appl Environ Microbiol 1978;35:1219–1220.
25. Kinney AB, Blount M. Effect of cranberry juice on urinary pH. Nurs Res 1979;28:287–290.
26. Blumenthal M, ed. *Popular Herbs in the U.S. Market*. Austin: American Botanical Council, 1997.

Chapter 13

Hawthorn

Timothy S. Tracy

SUMMARY

Hawthorn appears to be effective for the treatment of stage II congestive heart failure. The mechanism(s) by which hawthorn exerts this positive effect is still unclear as results regarding changes in particular cardiovascular parameters are mixed. Human clinical studies of the use of hawthorn in other cardiovascular conditions are lacking. Adverse effects of hawthorn therapy appear to be mild and no significant drug interactions have been reported (though, in theory it might potentiate the effect of vasodilators).

Key Words: *Crataegus oxyacantha*; heart failure; hypertension; vasodilation; digoxin.

1. HISTORY

Hawthorn is a spiny, small tree or bush with white flowers and red berries (haws), each containing one to three nuts, depending on the species *(1)*. Hybridization is common among individual species, making them difficult to identify *(2)*. Hawthorn is a member of the rose family and is found in Europe, North Africa, and western Asia *(3)*. It can reach heights of 25–30 ft and is used as a hedge *(1,4)*. The flowers grow in clusters and bloom from April to June, and the deciduous leaves are divided into three, four, or five lobes *(1)*. The use of hawthorn can be dated back to Dioscorides in the first century CE *(5)*.

Uses for the herb have included high and low blood pressure, tachycardia, arrhythmias, atherosclerosis, and angina pectoris *(1)*. Hawthorn is also purported to have spasmolytic and sedative effects *(1)*. Native Americans used it as a diuretic for kidney and bladder disorders and to treat stomach aches, stimulate appetite, and improve circulation *(4)*. The flowers and berries have

From Forensic Science and Medicine:
Herbal Products: Toxicology and Clinical Pharmacology, Second Edition
Edited by: T. S. Tracy and R. L. Kingston © Humana Press Inc., Totowa, NJ

astringent properties and have been used to treat sore throats in the form of haw jelly or haw marmalade *(5)*.

2. CURRENT PROMOTED USES

Hawthorn is promoted for use in heart failure, hypertension, arterioscle- rosis, angina pectoris, Buerger's disease, paroxysmal tachycardia *(6)*, heart valve murmurs, sore throat, skin sores, diarrhea, and abdominal distention *(7)*.

3. SOURCES AND CHEMICAL COMPOSITION

Crataegus oxyacantha (L.), *Crataegus laevigata*, *Crataegus monogyna Jacquin*, English hawthorn, haw, maybush, whitethorn *(1)*, may, mayblossom, hazels, gazels, halves, hagthorn, ladies' meat, bread and cheese tree *(3)*.

4. PRODUCTS AVAILABLE

Available products include tea, 1:5 tincture in 45% alcohol, 1:1 liquid extract in 25% alcohol *(6)*, and capsules of 250, 455, and 510 mg. The French Pharmacopoeia requires 45% ethanol for the fluid extract and 60% ethanol for the tincture *(8)*. It is recommended that 0.5–1 mL of liquid extract or 1–2 mL of tincture be taken three times a day *(6)*. The tea is made from 0.3–1 g of dried berries infused in hot water and taken three times a day *(4,6)*. A typical therapeutic dose of extract, standardized to contain 1.8% vitexin-4 rhamno- side, is 100–250 mg three times daily. A standardized extract containing 18% procyanidolic oligomers (oligomeric procyanidns) is dosed at 250–500 mg daily *(9)*.

5. PHARMACOLOGICAL/TOXICOLOGICAL EFFECTS

5.1. Cardiovascular Effects

Hawthorn extracts purportedly dilate coronary blood vessels, decrease blood pressure, increase myocardial contractility, and lower serum choles- terol *(9)*. Benefits have been demonstrated in patients with heart failure *(10)*. In patients with stage II New York Heart Association (NYHA) heart failure, doses of 160–900 mg/day of the aqueous-alcoholic extract for up to 56 days showed an increase in exercise tolerance, decrease in rate/pressure product, and increased ejection fraction *(11)*. Degenring and colleagues, in a random- ized, double-blind, placebo-controlled trial, studied a standardized extract of fresh *Crataegus* berries (Crataegisan®) for the treatment of patients with

NYHA II heart failure *(12)*. Using an intent-to-treat analysis, these investigators found that the hawthorn preparation significantly increased exercise tolerance as compared to placebo, but subjective symptoms of heart failure were unchanged. In a meta-analysis of 13 randomized trials of hawthorn extract in the treatment of heart failure, investigators noted that hawthorn produced a statistically significant increase in exercise tolerance over placebo *(13)*. In addition, symptoms such as dyspnea and fatigue improved significantly with hawthorn treatment. Adverse events were infrequent. These investigators concluded that hawthorn provides a significant benefit in the treatment of heart failure. The active principles are thought to be flavonoids, including hyperoside, vitexin, vitexin-rhamnose, rutin, and oligomeric procyanidins (dehydrocatechins, catechins, and/or epicatechins) *(4–6,11)*.

Two clinical trials have also been conducted to evaluate the ability of hawthorn to reduce blood pressure and treat hypertension. Asgary et al. studied the effect of Iranian *Crataegus curvisepala* hydroalcoholic extract in 92 men and women with primary mild hypertension *(14)*. These investigators found that treatment with the hawthorn extract for 4 months reduced both systolic (~13 mmHg decrease) and diastolic (~8 mmHg decrease) blood pressure as compared with placebo. The effect was progressive over the 4-month treatment period. In a similar study, a hawthorn extract was investigated for its ability to treat mild, essential hypertension *(15)*. Studying 36 subjects, these investigators found no difference in blood-pressure-lowering between hawthorn and placebo treatments (though both treatments did reduce blood pressure somewhat). Thus, results of hawthorn use in the treatment of hypertension are mixed.

Investigators attempted to elucidate the mechanism of action of the flavonoids hyperoside, luteolin-7-glucoside, rutin, vitexin, vitexin-rhamnoside, and monoacetyl-vitexin-rhamnoside in spontaneous-beating Langenhoff preparations of guinea pig hearts *(16)*. Dose-dependent effects on contractility, heart rate, and coronary blood flow similar to that of theophylline were exhibited by luteolin-7-glucoside, hyperoside, and rutin, whereas vitexin and its derivatives were less potent. These results were different from those of previous investigators, who found a decrease in coronary blood flow, contractility, and heart rate with hyperoside, whereas vitexin decreased contractility and increased heart rate and coronary blood flow. Vitexin-rhamnoside increased coronary blood flow, heart rate, and contractility in the previous study. These differences were attributed to differences in the experimental device. The investigators concluded that the mechanism behind the cardiac effects of these flavonoids involved phosphodiesterase inhibition, causing an increase in cyclic adenosine monophosphate concentration, as well as inhibition of thromboxane synthesis and enhancement of prostacyclin synthesis, as described by previous researchers. The authors also concluded that despite previous studies showing that vitexin-rhamnoside protected cultured heart cells from oxygen and glucose deprivation, the role of antioxidant activity as a mechanism behind the anti-ischemic effect of these flavonoids requires further study, given that vitexin-rhamnoside exhibited only minor effects in their study.

Because reactive oxygen species may play a role in the pathogenesis of atherosclerosis, angina, and cerebral ischemia, the antioxidant activity of dried hawthorn flowers and flowering tops, fluid extract, tincture, freeze-dried powder, and fresh plant extracts was investigated (8). Antioxidant activity, determined by the ability of the preparations to scavenge hydrogen peroxide, superoxide anion, and hypochlorous acid, was provided by all preparations, but was highest with the fresh young leaf, fresh floral buds, and dried flowers. The antioxidant activity was correlated to total phenolic proanthocyanidin and flavonoid content.

The effects of hawthorn extract LI 132 standardized to 2.2% flavonoids (Faros® 300, Lichtwer Pharma GmbH, Berlin, Germany) on contractility, oxygen consumption, and effective refractory period of isolated rat cardiac myocytes were studied (17). In addition, the effect of partially purified oligomeric procyanidins on contractility was also studied. The concentrations used in their study were chosen for their physiological plausibility, based on the assumption that the volume of distribution of both hawthorn extract and procyanidins in humans is 5 L, and that the daily dose is 900 mg and 5 mg, respectively. At concentrations of 30–180 μg/mL, the hawthorn extract increased myocardial contractility with a more favorable effect on oxygen consumption than β-1 agonists or cardiac glycosides. Hawthorn also prolonged the effective refractory period, indicating that it might be an effective antiarrhythmic agent. Oligomeric procyanidins at concentrations of 0.1–30 μg/mL had no detectable effect on contractility, suggesting that they are not responsible for the positive inotropic effect of hawthorn.

Tincture of Crataegus (TCR), made from hawthorn berries, was shown to have a hypocholesterolemic effect on rats fed 0.5 mL/100 g body weight for 6 weeks. These findings prompted a study that examined the ability of TCR to increase low-density lipoprotein (LDL) binding to liver plasma membranes in rats fed an atherogenic diet (18). The hypocholesterolemic effect of TCR appears to be caused by a 25% increase in LDL receptor activity, resulting in greater LDL uptake by the liver. This was caused by an increased number of receptors, not an increase in receptor binding affinity. In addition, TCR suppressed *de novo* cholesterol synthesis in the liver, and enhanced the use of liver cholesterol to make bile acids. Despite LDL receptor upregulation, the atherogenic diet fed to the rats offset the beneficial effects; LDL levels increased 104% and liver cholesterol increased by 231%. The investigators did not attempt to determine which TCR constituent was responsible for the hypocholesterolemic effect, but hypothesized that all contribute in some manner.

5.2. Neurological Effects

The flavonoids present in hawthorn purportedly have a sedative effect *(2,5)*.

5.3. Lethal Dose for 50% of Test Population

The Lethal Dose (LD_{50}) of an alcoholic extract of hawthorn leaves and fruit called Crataegutt® administered orally was 33.8 mL/kg in rats and 18.5 mL/kg in mice. This particular extract was manufactured by Schwabe and contained 2% or 10% oligomeric procyanidins. Death occurred after approx 30 minutes and was caused by sedation and apnea *(19)*.

5.4. Teratogenicity/Mutagenicity/Carcinogenicity

The German Commission E reports that hawthorn effects are unknown during pregnancy and lactation. No experimental data have been reported concerning toxicity in the embryo or fetus, or the effects on fertility or postnatal development. Commission E also reports the lack of experimental data concerning carcinogenicity. Despite experiential data that hawthorn may be mutagenic, Commission E feels that the amount of mutagenic substances ingested would not be sufficient to pose a risk to humans. Available information presents no indication of carcinogenic risk *(11)*.

6. PHARMACOKINETICS/TOXICOKINETICS

Although the investigators of one study *(17)* assumed a volume of distribution of 5 L (approximately plasma volume) for purposes of calculating a concentration to use in their in vitro study, there are no pharmacokinetic data to confirm this.

7. ADVERSE EFFECTS AND TOXICITY

7.1. Case Reports of Toxicity Caused By Commercially Available Products

Although several references mention that hawthorn in high doses may cause hypotension, arrhythmias, and sedation in humans *(1,2,4,5)*, no substantiative case reports can be located.

8. DRUG INTERACTIONS

Drug interactions with hawthorn are theoretically possible with cardioactive medications, but have not been documented *(2)*. In addition, the flavonoid constituents have been shown to have inhibitory and inducible effects on the cytochrome P-450 enzyme system, making other drug interactions possible *(20)*. However, an in vivo study of a potential pharmacokinetic interaction of digoxin and hawthorn demonstrated that concurrent administration had no effect on digoxin pharmacokinetics, suggesting that the two could be safely administered together from a pharmacokinetic point of view *(21)*. However, one must be mindful of additive effects and a potential pharmacodynamic interaction.

9. REGULATORY STATUS

Originally, all preparations of hawthorn were approved under one German Commission E monograph based on historical experience. However, in 1993, the preparations were reevaluated and it was concluded that sufficient scientific evidence was lacking to justify use of the flowers, leaves, and berries as individual compounds. As a result, there are currently four hawthorn monographs: three Unapproved monographs for the berry, flower, and leaf individually and an Approved monograph for the flower with leaves. In addition, the Approved monograph has only one approved indication: treatment of "decreasing cardiac output according to functional stage II of the NYHA *(11)*." In Canada, hawthorn carries new drug status and is not approved, as self-treatment of cardiovascular conditions is deemed inappropriate. Hawthorn is not on the General Sales List in the United Kingdom. In France, the flower and flowering top are permitted for oral use, and in Switzerland, the leaf and flower are permitted as herbal teas. In Sweden, hawthorn is classified as a natural product, whereas in the United States, it is considered a dietary supplement *(6)*.

REFERENCES

1. Anonymous. *Lawrence Review of Natural Products*. St. Louis: Facts and Comparisons, 1994.
2. Hamon NW. Herbal medicine: Hawthorns (*Genus crataegus*). Can Pharmaceut J 1988;121:708–709, 724.
3. Grieve M, ed. *A modern herbal*. New York: Dover, 1971.
4. Bigus A, Massengil D, Walker C. Hawthorn. www.unc.edu/~cebradsh/main.html. Date accessed: Oct 15, 1998.

5. Tyler VE. *The Honest Herbal, 3rd edition*. Binghamton: Pharmaceutical Products Press, 1993.
6. Blumenthal M, ed. *Popular Herbs in the U.S. Market*. Austin: American Botanical Council, 1997.
7. Williamson JS, Wyandt CM. Herbal therapies: the facts and the fiction. Drug Topics 1997;141:78–85.
8. Bahorun T, Gressier B, Trotin F, et al. Oxygen species scavenging activity of phenolic extracts from hawthorn fresh plant organs and pharmaceutical preparations. Arzneim Forsch 1996;46:1086–1089.
9. Anonymous. Hawthorn (*Crataegus monogyna*). Nat Med 1999;2:5.
10. Iwamoto M, Sato T, Ishizaki T. The clinical effect of *Crataegus* in heart disease of ischemic or hypertensive origin. A multicenter double-blind study. Planta Med 1981;42:1–16.
11. Blumenthal M, ed. *The Complete German Commission E Monographs*. Austin: American Botanical Council, 1998.
12. Degenring FH, Suter A, Weber M, Saller R. A randomized double blind placebo controlled trial of a standardized extract of fresh *Crataegus* berries (Crataegisan®) in the treatment of patients with congestive heart failure NYHA II. Phytomedicine 2003;10:363–369.
13. Pittler MH, Schmidt K, Ernst E. Hawthron extract for treating chronic heart failure: meta-analysis of randomized trials. Am J Med 2003;114:665–674.
14. Asgary S, Naderi GH, Sadeghi M, Kelishadi R, Amiri M. Antihypertensive effect of Iranian *Crataegus curvisepala* lind.: a randomized, double-blind study. Drugs Exp Clin Res 2004;5–6:221–225.
15. Walker AF, Marakis G, Morris AP, Robinson PA. Promising hypotensive effect of hawthorn extract: a randomized double-blind pilot study of mild, essential hypertension. Phytother Res 2002;15:48–54.
16. Schussler M, Holzl J, Fricke U. Myocardial effects of flavonoids from *Crataegus* species. Arzneim Forsch 1995;45:842–845.
17. Popping S, Rose H, Ionescu I, Fisher Y, Hammermeier H. Effect of a hawthorn extract on contraction and energy turnover of isolated rat caridiomyocytes. Arzneim Forsch 1995;45:1157–1160.
18. Rajendran S, Deepalakshmi PD, Parasakthy K, Devaraj H, Niranjali S. Effect of tincture of *Crataegus* on the LDL-receptor activity of hepatic plasma membrane of rats fed an atherogenic diet. Atherosclerosis 1996;123:235–241.
19. Ammon HO, Handel M. *Crataegus* toxicology and pharmacology. Part I: toxicity. Planta Med 1981;43:105–120.
20. Canivenc-Lavier M, Vernavaut M, Totis M, Siess M, Magdolou J, Suschetet M. Comparative effects of flavonoids and model inducers on drug-metabolizing enzymes in rat liver. Toxicology 1996;114:19–27.
21. Tankanow R, Tamer HR, Streetman DS, et al. Interaction study between digoxin and a preparation of hawthorn (*Crataegus oxyacantha*). J Clin Pharmacol 2003;43:637–642.

Chapter 14

Evening Primrose

Margaret B. Artz

SUMMARY

Evening primrose oil (OEP) is a dietary supplement that contains essential fatty acids (omega-3 and omega-6) and has been investigated in-depth for its effectiveness for conditions that are associated with a deficiency in essential fatty acids. OEP has a good safety profile with mild side effects and rare serious adverse events. OEP should not be taken during pregnancy, prior to surgery, in patients at risk for seizures or taking phenothiazine-related medications, antiplatelets, thrombolytics, low-molecular-weight heparins, or anticoagulants. There have been no reports of toxic ingestion, mortality, or teratogenicity with OEP supplementation, and usage during lactation is presumed to be safe. The German Commission E has not approved the use of OEP for any condition at this time. OEP is possibly effective for essential fatty acid deficiency and breast pain, and for rheumatoid arthritis after 6 months of treatment. Efficacy of OEP has not been clearly established for the following: atopic eczema, premenstrual syndrome, hot flashes, night sweats, preeclampsia, shortening duration of labor, attention-deficit/hyperactivity disorder, or chronic fatigue syndrome.

Key Words: *Oenothera* species; oil of evening primrose; essential fatty acids; linoleic acid; anti-inflammatory; eicosanoids.

1. HISTORY

Evening primrose is a botanical plant that has the following National Oceanographic Data Center Taxonomic Code (Kingdom: *Plantae*; Phylum: *Tracheobionta*; Class: *Magnoliopsida*; Order: *Myrtales*; Family: *Onagraceae*; Genus: *Oenothera* L.; Species: *Oenothera biennis* L.). A fragrant wildflower and biennial herb native to North America, the evening primrose reaches a height of 4 to 5 ft with flowers 2 to 3 cm long, and blossoms in the evening

From Forensic Science and Medicine:
Herbal Products: Toxicology and Clinical Pharmacology, Second Edition
Edited by: T. S. Tracy and R. L. Kingston © Humana Press Inc., Totowa, NJ

during June through September. A flower blooms only for one evening, thus the name "evening primrose" *(1)*. The evening primrose can be found in North America east of the Rocky Mountains, and was naturalized into Europe and Asia from North America in the early 17th century *(2,3)*. Its leaves are alternate, rough, hairy, lanceolate, 3 to 6 in. long, and lemon-scented. The fruit is a 1-in., oblong capsule that is approx 4 cm long, containing many tiny reddish seeds; seeds are 1.5 mm long, dark gray to black in color, and have irregular sharp edges *(2)*. The entire plant can be eaten (e.g., roots, leaves, flowers, buds, seedpods); leaves are cooked and eaten like spinach and the roots are boiled and taste sweet. Evening primrose was a staple food for many Native American tribes and a famine food for Chinese farmers *(3)*. European settlers and Native Americans used the whole plant to ameliorate ailments such as bruising, stomachaches, and shortness of breath *(4)*.

Evening primrose oil (OEP) is derived from the plant's small, dark seeds *(5)*. China is now the major grower of evening primrose seed in the world, supplying an estimated 90% of the world's crop *(3)*. A total of approx 400 t of seeds are processed each year in the United States and Canada. One major supplier of OEP derives the oil from specially selected and hybridized forms of *Oenothera* species *(6)*.

Today, the oil is used medicinally to treat a myriad of conditions related to essential fatty acid (EFA) deficiencies, low dietary intake of linoleic acid, and a variety of reproductive, cardiovascular, inflammatory, and neurological disorders. It is added to foods as a source of essential fatty acids and used in topical products such as soaps and cosmetics *(5–7)*.

2. CURRENT PROMOTED USES

Currently promoted uses of OEP include: EFA deficiency mastalgia, fibrocystic breast disease, endometriosis, menopause, premenstrual syndrome (PMS), and the prevention of preeclampsia, diabetic neuropathy, psoriasis, eczema/dermatitis, rheumatoid arthritis, cardiovascular disease, gastrointestinal disorders, attention deficit disorder in children, and hypercholesterolemia *(5,7)*. OEP is used topically as an ingredient in some soaps, cosmetics and medicinals.

3. SOURCES AND CHEMICAL COMPOSITION

The seeds of *O. biennis* contain approx 14–26% OEP, which is a fixed oil. Within this oil, several important fatty acids are present:

- 50–85% of *cis*-linoleic acid

- 2–16% of *cis*-γ-linolenic acid
- 6–11% of *cis* 6,9,12-octadecatrienoic acid; oleic acid
- 7–10% of palmitic acid
- Miscellaneous components: stearic acids, steroids, campesterol, vitamin E, and β-sitosterol.

A descriptive report of OEP produced in China states the following information about OEP: refractive index (20°C) of 1.48, specific gravity (20°C) of 0.93, iodine value of 140, saponification value of 188, thiocyanogen value of 84, and unsaponifiable matter of 1% *(3)*.

In other parts of the plant, mucilage and tannin are present. OEP contains the highest amount of γ-linolenic acid (an EFA) of any food substance.

4. PRODUCTS AVAILABLE

A total of approx 400 t of seeds are processed each year in the United States and Canada. One major supplier of OEP derives the oil from specially selected and hybridized forms of *Oenothera* species *(6)*. OEP in oral tablets or capsules usually range from 500 to 1300 mg. Most commercial products are standardized for a γ-linoleic acid content of 9% *(2)*. Dosage forms include oral formulations (capsules, tablets, oil swallowed directly or mixed with another liquid/food) and topical *(7)*.

Examples of products containing OEP include Efamol® Pure Evening Primrose Oil, Efamol PMS Control, Efamol Fortify, Efalex® capsules, Efalex® liquid, and Efanatal®. Efamol Pure Evening Primrose Oil is a natural colored, oval, soft gelatin capsule and a 500-mg capsule contains Efamol Pure Evening Primrose Oil 500 mg (linoleic acid 165 mg and γ-linolenic acid 40 mg). Efamol PMS Control is an opaque, pink, oval, soft gelatin capsule and a 695-mg capsule contains Efamol Pure Evening Primrose Oil 250 mg (linoleic acid 320 mg and γ-linolenic acid 20 mg), vitamin C (as ascorbic acid) 30 mg, magnesium (as heavy magnesium oxide) 20 mg, vitamin B6 (as pyridoxine HCl) 20 mg, niacin (as niacinamide) 6 mg, zinc (as zinc sulfate monohydrate) 2 mg, vitamin E (as *d*-α-tocopheryl acetate) 15 IU, and *d*-biotin 40 μg. Efamol Fortify is a white, oblong, soft gelatin capsule and a 750-mg capsule contains calcium (as calcium carbonate) 100 mg, Efamol Pure Evening Primrose Oil 400 mg (255 mg linoleic acid and 32 mg γ-linolenic acid), marine fish oil 44 mg with an eicosapentaenoic acid content of 7 mg, and vitamin E (as *d*-α-tocopheryl acetate) 15 IU. Efalex is a clear, oblong, soft gelatin capsule and a 450-mg capsule contains a docosahexaenoic acid-rich fish oil 294 mg with a docosahexaenoic acid content of 60 mg, Efamol Pure Evening Primrose Oil 140 mg (γ-linolenic acid 12 mg and arachidonic acid 5.25 mg), vitamin E (as *d*-α-tocopheryl acetate) 15 IU, and thyme oil 1 mg. Efalex Liquid is a pale yellow-green, lemon-lime flavored, free-flowing oil. One teaspoon (5 mL) contains sunflower oil 3584 mg, docosahexaenoic acid-rich fish oil 520 mg with a docosahexaenoic acid content of 120 mg, Efamol Pure Evening Primrose Oil 300 mg (γ-linolenic acid 24 mg and arachidonic acid content of 10.5 mg), vitamin E (as *d*-α-tocopheryl acetate) 3.7 IU, and thyme oil 2 mg. Efanatal is a pink, oval, soft gelatin capsule and a 517-mg capsule contains Efamol Pure Evening Primrose Oil 140 mg (linoleic acid 162 mg, γ-linolenic acid 20 mg, and arachidonic acid 4.3 mg), fish oil 250 mg, docosahexaenoic acid 62.5 mg, and vitamin E (as *d*-α-tocopheryl acetate) 7.5 IU. Other products containing OEP are available from manufacturers such as Jamieson, Holista, and Nutrilite. Some OEP products on the market may contain other oils in their formulations including borage oil and black current oil.

4.1. Dosage

Use of more than 4 g OEP daily (300–600 mg γ-linolenic acid) is not recommended *(6)*. However, adult doses of 3–8 g/day have been used for various conditions *(2)*. For a OEP product with a standardized γ-linolenic acid content of 8%, the following dosages are recommended for the following conditions *(7)*: atopic eczema, 4 to 8 g daily for adults, 2 to 4 g daily for children; cyclical and noncyclical mastalgia, 3 to 4 g daily; PMS, 3 g daily.

5. PHARMACOLOGICAL/TOXICOLOGICAL EFFECTS

5.1. Dermatological Effects

Clinical evidence of nutritional supplementation with OEP to correct dermal conditions is mixed. One theory for the mixed results is that in some persons, once sensitized, immunological factors may override what help OEP can offer. Very high doses of OEP or linoleic acid, or modest doses of γ-linolenic acid, with corresponding correction of plasma EFA levels, produce some clinical improvement *(8)*.

A defect in the capability of the enzyme δ-6-desaturase to convert linoleic acid to γ-linolenic acid is known to occur in patients with atopic dermatitis *(9)*. Patients with atopic eczema have a dietary deficiency in metabolites of linoleic: γ-linolenic acid, dihomo-γ-linolenic acid, arachidonic acid, adrenic acid, and docosapentaenoic acid caused by a reduced rate of activity in the δ-6-desturase enzyme *(8)*. Galli et al. compared blood samples from babies born to parents who suffered from atopic eczema. Results showed that dihomo-γ-linolenic acid and arachidonic acid were consistently and significantly lower in children who later had atopic eczema *(10)*.

Some studies have shown that OEP administration can improve the percentage of body surface involvement, itch, dryness, scaling, and inflammation associated with atopic eczema. A meta-analysis *(11)* of nine controlled trials involving OEP in the treatment of atopic eczema showed a highly significant improvement in the symptom of itch over placebo ($p < 0.0001$). In 1993, Berth-Jones et al. conducted a randomized, double-blind, parallel-group–designed study to investigate whether supplementation with OEP alone or a combination of OEP and fish oil helped with clinical symptoms of atopic dermatitis. A total of 133 patients (adults and children were evenly distributed) with chronic hand dermatitis enrolled and were randomized to receive either Epogam® (per 500 mg capsule of OEP [321 mg linoleic acid and 40 mg γ-linolenic acid]), Efamol Marine (per 430 mg capsule of OEP/fish oil [17 mg eicosapentaenoic acid and 11 mg docosahexaenoic acid]), or placebo (paraffin/olive oil). No improvement with OEP was found *(12)*.

In 1996, Whitaker et al. conducted a clinical trial to test if OEP supplementation affected the changes in lamellar bodies and lipid layers of the stratum corneum in patients with chronic (longer than 12 months) hand dermatitis. This parallel, double-blind, placebo-controlled trial had 39 patients with chronic hand dermatitis or eczema and 10 age- and sex-matched healthy controls for statistical comparison. Treatment lasted for 16 weeks, with the active group taking OEP (twelve 500-mg Epogam capsules daily for a total dose of 600 mg γ-linolenic acid) and the placebo group taking placebo (twelve 500-mg sunflower capsules daily), after which there was an 8-week washout period. Although the Epogam group improved in the clinical impressions of dermatitis, there was no statistical difference between groups and no structural change in skin specimens was seen *(13)*.

5.2. Anti-Inflammatory Effects

Increased concentrations of eicosanoids (leukotriene B_4, prostaglandin E_2 [PGE_2] and thromboxane A_2) have been reported to exist in the colon mucosa and rectal areas of patients with ulcerative colitis *(14)*. In 1993, a randomized, placebo-controlled study was conducted by Greenfield et al. examining the effect of OEP and fish oil supplementation on cell membranes and symptom control in 43 patients diagnosed with stable ulcerative colitis. Treatment with OEP increased red-cell membrane concentrations of dihomo-γ-linolenic acid by 40% at 6 months ($p < 0.05$), and compared to MaxEPA® and placebo, OEP significantly improved stool consistency at 6 months, with this difference being maintained 3 months after OEP treatment was discontinued ($p < 0.05$). There was no difference in stool frequency, rectal bleeding, relapse rates, sigmoidoscopic findings between the three groups. OEP appeared to be of minimum benefit over placebo and fish oils *(14)*.

5.3. Autoimmune

In a randomized, double-blind, placebo-controlled, three-arm, parallel study that used OEP treatment, 49 adult participants with a diagnosis of rheumatoid arthritis that required nonsteroidal anti-inflammatory medication (NSAID) but not second-line therapy were included. Treatment groups consisted of the control group (liquid paraffin placebo), OEP group (OEP daily dose containing 540 mg γ-linolenic acid), and OEP/fish oil group (dose not recorded). At 12 months, both active treatment groups reported significant subjective improvement compared to the placebo group and had significantly reduced their NSAID use. However, at 15 months, both treatment groups had relapsed. There was no evidence that OEP or OEP with fish oil had modified the disease process in any way. The reviewers stated that insufficient data was given by the study *(15,16)*.

Researchers conducted a 6-month, double-blind, placebo-controlled study involving 40 patients (male and female) with rheumatoid arthritis and upper gastrointestinal lesions caused by NSAIDs *(17)*. For 6 months, 19 patients (17 females and 2 males) received 6 g of OEP daily (total daily dose of γ-linolenic acid 540 mg) and 21 patients (15 females and 6 males) received 6 g of olive oil daily. The results of this study found that there was a significant reduction in morning stiffness after 3 months. No patient was able to stop NSAID medication after completing OEP treatment and only 23 percent of patients could reduce their dose. These results were similar to that of the placebo (olive oil) group. Jäntti et al. examined the effect of OEP on clinical symptoms and plasma prostaglandin levels of patients with rheumatoid arthritis. Study results showed that plasma concentrations of PGE_2 decreased and thromboxane B_2 increased in both groups in addition to no clinical benefits reported *(18)*. Other experimental studies have also concluded that oral OEP has no general therapeutic effect in patients with rheumatoid arthritis.

Manthorpe et al. examined the effect of OEP supplementation on lacrimal gland function in patients with primary Sjögren's syndrome using a double-blind, crossover design. The active treatment group received the following twice daily: Efamol (1500 mg [9% γ-linolenic acid, 73% *cis*-linoleic acid]), Efavit® (375 mg vitamin C, 75 mg pyridoxine, 75 mg niacin, 15 mg zinc sulfate), and vitamin E (40.8 IU). The control group received the same number of placebo capsules and directions. Study results were mixed, with benefit seen in the lacrimal film function (Schirmer's I-test, $p < 0.03$) and nonsignificant findings for the remaining outcome measures. Because of the vitamins given with the OEP, it is difficult to say whether the OEP was of any benefit *(19)*. In a randomized, placebo-controlled trial by Theander et al., 90 patients diagnosed with primary Sjögren's syndrome (with or without signs of autoimmunity) were given either OEP or corn oil (placebo) for 6 months and their symptom levels recorded. Patients were evaluated at baseline, 3, and 6 months. No significant improvements were found for any of the outcomes with OEP supplementation *(20)*.

Increased levels of arachidonic acid and 4-series leukotriene have been reported in the skin plaques of patients with psoriasis *(21)*. OEP is rich in three fatty acids (eicosapentaenoic, γ-linolenic, and docosahexaenoic acid), and eicosapentaenoic acid may help improve psoriasis by inhibiting the formation of 4-series leukotrienes by forming the 5-series leukotrienes, which are considered biologically less active than the 4-series.

In a double-blind, placebo-controlled trial by Veale et al., the effect of OEP supplementation on the improvement of psoriatic arthritis was examined in 38 patients with chronic stable plaque psoriasis and inflammatory arthritis. Results reported no changes in outcome measurements except a decrease in leukotriene B_4 production during the active phase in the Efamol Marine group compared to baseline ($p < 0.03$) and a rebounding increase in thromboxane B_2 during the group's placebo phase run-out phase. The authors suggest that the dose used in this study was sufficient to show some competition with arachidonic acid in its metabolic pathways but not high enough to show improvement in clinical outcomes *(21)*.

5.4. Neurological System Effects

In diabetes, the δ-6-desaturation of linoleic acid into γ-linolenic acid is impaired. With evidence that a high intake of linoleic acid may have some benefit in cardiovascular problems in diabetics, there are hypotheses that supplementation with products high in γ-linolenic acid might benefit diabetic neuropathy. The 1993 study by the Gamma-Linolenic Acid Multicenter Trial Group specifically investigated the effects of OEP (12 capsules daily of a product identical to Epogam, 480 mg total daily dose of γ-linolenic acid) on the clinical outcomes of 111 patients with mild diabetic neuropathy over 1 year using a randomized, double-blind, placebo-controlled, parallel design. Participants were evaluated at baseline, 3, 6, and 12 months. Compared to placebo, OEP supplementation improved 8 out of 10 neurophysiological measures and 5 out of 6 neurological assessments ($p < 0.05$). The authors noted that the changes seen were of a magnitude that was clinically meaningful and consistent among the clinical, thermal, and neurophysical assessments *(22)*. Using a double-blind, placebo-controlled study design, Jamal and Carmichael investigated the effect of 6-month OEP supplementation in clinical improvement in 22 patients with either type I or II diabetes mellitus with distal diabetic polyneuropathy. OEP (a total daily dose of 4 g containing 360 mg γ-linolenic acid) was given to 12 patients, and placebo was given to 10. Compared with the placebo group, the OEP group showed statistically significant improvement ($p < 0.05$) in neuropathy symptom scores, median nerve motor conduction velocity/compound muscle action/potential amplitude, peroneal nerve motor conduction velocity/compound muscle action/ potential amplitude, median sensory nerve action potential amplitude, ankle heat threshold, and cold threshold *(23)*.

Two major controlled trials have examined supplementation of OEP and its effect on attention-deficit/hyperactivity disorder (ADHD) with mixed results. In a double-blind, placebo-controlled, crossover study, Aman et al. examined the effect of OEP supplementation on ADHD in 31 children with ADHD (4 girls, 27 boys). A total of 26 children (mean age of 9 years [age range not reported]) fulfilled the following inclusion criteria: 90th percentile or greater scores on both the Attention Problem subscale III of the Revised Behavior Problem Checklist and the Inattention subscale II of the Teacher Questionnaire *(24,25)*. Each child received either OEP supplementation (6 Efamol capsules daily containing a total daily content of 2.16 g linoleic acid and 270 mg γ-linolenic acid) or placebo (500 mg liquid paraffin) for 4 weeks each, then switched to the other treatment with a 1-week washout period between crossovers. When the experiment-wise probability level was set at 0.05, the authors concluded that OEP supplementation showed no effect in hyperactive children *(26)*. In the second trial, Arnold et al. compared *d*-amphetamine to OEP treatments using a double-blind, placebo-controlled, crossover treatment of 18 boys suffering from ADHD. Outcomes were parent and teacher ratings using standardized hyperactivity scales at screening, baseline, and every 2 weeks during the 3-month study period. Parent ratings showed no effect regarding OEP treatment. Teachers' ratings showed a trend of OEP effect between placebo and *d*-amphetamine with the only statistically significance ($p < 0.05$) demonstrated on the Conners Hyperactivity Factor *(27)*.

Only one double-blind, crossover study has been conducted to examine the effect of OEP supplementation in 13 patients (8 men, 5 women) diagnosed with schizophrenia. Active treatment (4 g OEP, 40 mg vitamin E, 1000 mg vitamin C, 200 mg vitamin B_6, 300 mg vitamin B_3, and 40 mg zinc sulfate) lasted 4 months, with a 2-month washout period before crossing over to placebo treatment (content not described) for 4 months. Nonsignificant results were obtained when comparing mean scores during the active and placebo treatment periods *(28)*.

5.5. Endocrine System Effects

Because levels of γ-linolenic acid are lower in women with PMS compared to non-PMS women *(29)*, it is thought that a defect in converting linoleic acid to γ-linolenic acid may contribute to the sensitivity to normal changes in prolactin that happen during the menstrual cycle *(6)*. Unfortunately, most studies of OEP in treatment of PMS have been open-label, nonplacebo-controlled studies. Only two small studies are considered well-designed and are summarized below. Khoo et al. examined the effect of OEP on PMS symptoms of 38 women, aged 20 to 40 years, using a randomized, double-blind, placebo-controlled, crossover study design. Treatment duration for the first phase lasted 3 cycles, after which women crossed over into the other group for the next 3 months. Active treatment was OEP supplementation (8 Efamol Vita-Glow® capsules daily, each capsule containing 72% linoleic acid, 9% γ-linolenic acid, and 12% oleic acid). Content of placebo capsule was not stated. Results reported showed that over 6 months, there were no significant differences in the symptom scoring between the active and placebo groups. The authors determined that the slight improvements noted by women with moderate PMS were caused by a placebo effect *(30)*. Collins et al. conducted a 10-month randomized, double-blind crossover trial to determine the effect of OEP supplementation in 27 women diagnosed with PMS using *Diagnostic Manual of Mental Disorders, 3rd Edition, Revised* criteria. All women were given placebo in the first month, which was considered the baseline month. All women with PMS received placebo in the second cycle to reduce placebo effects. For the OEP phase, total daily dose was 12 Efamol capsules (each capsule contained 4.32 g linoleic acid, 540 mg γ-linolenic acid). For the placebo phase, the same number of capsules was given and contained paraffin. After starting the treatment in the third cycle, women in one group crossed over to the other group in the seventh cycle. Of the 68 women who participated, 38 completed the study. Analyses of all outcome measures showed that OEP did not improve PMS symptoms or their cyclic nature. All women had improvements in their PMS symptoms over time. The study investigators stated that this result came from a placebo or study participation effect *(31)*.

Chenoy et al. conducted a randomized, double-blind, placebo-controlled study of 56 women that examined the effects of OEP on menopausal flushing (hot flashes). Inclusion criteria were that women had (1) suffered hot flashes at least 3 times a day, and (2) had increased follicle-stimulating hormone and luteinizing hormone levels and/or amenorrhea for at least 6 months. Baseline levels were taken for 1 month when no treatment was given. At the beginning of the second month, 6 months of treatment began with the OEP group taking 4 g of OEP daily (4 capsules twice a day, each capsule containing 500 mg of OEP with 10 mg of vitamin E) and the control group taking the same regimen of placebo capsules. Women recorded the number and severity of flushing and sweating episodes in daily diaries. Assessments were conducted at baseline, 1, 4, and 7 months. Of the 35 women who finished the study, improvements between control cycle and last treatment cycle were statistically significant for the placebo group but not for the OEP group, showing that OEP offered no benefit over placebo in treating menopausal flushing (32).

Although there has been no direct evidence, hormone imbalances (e.g., progesterone deficiency in the luteal phase, high estrogen levels or increased sensitivity to estrogen, higher-than-normal basal prolactin levels) have been associated with mastalgia (33,34). It is suggested that women with breast pain have low levels of γ-linolenic acid, possibly caused by the competition from high levels of saturated fatty acids that make the woman less able to convert linoleic acid to γ-linolenic acid. Another proposed mechanism suggests that OEP may help reduce pain through decreased peripheral prolactin via 1-series prostaglandins that is made from an OEP constituent, dihomo-γ-linolenic acid (34).

OEP was found to have a favorable response in 45% of patients treated at the Cardiff Mastalgia Clinic for cyclical mastalgia (33). In this study, results from clinical trials (ranging in design from randomized and placebo-controlled to open-label) of drug treatment for mastalgia were grouped and descriptively analyzed. Four drugs were compared: bromocriptine, danazole, OEP, and progestins (dydrogesterone and norethisterone). Typical doses and durations were different for each drug. The usual dose of OEP used was six capsules daily for 3 to 6 months; milligram, product name, or percent of fatty acids was not disclosed. Study results report that of the 291 women who received medications, 45% of those with cyclical mastalgia reported good responses using OEP compared to 70% using danazol. Women with noncyclical mastalgia had less impressive good responses from all the drugs (31% danazol vs 27% OEP). No statistical analyses were performed to control for study biases, so study results are suspect (33). In a later study by Wetzig, 170 women with severe mastalgia who were treated at a single clinic were followed for 3 years, with assessments performed on their responses to various medications. Sequence of drugs given depended on previous responses: vitamin B6 (50–100 mg twice daily), OEP (1 g two to three times daily), danazol (100 mg twice daily tapering to 100 mg daily when pain was controlled). In some cases, progesterone, tamoxifen, or NSAIDs were also prescribed. Results regarding the effect of OEP supplementation on mastalgia were similar to placebo *(35)*. A more recent study by Blommers et al. examined the effect of OEP and fish oil supplementation on severe mastalgia using a randomized, double-blind, controlled design. A total of 120 women were randomly placed into four groups. Group 1 took fish oil and one control oil, group 2 took OEP and a control oil, group 3 took fish oil and OEP, and group 4 took two control oils. Duration of therapy was 6 months. Results showed that neither OEP or fish oil were better than placebo in decreasing the number of days with pain *(36)*.

Prostaglandins may be an important factor in the cervical ripening process of labor *(37)*. Topical or oral OEP is promoted as an agent to speed cervical ripening *(38)*. For this particular condition, continuing OEP use is typically reevaluated after 1 week. If no cervical change is noted, the woman may choose to continue for another week or change to another ripening agent *(39)*. A 1999 retrospective cohort study investigated the effect of oral OEP on pregnancy length, duration of labor, incidence of postdates induction, incidence of prolonged rupture of membranes, occurrence of abnormal labor patterns, and cesarean delivery in low-risk nulliparous women. The sample (54 subjects receiving OEP and 54 random controls) was drawn from records of all nulliparous women registered for care at one birth center in the northeastern region of the United States from 1991 to 1998. Results of this study showed that there were no apparent benefits from taking oral OEP with regards to reducing the incidence of adverse labor outcomes or decreasing the overall length of labor. In fact, the study reported a trend of increased incidence of birthing problems such as prolonged rupture of membranes and the need for oxytocin augmentation or vacuum extraction *(37)*.

5.6. Cardiovascular System Effects

Low levels of linoleic acid and dihomo-γ-linolenic acid may predispose coronary heart disease *(6,40)*. For this benefit, linoleic acid needs to be converted by the enzyme δ-6-desaturase to other highly unsaturated, long-change fatty acids, such as γ-linolenic acid. OEP contains both linoleic and γ-linolenic acid, and OEP has been reported to reduce elevated serum cholesterol levels, with γ-linolenic acid having a more dramatic effect of the two *(6)*.

In 1986, Boberg et al. examined the effects of n-3 and n-6 long-chain polyunsaturated fatty acid supplementation on serum lipoproteins and platelet function in 28 adults with high triglyceride levels. Using a placebo-controlled, double-blind, crossover design, 14 adults were randomized to either Efamol supplement group (total daily dose of 4 g had a content of 2.88 g of linoleic acid and 0.36 g of γ-linolenic acid) or the placebo group for 8 weeks, then switched to the other group for another 8 weeks. Another 14 adults were randomized to either MaxEPA supplement group (total daily dose of 10 g had a content of 1.8 g of eicosapentaenoic acid and 1.2 g of docosahexaenoic acid) or the placebo group for 8 weeks, with the same group switching noted previously. Results showed that although OEP supplementation increased γ-linolenic and dihomogammalinolenic acid content in plasma triglycerides and cholesterol esters, compared to placebo, no statistically significant changes were demonstrated in serum lipoprotein lipids or apolipoproteins, triglyceride levels, platelet aggregation, or plasma β-thromboglobulin levels *(41)*. In a randomized, placebo-controlled study of healthy men aged 35 to 54 years who had low levels of dihomo-γ-linolenic acid, the effect of OEP on the fatty acid composition of their adipose tissue and serum lipids was investigated. A total of 35 subjects were enrolled in the study and randomized into four groups. Group 1 ($n = 9$) received 10 mL OEP daily, Group 2 ($n = 8$) received 20 mL OEP daily, Group 3 ($n = 9$) received 30 mL OEP daily, and Group 4 ($n = 9$) received 20 mL of safflower oil daily for 4 months. Results showed that whereas 20 mL daily OEP supplementation increased adipose dihomo-γ-linolenic acid levels ($p < 0.01$), no effects were seen on serum cholesterol, low-density lipoprotein (LDL) cholesterol, or high-density lipoprotein (HDL) cholesterol. OEP was deemed to be an ineffective cholesterol-lowering agent *(40)*.

In a randomized, blinded, crossover study, 12 males with hyperlipidemia took 3 g of OEP daily (containing linoleic acid 2200 mg and γ-linolenic acid 240 mg). After a receiving a placebo for 4 weeks, participants were randomly divided into the treatment or placebo group. For 4 months, the treatment group received OEP (total daily dose of 3 g containing 2.2 g linoleic acid and 240 mg γ-linolenic acid) and placebo group received liquid paraffin. After 4 months, each group received 4 weeks of placebo (washout period) then crossed over to the alternate group for 4 more months of supplementation. Comparing blood samples taken at placebo phase and after 4 months of OEP use, serum triglyceride levels, serum cholesterol, and LDL cholesterol had decreased 48, 32, and 49%, respectively, and HDL cholesterol had increased by 22% ($p < 0.01$). Adenosine diphosphate- and adrenaline-induced platelet aggregation were reduced 50 and 60%, respectively, after 2 and 4 months of OEP use (p-values not reported), and platelet production of thromboxane B_2 went from 26 ± 1.8

to 11.8 ± 3.8 ng/mL plasma ($p < 0.001$) after OEP use. Compared with placebo, after 3 or 4 months of OEP use, bleeding time increased 40% (from 6.8 ± 0.3 minutes to 12.0 ± 0.8 percent [$p < 0.001$]) *(42)*.

Using a randomized, placebo-controlled design, Leng et al. examined the effect of γ-linolenic acid on cholesterol and lipoprotein levels in patients with lower limb atherosclerosis. A total of 120 adults with stable intermittent claudication (ankle brachial pressure index of ≤ 0.9 in at least one limb) were given either active treatment or placebo for 2 years. Active treatment was three capsules twice daily of polyunsaturated fatty acid (one capsule contained 280 mg γ-linolenic acid and 45 mg eicosapentaenoic acid) or the same doses of placebo capsules (one capsule contained 500 mg sunflower oil) for 2 years. Of the 120 participants, 39 (65%) taking active treatment and 36 (60%) taking placebo completed the trial. Lipid concentrations and walking distance were not different between groups. However, hematocrit was higher ($p < 0.01$) and systolic blood pressure lower ($p < 0.05$) in the γ-linolenic acid groups *(43)*.

Dietary supplementation with OEP, which contains the fatty acid prostaglandin precursors linoleic and γ-linolenic acid, may enhance the synthesis of prostaglandins, which might help lower vascular sensitivity to increased levels of angiotensin II in pregnancy *(44)*. To examine the effect that OEP supplementation has in hypertension during pregnancy, randomized, placebo-controlled studies have investigated its use in women diagnosed with preeclampsia. Unfortunately, OEP did not lower blood pressure in women suffering from hypertension in pregnancy *(44,45)*.

5.7. Cytotoxic Effects

Mansel et al. conducted a randomized, double-blind, placebo-controlled clinical trial in 200 women with proven recurrent breast cysts that could be aspirated and were determined to be noncancerous by mammography and biopsy. For 1 year, one group took six Efamol capsules per day (total daily OEP dose of 3 g containing 9% γ-linolenic acid) and the other group took placebo. After 1 year, only 15 women had dropped out of the study (eight from the placebo group, seven from the OEP group). The overall cyst recurrence rate was 44% (46% for the placebo group, and 43% for the OEP group) and statistically nonsignificant. Cyst fluid electrolyte ratio was unremarkable. Fatty acid analysis results were not stated *(46)*.

5.8. Miscellaneous

Chronic fatigue syndrome (CFS) is a condition in which the etiology is unclear. The illness is characterized by persistent and relapsing fatigue, and other varying symptoms throughout the body. In 2000, a critique of the literature on all randomized, controlled trials (RCTs) regarding treatment for CFS reported mixed results for OEP therapy *(47)*. The reviewers reported two RCTs comparing OEP with placebo in patients with either a diagnosis of postviral fatigue syndrome or CFS.

In the first double-blind, randomized, placebo-controlled study, Behan et al. examined the physical and psychological effects of high doses of OEP/ fish oil supplementation on 63 adults (27 men, 36 women) with postviral fatigue syndrome. Participants were randomly assigned to placebo group (liquid paraffin containing a total daily dose of 400 mg linoleic acid and 80 IU of vitamin E) or treatment group (total daily OEP/fish oil preparation [Efamol Marine] containing 288 mg γ-linolenic acid, 136 mg eicosapentaenoic acid, 88 mg docosahexaenoic acid, 2.04 g linoleic acid, and 80 IU of vitamin E) for 3 months and were evaluated at baseline, 1, and 3 months. At 3 months, 85% of the OEP group showed improvement compared to 17% of the placebo group (scale of better, worse, unchanged; $p < 0.0001$). The EFA levels were abnormal at the baseline and were corrected in the OEP group by study end ($p < 0.05$) *(48)*. The second study was a double-blind, placebo-controlled, randomized design in which Warren et al. looked at the effect of 3 months of OEP supplementation on physical and psychological outcome measures of 50 patients (21 men, 29 women) who were diagnosed with CFS using the Oxford Criteria. Patients were randomized into the placebo group (sunflower oil) or treatment group (eight Efamol Marine 500 mg daily, which contained, in the OEP and fish oil components, a total daily dose of 288 mg γ-linolenic acid, 136 mg eicosapentanoic acid, 88 mg of docosahexanoic acid, and 2.04 g of linoleic acid) or placebo group (sunflower oil). At the end of the 3 months, seven participants from the treatment group and five from the placebo group had quit the study because of lack of clinical response. Results showed that although both physical and psychological symptoms improved with time, the differences before and after treatment between groups were not statistically significant *(49)*.

Khan et al. examined the effects of n-3 and n-6 fatty acid supplements (two of which contained OEP) on the microvascular blood flow and endothelial function in 173 healthy men and women aged 40 to 65 years in an 8-month, double-blind, randomized, placebo-controlled study. For the single OEP supplementation, the group received a total daily OEP of 5 g (which contained 400 mg/day of γ-linolenic acid). For the tuna oil/OEP supplementation, the group received a total daily tuna oil of 5 g (which contained 6% of eicosapentaenoic acid and 27% of docosahexaenoic acid per day) and OEP of 5 g. Results showed that there although there were significant improvements in the tuna oil supplementation group, there were no significant changes in any of the outcome measures with either single OEP or tuna oil/OEP supplementation group *(50)*.

6. PHARMACOKINETICS

The pharmacokinetic parameters of γ-linolenic acid and its metabolic products were studied in six healthy volunteers (three males, three females) following the administration of OEP. Serum level-time courses of eight fatty acids (γ-linolenic, palmitic, linoleic, linolenic, oleic stearic, arachidonic, dihomo-γ-linolenic acids) were profiled twice over a 24-hour period after receiving six capsules of Epogam (total dose being 240 mg of γ-linolenic acid) at 7:00 AM and 7:00 PM (γ-linolenic acid total daily dose being 480 mg). The following mean pharmacokinetic parameters were obtained for γ-linolenic acid: t_{max}^{am} (hour) = 4.4 (significantly higher than t_{max}^{pm} [$p < 0.05$]); C_{max}^{am} (μg/mL) = 22.6; t_{max}^{pm} (hour) = 2.7; C_{max}^{pm} (μg/mL) = 20.7; area under the curve $(AUC)_{12h}^{am}$ ([μg · hour]/mL) = 119.0; AUC_{12h}^{pm} ([mg · hour]/mL) = 155.1; and AUC_{24h} ([mg · hour]/mL) = 274.1 (significantly higher than AUC_{24h} at baseline). This study found that the absorption of γ-linolenic acid from the gastrointestinal tract is much lower than in the morning than in the evening. The concentrations of the metabolites of γ-linolenic acid, arachidonic acid, and dihomo-γ-linolenic acid could not be established in these volunteers *(51)*.

7. ADVERSE EFFECTS AND TOXICITY

When taken within the recommended dosage range, the γ-linolenic acid content of OEP is equivalent to that present in a normal diet *(6)*. Thus, although adverse effects are rare at recommended doses, occasionally, mild gastrointestinal effects and headache may occur with oral use of OEP. The World Health Organization Programme for International Drug Monitoring reported that, in the period between 1968 and 1997, there were 193 adverse reactions reported mentioning OEP. The most critical of these OEP reports mentioned convulsions, aggravated convulsion, face edema, and asthma. The most noncritical OEP adverse effects included headache, nausea, itching, abdominal pain, and diarrhea *(52)*. In the study by Guivernau et al. summarized in Subheading 5.6, they reported that OEP inhibited platelet aggregation and prolonged bleeding time in 12 males with hyperlipidemia taking 3 g of OEP daily (containing linoleic acid 2200 mg and γ-linolenic acid 240 mg). Compared to placebo, bleeding time at 3 and 4 months increased 40%, with the group's mean rising from 6 to 12 minutes ($p < 0.001$) *(42)*.

8. INTERACTIONS

Current references advise caution with concurrent use of OEP and several classes of medications (anticoagulants, antipsychotics/anticonvulsants) because of possible serious side effects *(53–55)*.

The use of anticonvulsants and OEP may result in a delayed reduction in the effectiveness of the anticonvulsant because OEP may lower the seizure threshold *(56,57)*. In 1981, three patients on phenothiazine therapy for schizophrenia were given OEP (around this time, OEP was being considered a possible adjunct to therapy for this mental illness). The patients suffered seizures, OEP was discontinued, and electroencephalograms were performed. The tests showed temporal lobe epileptic disorders and phenothiazine therapy was discontinued or reduced in the patients and carbamazepine started. The authors suggested that OEP supplementation be used with caution in patients taking phenothiazines *(58)*. In 1983, Holman et al. reported an incident involving a 43-year old man with schizophrenia on fluphenazine decanoate therapy who suffered a grand mal seizure after using 4 g of OEP daily for 3 months. Once the OEP was discontinued, no seizures occurred in the next 7 months *(28)*.

The use of OEP with antiplatelets, thrombolytics, low-molecular-weight heparins, or anticoagulants may result in a delayed but increased risk of bleeding, and concomitant use is not advised *(54,59)*. The mechanism of action is theorized to be decreased thromboxane B_2 synthesis and increased prostacyclin production caused by γ-linolenic acid, resulting in inhibition of platelet aggregation and prolonged bleeding time *(42)*.

In vitro experiments by Zou et al. in 2002 examined the effect of the *cis*-linoleic acid component of OEP on the catalytic activity of cDNA-expressed cytochrome P450 isoforms (CYP1A2, CYP2C9, CYP2C19, CYP2D6, and CYP3A4). The highest concentration of *cis*-linoleic acid tested was 179 μM. In these assays, inhibitory concentration of 50% (IC_{50}) values no greater than 10 μM were considered potent inhibitors and IC_{50} values between 10 and 50 μM were labeled moderate inhibitors. *cis*-Linoleic acid tested as a moderate inhibitor of all the cDNA-derived enzymes except CYP3A4 (resorufin benzyl ether substrate) *(60)*. However, there have been no case reports of toxicity or adverse reactions with OEP and drugs metabolized by this enzyme.

9. REPRODUCTION

There are no known reports of OEP causing teratogenicity in humans. No effects of OEP administration on reproduction were found in 2-year teratological investigations *(6)*. Specifically, reproductive investigations with OEP supplementation have been performed in mice and rats, mink, and the blue fox. In a study examining the teratogenicity of OEP, mice and rats were fed a diet containing 10% of oxidized linoleic acid. The rats, but not the mice, had babies with an increase in urogenital anomalies *(61)*. Tauson et al. conducted reproduction studies in the male and female mink. The results of these studies suggested there was a tendency (nonsignificant) for a decreased rate of stillbirths and deaths during the first 21 days of life when the male was administered OEP. OEP administration did not affect reproductive performance *(62)*.

10. REGULATORY STATUS

OEP is regulated as a dietary supplement in the United States. It is approved in Canada as an over-the-counter product for use in EFA-deficiency conditions and as a dietary supplement to increase EFA intake. In the United Kingdom, it is on the General Sales List. In Germany, OEP is approved for use as food and is approved there in the treatment and symptomatic relief of atopic eczema. In Sweden, OEP is classified as a natural product. OEP has a Class 1 Safety Rating with the American Herbal Product Association *(1,2,7)*.

REFERENCES

1. Chen, JK. Evening Primrose Oil. Continuing Education Module. Continuing Pharmaceutical Education Credit. University of Southern California School of Pharmacy. ACPE#007-999-99-049-H01. March 1999.
2. Anonymous. Evening primrose. In: *PDR for Herbal Medicines*. Montvale: Medical Economics Company, 2000.
3. Deng Y, Hua H, Li J, Lapinskas P. Studies on cultivation and uses of Evening Primrose (*oenothera* spp.) in China. Econ Bot 2001;55:83–92.
4. Tyler VE. *A Sensible Guide to the Use of Herbs and Related Remedies, 3rd edition*. New York: Pharmaceutical Press, 1993.
5. Evening Primrose oil in: Natural Medicines Comprehensive Database (intranet database). University of Minnesota Biomedical Library. http://www.natural databse.com. Last accessed: Jan. 2004.
6. Anonymous. *Facts and comparisons. The Review of Natural Products*. St. Louis: Wolters Kluwer Company, 1997.

7. Blumenthal M, ed. *Popular Herbs in the US Market*. Austin: American Botanical Council, 1997.

8. Horrobin D. Essential fatty acid metabolism and its modification in atopic eczema. Am J Clin Nutr 2000; 71:367S–372S.

9. Kerscher MJ, Korting HC. Treatment of atopic eczema with evening primrose oil: rationale and clinical results. Clin Investig 1992;70:167–171.

10. Galli E, Picardo M, Chini L, et al. Analysis of polyunsaturated fatty acids in newborn sera: a screening tool for atopic disease? Br J Dermatol 1994;130:752–756.

11. Morse P, Horrobin D, Manku M, et al. Meta-analysis of placebo-controlled studies of the efficacy of OEPgam in the treatment of atopic eczema. Relationship between plasma essential fatty acid changes and clinical response. Br J Dermatol 1989;121:75–90.

12. Berth-Jones J, Graham-Brown RAC. Placebo-controlled trial of essential fatty acid supplementation in atopic dermatitis. Lancet 1993;341:1557–1560.

13. Whitaker D, Cilliers J, de Beer C. Evening Primrose Oil (OEPgam) in the treatment of chronic hand dermatitis: disappointing therapeutic results. Dermatology 1996;193:115–120.

14. Greenfield SM, Green AT, Teare JP, et al. A randomized controlled study of evening primrose oil and fish oil in ulcerative colitis. Aliment Pharmacol Ther 1993;7:159–166.

15. Belch JJF, Ansell D, Madhok R, O'Dowd A, Sturrock D. Effects of altering dietary essential fatty acids on requirements for non-steroidal anti-inflammatory drugs in patients with rheumatoid arthritis: a double blind placebo controlled study. Ann Rheum Dis 1988;47:96–104.

16. Belch J, Hill A. Evening primrose oil and borage oil in rheumatologic conditions. Am J Clin Nutr 2000;71:352S–356S.

17. Brzeski M, Madhok R, Capell A. Evening Primrose oil in patients with rheumatoid arthritis and side effects of non-steroidal anti-inflammatory drugs. Brit J Rheumatol 1991;30:370–372.

18. Jäntti J, Seppälä E, Vapaatalo H, Isomäki H. Evening primrose oil and olive oil in treatment of rheumatoid arthritis. Clin Rheumatol 1989;8:238–244.

19. Manthorpe R, Petersen SH, Prause JU. Primary Sjögren's syndrome treated with Efamol/Efavit. A double-blind cross-over investigation. Rheumatol Int 1984;4:165–167.

20. Theander E, Horrobin DF, Jacobsson LT, Manthorpe R. Gammalinolenic acid treatment of fatigue associated with primary sjögren's syndrome. Scand. J Rheumatol 2002;31:72–79.

21. Veale DJ, Torley HI, Richards IM, et al. A double-blind placebo controlled trial of Efamol Marine on skin and joint symptoms of psoriatic arthritis. Br J Rheumatol 1994;33:954–958.

22. Keen H, Payan J, Allawi J, et al. Treatment of diabetic neuropathy with gamma-linolenic acid. The gamma-Linolenic Acid Multicenter Trial Group. Diabetes Care 1993;16:8–15.

23. Jamal GA, Carmichael H. The effect of gamma-linolenic acid on human diabetic peripheral neuropathy: a double-blind placebo-controlled trial. Diabet Med 1990;7:319–323.
24. Werry JS, Hawthorne D. Connors Teacher Questionnaire—Norms and validity. Aust N Z J Psychiatry 1976;10:257–262.
25. Quay HC. A dimensional approach to children's behavior disorders: The Revised Behavior Problem Checklist. School Psych Rev 1983;12:244–249.
26. Aman MG, Mitchell EA, Turbott SH. The effects of essential fatty acid supplementation by Efamol in hyperactive children. J Abnorm Child Psychol 1987;15(1):75–90.
27. Arnold LE, Kleykamp D, Votolato NA, Taylor WA, Kontras SB, Tobin K. Gamma-linolenic acid for attention-deficit hyperactivity disorder: placebo-controlled comparison to D-amphetamine. Biol Psychiatry 1989;25(2):222–228.
28. Holman CP, Bell AFJ. A trial of evening primrose oil in the treatment of chronic schizophrenia. J Orthomol Psychiatr 1983;12:302–304.
29. Bendich A. The potential for dietary supplements to reduce premenstrual syndrome (PMS) symptoms. J Am Coll Nutr 2000;19:3–12.
30. Khoo SK, Munro C, Battistutta D. Evening primrose oil and treatment of premenstrual syndrome. Med J 1990;153:189–192.
31. Collins A, Cerin A, Coleman G, Landgren B. Essential fatty acids in the treatment of premenstrual syndrome. Obstet Gynecol 1993;81:93–98.
32. Chenoy R, Hussain S, Tayob Y, O'Brien P, Moss M, Morse P. Effect of oral gamalenic acid from evening primrose oil on menopausal flushing. BMJ 1994;308:501–503.
33. Pye J, Mansel R, Hughes L. Clinical experience of drug treatments for mastalgia. Lancet 1985;2(8451):373–376.
34. Horner NK, Lampe JW. Potential mechanisms of diet therapy for fibrocystic breast conditions show inadequate evidence of effectiveness. J Am Diet Assoc 2000;100:1368–1380.
35. Wetzig N. Mastalgia: a 3 year Australian study. Aust N Z J Surg 1994;64:329–331.
36. Blommers J, de Lange-de Klerk ESM, Kuik DJ, Bezemer PD, Meijer S. Evening primrose oil and fish oil for severe chronic mastalgia: a randomized, double-blind, controlled trial. Am J Obstet Gynecol 2002;187:1389–1394.
37. Dove D, Johnson P. Oral evening primrose oil: its effect on length of pregnancy selected intrapartum outcomes in low-risk nulliparous women. J Nurse Midwifery 1999;44:320–323.
38. Evening primrose oil. In: REPRORISK® System (intranet database) version 5.1. Greenwood village, CO, Thomson Micromedex.
39. Adair, CD. Nonpharmocologic approaches to cervical priming and labor induction. Clin Obstet Gynecol 2000;43:447–454.
40. Abraham RD, Riemersma RA, Elton RA, Macintyre C, Oliver MF. Effects of safflower oil and evening primrose oil in men with a low dihomo-γ-linolenic level. Atherosclerosis 1990;81:199–208.

41. Boberg M, Vessby B, Selinus I. Effects of dietary supplementation with n-6 and n-3 long-chain polyunsaturated fatty acids on serum lipoproteins and platelet function in hypertriglyceridaemic patients. Acta Med Scand 1986;220(2):153–160.
42. Guivernau M, Meza M, Barja P, Roman O. Clinical and experimental study on the long-term effect of dietary gamma-linolenic acid on plasma lipids, platelet aggregation, thromboxane formation, and prostacyclin production. Prostaglandins Leukot Essent Fatty Acids 1994;51:311–316.
43. Leng GC, Lee AJ, Fowkes FG, et al. Randomized controlled trial of gamma-linolenic acid and eicosapentaenoic acid in peripheral arterial disease. Clin Nutr 1998;17:265–271.
44. Moodley J, Norman R. Attempts at dietary alteration of prostaglandin pathways in the management of pre-eclampsia. Prostaglandins Leukot Essent Fatty Acids 1989;37:145–147.
45. Laivouri H, Horatta O, Viinikka L, Ylikorkala O. Dietary supplementation with primrose oil or fish oil does not change urinary excretion of prostacyclin and thromboxane metabolites in pre-eclamptic women. Prostaglandins Leukot Essent Fatty Acids 1993;49:619–694.
46. Mansel RE, Harrison BJ, Melhuish J, et al. A randomized trial of dietary intervention with essential fatty acids in patients with categorized cysts. Ann NY Acad Sci 1990;586:288–294.
47. Reid S, Chalder T, Cleare A, Hotopf M, Wessely S. Chronic fatigue syndrome. Br Med J 2000;320:292–296.
48. Behan PO, Behan WMH, Horrobin D. Effect of high doses of essential fatty acids on the postviral fatigue syndrome. Acta Neurol Scand 1990;82:209–216.
49. Warren G, McKendrick M, Peet M. The role of essential fatty acids in chronic fatigue syndrome: a case-controlled study of red-cell membrane essential fatty acids (EFA) and a placebo-controlled treatment study with high dose of EFA. Acta Neurol Scand 1999;99:112–116.
50. Khan F, Elherik K, Bolton-Smith C, et al. The effects of dietary fatty acid supplementation on endothelial function and vascular tone in healthy subjects. Cardiovasc Res 2003;59(4):955–962.
51. Martens-Lobenhoffer J, Meyer F. Pharmacokinetic data of gamma-linolenic acid in healthy volunteers after the administration of evening primrose oil (Epogam). Int J Clin Pharmacol Ther 1998;36:363–366.
52. Farah MH, Edwards R, Lindquist M, Leon C, Shaw D. International monitoring of adverse health effects associated with herbal medicines. Pharmacoepidemiol Drug Saf 2000;9:105–112.
53. DRUG-REAX®, system (intranet database). Version 5.1. Greenwood Village, CO, Thomsom Micromedex.
54. Ernst E. Possible interactions between synthetic and herbal medicinal products. Part I: a systematic review of the evidence. Perfusion 2000;13:4–15.
55. Miller LG. Herbal medicinals: selected clinical considerations focusing on known or potential drug–herb interactions. Arch Intern Med 1989;158:2200–2211.

56. Newall CA, Anderson LA, Phillipson JD, eds. Herbal medicines: a guide for health-care professionals. London: The Pharmaceutical Press, 1996.
57. Barber AJ. Evening primrose oil: a panacea? Pharm J 1988;240:723–725.
58. Vaddadi KS. The use of gamma-linolenic acid and linoleic acid to differentiate between temporal lobe epilepsy and schizophrenia. Prostaglandins Med 1981;6:375–379.
59. Norred CL, Brinker F. Potential coagulation effects of preoperative complementary and alternative medicines. Alt Ther 2001;7(6):58–67.
60. Zou L, Harkey MR, Henderson GL. Effects of herbal components on cDNA-expressed cytochrome P450 enzyme catalytic activity. Life Sci 2002;71:1579–1589.
61. Cutler MG, Schneider R. Malformations produced in mice and rats by oxidized linoleate. Food Cosmet Toxicol 1973;11:935–942.
62. Tausen A, Neil M, Forsberg M. Effect of evening primrose oil as a food supplement on reproduction in the milk. Acta Vet Scand 1991;32:337–344.

Chapter 15

Citrus aurantium

Anders Westanmo

SUMMARY

Citrus aurantium has enjoyed a rich history of uses in food, cosmetics, and medicine. Recent misuse of this product for weight loss, however, is threatening to tarnish the holistic reputation of this fruit. Manufacturers are isolating and concentrating the synephrine content from the 0.33 mg/g contained in the pulp of whole fruit to 20 mg/g in some dietary supplements, and over 100-fold increase to 35 mg/g in extracts. With the known cardiovascular effects of synephrine, this may be creating a potentially dangerous or abuseable supplement out of what people once safely enjoyed. The use of *C. aurantium* for weight loss has little support in the literature, but this has not stopped producers from marketing the drug for this purpose since the void left after the ban of ephedra. The increased frequency in which case reports of toxicity have emerged since this product has started being used for weight loss should serve as a cautionary note for more vigilant monitoring of safety.

Key Words: *Citrus aurantium*; bitter orange; synephrine; ephedra substitute; weight loss; adrenergic amines.

1. HISTORY

It is generally thought that the Citrus originated in Southeast Asia *(1)*. There are a number of Citrus types, but within oranges the principal members are the sweet orange and the bitter orange *(2)*. The horticultural mapping of Citrus is notoriously difficult, with large discrepancies between writings of number of variants of all types of Citrus. Despite this, there are three properly recognized bitter orange fruits and two other closely related fruits. One of these relatives is the bergamot, the oil of which is famous for giving the distinct taste in Earl Gray tea. Bitter orange fruit is too sour for general con-

From Forensic Science and Medicine:
Herbal Products: Toxicology and Clinical Pharmacology, Second Edition
Edited by: T. S. Tracy and R. L. Kingston © Humana Press Inc., Totowa, NJ

sumption, although it is eaten with salt and chili in Mexico *(3)* and raw in Iran *(4)*. The peel of the fruit has a distinctive taste that is highly valued in marmalade, its most widespread culinary use *(1)*. *Citrus aurantium* is used in several alcoholic beverages. When dried, the peel is used in a distinctive Belgian beer called Orange Muscat *(5)*. The oil from *C. aurantium* is a standard ingredient in other liquors such as Triple Sec, Cointreau, and Curacao *(5)*. The oil is also common in perfumes and as a flavoring agent in sweets *(5)*.

The bulk primary usage of *C. aurantium* is for medicinal purposes. In China, Japan, and Korea, when dried, the entire unripe fruit is used to treat digestive problems. The dried fruit is used to stimulate gastric acid secretion and appetite in Western countries.

2. CURRENT PROMOTED USES

C. aurantium is still used in traditional culinary ways as described previously. It has also been investigated for a number of other medicinal uses in addition to gastrointestinal disturbances. As a dermatological agent, it has been used as an antifungal and exhibits some evidence of fungicidal and fungistatic activity *(6)*. Other uses cited are as a stimulant, a sedative, and for the treatment of anemia, frostbite, general feebleness, retinal hemorrhage, bloody stools, duodenal ulcers, and prolapsed uterus or anus, among other things *(7)*.

Recently, the most predominant use of *C. aurantium* has been as a weight loss aid or as an energy-boosting supplement. After the ban of phenylpropanolamine owing to increased risk of hemorrhagic stroke *(8)*, ephedra use in supplements increased substantially. In April of 2004, the Food and Drug Administration (FDA) banned the sale of ephedra *(9)* because of the dangers of this alkaloid *(10–12)*. Since then, alternatives to this popular supplement have been sought and bitter orange has emerged as the mostly predominantly used substitute *(13)*. Although *C. aurantium* has been reviewed for this use *(14,15)*, lack of studies make this difficult—as evidenced by a 2004 Cochrane Collaboration Database that identified only one eligible randomized placebo controlled trial *(16)*. Despite this relative lack of data, since the banning of ephedra, *C. aurantium* has been used extensively in a variety of products—from weight loss pills to two patents for weight-loss toothpaste *(17)*.

3. SOURCES AND CHEMICAL COMPOSITION

C. aurantium, synonyms *Citrus amara*, *Citrus bigarradia*, *Citrus vulgaris*. Also known as *Aurantii pericarpium*, Chisil, *Fructus aurantii*, Green Orange, Kijitsu, Neroli Oil, Seville Orange, Shangzhou Zhiqiao, Sour Orange, Synephrine, Zhi Qiao, Zhi Shi.

4. PRODUCTS AVAILABLE

C. aurantium is available in a wide variety of products in multiple dosage forms.

5. PHARMACOLOGICAL/TOXICOLOGICAL EFFECTS

5.1. Pharmacology

C. aurantium contains the adrenergic amines (18) synephrine, octopamine, and tyramine, and many flavones and glycosylated flavanones (19). These sympathomimetic molecules can be found in normal human plasma (20) and are known to be involved in alternate metabolic pathways of common biogenic amines (21). Of the three active adrenergic molecules in C. aurantium, synephrine is present in amounts at least 100-fold greater than either octopamine or tyramine found in the fresh fruit, dried extracts, or herbal medicines (18).

The locations of the substituents on the phenylethylamine backbone play an integral role in determining the observed pharmacological effects of sympathomimetic molecules (Table 1). Substitution of a hydroxyl group on the β-carbon tends to increase activity toward both α and β receptors, but decreases activity in the central nervous system (CNS) (23). Substitution of hydroxyl groups in any place in the phenylethylamine structure increases the hydrophilicity, and thus decreases the propensity of the molecule to enter the CNS (23). Ephedrine, for example, is a weaker CNS stimulant than amphetamine but is a stronger bronchodilator and has greater effect on increasing heart rate and blood pressure (23). The relatively polar epinephrine is essentially devoid of CNS activity aside from anxiety related to other systemic effects (23). Hydroxyl groups at both the 3 and 4 position provides the most α and β activity (23). Also, substitution at the amino position generally enhances the effect on β receptors (23). This accounts for the strength of epinephrine for β_2 receptor subtype relative to that of norepinephrine and is probably important in interpreting the potential cardiovascular effects of C. aurantium (23).

Although the effects of synephrine and C. aurantium differ slightly in hemodynamic studies, the relative content of synephrine compared to octopamine and tyramine (at least 100 times more synephrine) in C. aurantium products substantially outweighs the effects of octopamine and tyramine (18). Although synephrine is found endogenously in the adrenal glands (24), the function is still unclear (20). The binding of synephrine to various adrenergic receptors is shown in Table 2.

Table 1
Chemical Structures of Selected Sympathomimetic Molecules

Epinephrine	3-OH,4-OH	OH	H	CH_3
Norepinephrine	3-OH,4-OH	OH	H	H
Synephrine	4-OH	OH	H	CH_3
Octopamine	4-OH	OH	H	H
Tyramine	4-OH	H	H	H
Ephedrine	OH	CH_3	CH_3	
Amphetamine	H	CH_3	H	
Phenylpropanolamine	OH	H	H	

Adapted from ref. *23*.

Table 2
Synephrine Binding and Activity in Adrenergic Receptors

Receptor	Activated by synephrine	Binding relative to norephinephrine	Reference	Clinical hemodynamic response to receptor activation (23)
α_1	Yes	↓15X	25,26	Constrict vessels through-out body
α_2	Yes	?	26,27	Constrict coronary/renal arteries and systemic veins
β_1	No	?	28	Increase heart reate, contractility, conduction velocity
β_2	No, yes	↓100X	28,29	Dilate coronary/renal/ skeletal/pulmonary arties and systemic veins. Also same heart effects as β_1
β_3	Yes	?	30	Lypolysis and thermogenesis

There seems to be general agreement that synephrine is an active agonist of β_1, β_2, and β_3 adrenergic receptors. There are conflicting reports, however, as to whether synephrine possesses any β_1- or β_2-receptor-agonist activity *(28,29)*. The effects observed in studies and case reports of *C. aurantium* use are consistent with that of both α- and β-receptor activation. Also, the published study demonstrating lack of β-adrenergic activation *(28)* was conducted in guinea pig atria/trachea, whereas the study demonstrating activation of β receptors was conducted in cloned human β_2 adrenergic receptors *(29)*. Drugs such as phenylephrine, which primarily exhibit α-agonist activity, induce a rise in blood pressure and a dose-related increase in peripheral vascular resistance *(23)*. Associated with these actions is a sinus bradycardia caused by vagal reflex *(23)*. Epinephrine, on the other hand, exhibits both α- and β-agonist activity. Effects seen with norepinephrine infusion are tachycardia, a moderate increase in systolic blood pressure, a rise in cardiac output, and shorter and more powerful cardiac systole *(23)*. The β_2 effect relaxes skeletal muscle vasculature, leading to a lowering of peripheral resistance and diastolic blood pressure *(23)*.

5.2. Cardiovascular Effect

There are limited animal and human data on the cardiovascular effects of *C. aurantium*. When administered to rats, *C. aurantium* and synephrine both raised blood pressure in a dose-dependent manner *(31)*. In another study in rats, repeated oral *C. aurantium* extract led to dose-dependent cardiovascular toxicity and mortality *(32)*. Two studies using synephrine or *C. aurantium* in rats with induced portal hypertension (by portal vein ligation) have been conducted *(31,33)*. In these studies, both synephrine and *C. aurantium* significantly reduced portal venous pressure. Interestingly, *C. aurantium* had a greater effect on portal hypertension than synephrine alone.

In one of two human clinical studies to date, intravenous synephrine was injected into 12 healthy male volunteers *(34)*. The parameters recorded were transthoracic echocardiography (ECG), cardiac index, arterial blood pressure, and peripheral vascular resistance. Volunteers were given a continuous infusion of 4 mg/minute synephrine. Systolic blood pressure increased by a mean of 27 mmHg ($p < 0.005$) and mean arterial blood pressure increased by 9 mmHg ($p < 0.005$). The cardiac index increased from 3.6 to 4.6 L/(minute $\cdot$ μ_2) ($p < 0.001$) whereas peripheral vascular resistance was decreased ($p < 0.01$). The heart rate and diastolic pressure remained unchanged. By ECG, left ventricular contractility parameters also increased significantly—as shown by systolic shortening fraction ($p = 0.001$) and maximal velocity of shortening of the left ventricular diameter (p = 0.005). These data are consistent with predominant stimulation of α-adrenergic receptors, but also stimulation of both β_1/β_2 receptors.

In another study, the effects of ingestion of Seville orange (*C. aurantium*) juice on blood pressure were studied *(35)*. A group of 12 normotensive adults were given an 8-oz glass of juice (approx 13–14 mg synephrine) 8 hours apart. Blood pressure and heart rate were measured every hour for 5 hours after the second glass. The study was conducted in a crossover design and subjects returned in 1 week to repeat the test with water control. The administration of *C. aurantium* juice did not result in the change of any hemodynamic parameters.

5.3. Thermogenic/Lypolytic Effects

Synephrine and octopamine have been shown to stimulate β_3 receptors *(30)*, the stimulation of which is thought to contribute to lypolysis and thermogenesis. Stimulation of the α_2-adrenergic receptor is also reported to have this effect *(23)*. To date, one placebo-controlled, randomized, double-blind study evaluating the effects of a *C. aurantium*-containing weight loss product has been conducted *(36)*. A total of 20 healthy adults received either the active pill (n = 9), a placebo (n = 7), or nothing (n = 4). The treatment group was given a daily pill for 6 weeks containing 975 mg *C. aurantium* extract (6% synephrine alkaloids), 528 mg of caffeine, and 900 mg of St. John's wort. Subjects followed an 1800 kcal/day diet and engaged in a 3-day/week training program under the guidance of an exercise physiologist. Measurements of weight, blood pressure, fat loss, and mood were taken at baseline, 3 weeks, and 6 weeks. At week 6, the treatment group lost significantly more fat (average of 3.1 kg) compared with other groups and also lost a significantly greater amount of body weight (average of 1.4 kg). No significant changes in blood pressure were seen at weeks 3 or 6, although it is unclear whether or not the treatment group took the medication prior to measurement on those days.

5.4. Dermatological Effects

Oil of bergamot (*C. aurantium* spp. *bergamia*) was a relatively frequent cause of dermatitis prior to the banning of this aromatic substance from cosmetics *(37)*. The phenomena of "Berloque dermatitis" has been extensively reviewed *(38,39)* and is thought to be caused mainly by the furocoumarin bergapten (5-methoxypsoralen) *(40–42)*. This photomutagenic and photocarcinogenic substance has been largely removed from the cosmetic industry, but the extract is still prevalent in aromatherapy *(43)*. Recently cases have been reported of bullous reactions after contact or aerosolization exposure followed by sun exposure or tanning *(37)*.

6. ADVERSE EFFECTS AND TOXICITY

6.1. Case Reports of Toxicity Caused By Commercially Available Products or Traditional Uses by Various Specialty Populations

One case of an acute lateral wall myocardial infarction (MI) was reported in a woman after daily ingestion of Edita's Skinny Pill (containing 300 mg bitter orange plus caffeine and guarana) for 1 year *(44)*. The 55-year-old Caucasian woman developed chest discomfort after eating Chinese food. After workup at the hospital, the woman was diagnosed with acute lateral-wall MI and smoking addiction. Her ejection fraction was 0.45. Prior to this incident, she had no known coronary artery disease, hypertension, or hyperlipidemia.

In another case, a 52-year-old woman experienced tachycardia shortly after taking a dry herbal extract of unripe *C. aurantium* fruit *(45)*. The patient took no medications except for a 10-year history of thyroxine (50 µg/day) treatment. The woman had ingested a dietary supplement for weight loss and consumed 500 mg of *C. aurantium* titrated at 6% synephrine (30 mg). Later in the same day of her first dose, she experienced unrelenting tachycardia and was admitted to the emergency room. She stated that she had never experienced prolonged tachycardia in the past. She was released from the ER after the tachycardia subsided, and felt well. After approx 1 month of feeling well, the woman took another dose of the supplement. Again, later in the day after ingestion of the supplement, she experienced a new episode of prolonged tachycardia. She was seen at the same hospital and released without incident. At the time of publication, she had not taken any more of the supplement and had no reports of tachycardia or other medical problems.

An article in the Canadian Adverse Reaction Newsletter published their reporting of adverse effects caused by products containing *C. aurantium* from January 1, 1998 to February 28, 2004 *(46)*. The article lists 16 reports of synephrine associated with cardiovascular events including tachycardia, cardiac arrest, ventricular fibrillation, transient collapse, and blackout. In one case, bitter orange was the sole suspected culprit. In seven others the products also contained caffeine, and in eight cases the product contained both caffeine and ephedrine. Health Canada has issued an advisory stating that synephrine may have effects similar to ephedrine and caution should be used if taking it *(47)*.

7. Pharmacokinetics

No current information exists on the pharmacokinetics of exogenously administered synephrine.

8. Interactions

Juice of the Seville orange is known to inhibit intestinal CYP3A4 *(48,49)* and has been used for this purpose experimentally *(50)*. This effect is similar to that observed with grapefruit juice in that it only affects intestinal CYP3A4 and has no effect on hepatic CYP3A4. Many drugs may be potentially affected by coadministration with Seville orange juice, including antifungals (ketoconazole, itraconazole), calcium channel blockers (diltiazem, nicardipine, verapamil), chemotherapeutic agents (etoposide, paclitaxel, vinblastine, vincristine, vindesine), dextromethorphan, felodipine, fexofenadine, glucocorticoids, losartan, midazolam, and others *(7)*. The effects of Seville orange juice on cyclosporine disposition has been evaluated in several studies with mixed results. In one study, the C_{max} of cyclosporine was increased by 64% after ingestion of *C. aurantium (51)*. Another study demonstrated no effect of *C. aurantium* on cyclosporine concentrations *(52)*. A third study looked at the effect of long-term administration of *C. aurantium* extract on CYP3A4 activity *(53)*. The study found no effect of the extract on CYP3A4 activity. The authors proposed a novel explanation for the discrepancy in results from that of Seville orange juice. They suggest that although juice retains the poorly soluble furanocoumarins, most hot-water prepared extracts (including those used in the study) do not.

There are theoretical drug interactions with caffeine and monoamine oxidase inhibitors. Caffeine could increase risk of cardiovascular events when taken with *C. aurantium (54)*. The case report of MI *(45)* and several of the Canadian reported adverse events included caffeine *(46)*. Synephrine, tyramine, and octopamine are all substrates of monoamine oxidase *(55)*. Taking a monoamine oxidase inhibitor with *C. aurantium* could increase concentrations of these sympathomimetics, and thus should be avoided.

9. Reproduction

Effects of *C. aurantium* on reproduction are unknown.

10. Regulatory Status

Citrus oils are found on the list of items on the FDA's Generally Recognized as Safe list. Because of this, manufacturers are only required to include the term, "natural flavoring" on the package label when referring to *C. aurantium*. Canada has issued warnings about synephrine and has not allowed cross-border shipment of one synephrine-containing supplement ("Thermonex") *(47)*.

REFERENCES

1. Webber HJ. History and development of the citrus industry. In: Reuther W, Webber HJ, Batchelor LD, eds., *The Citrus Industry, vol. 1*. Berkeley: University of California Press, 1967, pp. 1–39.
2. Webber HJ. Cultivated varieties of the citrus. In: Webber HJ, Batchelor DL, eds., *The Citrus Industry, vol. 1*. Berkeley: University of California Press, 1943, pp. 475–668.
3. Facciola S, Cornucopia II, eds. *A Source Book of Edible Plants*. Vista: Kampong Publications, 1998.
4. Hosseinimehr SJ, Tavakoli H, Pourheirdari G, Sobhani A, Shafiee A. Radioprotective effects of citrus extract agains gamma-irradiation in mouse bone marrow cells. J Radiat Res 2003;44:237–241.
5. Kiple KF, Ornelas KC, eds. *The Cambridge World History of Food*. Cambridge: Cambridge University Press, Vol. 2, 2000, pp. 1822–1826.
6. Ramadan W, Mourad B, Ibrahim S, Sonbol F. Oil of bitter orange: new topical antifungal agent. Int J Dermatol 1996;35:448–449.
7. Natural Medicines Comprehensive Database online. www.naturaldatabase.com. Date accessed: June 15, 2006.
8. FDA Drug Information. www.fda.gov/cder/drug/infopage/ppa/. Date accessed June 15 2006.
9. Department of Health and Human Services, Food and Drug Administration 2004;1–363. Final rule declaring dietary supplements containing ephedrine alkaloids adulterated because they present an unreasonable risk. www.fda.gov/OHRMS/DOCKETS/98fr/1995n-0304-nfr0001.pdf. Date accessed: June 15, 2006.
10. Shekelle PG, Hardy ML, Morton SC, et al. Efficacy and safety of ephedra and ephedrine for weight loss and athletic performance: a meta-analysis. JAMA 2003;289:1537–1545.
11. Haller CA, Benowitz NL. Adverse cardiovascular and central nervous system events associated with dietary supplements containing ephedra alkaloids. N Engl J Med 2000;343:1833–1838.
12. Bent S, Tiedt TN, Odden MC, Shlipak MG. The relative safety of ephedra compared with other herbal products. Ann Intern Med 2003;138:468–471.
13. Marcus DM, Grollman AP. Ephedra-free is not danger free. Science 2003;301:1667–1671.
14. Preuss HG, DiFerdinando D, Bagchi M, Bagchi D. Citrus aurantium as a thermogenic, weight-reduction replacement for ephedra: an overview. J Med 2002;33(1–4):247–264.
15. Fugh-Berman A, Myers A. Citrus aurantium, an ingredient of dietary supplements marketed for weight loss: current status of clinical and basic research. Exp Biol Med 2004;229:698–704.

16. Bent S, Padula A, Neuhaus J. Safety and efficacy of Citrus aurantium for weight loss. Am J Cardiol 2004;94:1359–1361.
17. United States Patent and Trademark Office. www.uspto.gov/patft. Search terms: citrus and aurantium. Date accessed: June 15, 2006.
18. Pellati F, Benvenuti S, Melegari M, Firenzuoli F. Determination of adrenergic agonists from extracts and herbal products of Citrus aurantium L. var by LC. J Pharm Biomed Anal 2002;29:1113–1119.
19. Pellati F, Benvenuti S, Melegari M. High-performance liquid chromatography methods for the analysis of adrenergic amines and flavanones in Citrus aurantium L. var. amara. Phytochem Anal 2004;15:220–225.
20. Andrea GD, Terrazzino S, Fortin D, Farruggio A, Rinaldi L, Leon A. HPLC electrochemical detection of trace amines in human plasma and platelets and expression of mRNA transcripts of trace amine receptors in circulating leukocytes. Neurosci Lett 2003;346:89–92.
21. Boulton AA. Trace amines: comparative and clinical neurobiology. In: Juorio AV, Downer RGH, eds. *Experimental and Clinical Neuroscience*. Totowa, NJ: Humana, 1988.
22. Pellati F, Benvenuti S, Melegari M. Enantioselective LC analysis of synephrine in natural products on a protein-based chiral stationary phase. J Pharm Biomed Anal 2005;37(5):839–849.
23. Hoffman BB. Catecholamines, sympathomimetic drugs, and adrenergic receptor antagonists. In: Hardman JG, Limbird LE, eds., *Goodman and Gilman's the Pharmacologic Basis for Therapeutics, 10th edition*. New York, McGraw Hill: 2001, pp. 215–268.
24. Williams CM, Couch MW, Thonoor CM, Midgley JM. Isomeric octopamines: their occurrence and functions. J Pharm Pharmacol 1987;39:153–157.
25. Hwa J, Perez DM. The unique nature of the serine interactions for a1-adrenergic receptor agonist binding and activation. J Biol Chem 1996;271(11):6322–6327.
26. Brown CM, McGrath JC, Midgley JM, et al. Activities of octopamine and synephrine stereoisomers on alpha-adrenoceptors. Br J Pharmacol 1988;93:417–429.
27. Airriess CN, Rudling JE, Midgley JM, Evans PD. Selective inhibition of adenylyl cyclase by octopamine via a human cloned alpha 2A-adrenoreceptor. Br J Pharmacol 1997;12(2):191–198.
28. Jordan R, Midgley JM, Thonoor CM, Williams CM. Beta-adrenergic activities of octopamine and synephrine stereoisomers on guinea-pig atria and trachea. J Pharm Pharmacol 1987;39(9):752–754.
29. Liapakis G, Balesteros JA, Papachristou S, Chan WC, Chen X, Javitch JA. The forgotten serine: a critical role for Ser-203 in ligand binding to and activation of the β2-adrenergic receptor. J Biol Chem 2000;275(48):37,779–37,788.
30. Carpene C, Galitzky J, Fontana E, Atgie C, Lafontan MB. Selective activation of β3-adrenoceptors by octopamine: comparative studies in mammalian fat cells. Arch Pharmacol 1999;359(4):310–321.
31. Huang YT, Wang GF, Chen CF, Chen CC, Hong CY, Yang MCM. Fructus aurantii reduced portal pressure in portal hypertensive rats. Life Sci 1995;57:2011–2020.

32. Calapai G, Firenzuoli F, Saitta A, et al. Antiobesity and cardiovascular effects of Citrus aurantium extraces in the rat: a preliminary report. Fitoterapia 1999;70:586–592.
33. Huang YT, Lin HC, Chang YY, Yang YY, Lee SD, Hong CY. Hemodynamic effects of synephrine treatment in portal hypertensive rats. Jpn J Pharmacol 2001;85:183–188.
34. Hofstetter R, Kreuder J, von Bernuth G. The effect of oxedrine on the left ventricle and peripheral vascular resistance. Arzneimittelforschung 1985;12:1844–1846.
35. Penzak SR, Jann MW, Cold A, Hon YY, Desai HD, Gurley BJ. Seville (sour) orange juice: synephrine content and cardiovascular effects in normotensive adults. J Clin Pharmacol 2001;41:1059–1063.
36. Colker CM, Kalman DS, Torina GC, Perlis T, Street C. Effects of Citrus aurantium extract, caffeine, and St. John's wort on body fat loss, lipid levels, and mood states in overweight healthy adults. Curr Ther Res 1999;60:145–153.
37. Kaddu S, Kerl H, Wolf P. Accidental bullous phototoxic reactions to bergamot aromatherapy oil. J Am Acad Dermatol 2001;45(3):458–461.
38. Zaynoun ST, Aftimos BA, Tenekjian KK, Kurban AK. Berloque dermatitis—a continuing cosmetic problem. Contact Dermatitis 1981;7(2):111–116.
39. Chew A, Maibach H. Berloque Dermatitison. eMedicine 2001. www.emedicine.com/derm/topic52.htm. Date accessed: June 15, 2006.
40. Makki S, Treffel P, Humbert P, Agache P. High-performance liquid chromatographic determination of citropten and bergapten in suction blister fluid after solar product application in humans. J Chromatogr 1991;563:407–413.
41. Clark SM, Wilkinson SM. Phototoxic contact dermatitis from 5-methoxypsoralen in aromatherapy oil. Contact Dermatitis 1998;38:289–290.
42. Levine N, Don S, Owens C, Rogers DT, Kligman AM, Forlot P. The effects of bergapten and sunlight on cutaneous pigmentation. Arch Dermatol 1989;125:1225–1230.
43. Ashwood-Smith MJ, Poulton GA, Barker M, Mildenberger M. 5-Methoxypsoralen, an ingredient in several suntan preparations, has lethal mutagenic and clastogenic propertics. Nature 1980;285:407–409.
44. Nykamp DL, Fackih MN, Compton AL. Possible association of acute lateral-wall myocardial infarction and bitter orange supplement. Ann Pharmacother 2004;38:812–816.
45. Firenzuoli F, Gori L, Galapai C. Adverse reaction to an adrenergic herbal extract (Citrus aurantium). Phytomedicine 2005;12(3):247–248.
46. Jordan S, Murty M, Pilon K. Products containing bitter orange or synephrine: suspected cardiovascular adverse reactions. Can Adverse React Newsl 2004;14(4):3–4.
47. Health Canada warns Canadians not to use "Thermonex." Ottawa: Health Canada, 2004. www.hc-sc.gc.ca/english/protection/warnings/2004/2004_30.htm. Date accessed: November 28, 2004.
48. Malhotra S, Bailey DG, Paine MF, Watkins PB. Seville orange juice-felodipine interaction: comparison with dilute grapefruit juice and involvement of furocoumarins. Clin Pharmacol Ther 2001;69:14–23.

49. Di Marco MP, Edwards DJ, Wainer IW, Ducharme MP. The effect of grapefruit juice and Seville orange juice on the pharmacokinetics of dextromethorphan: the role of gut CYP3A4 and p-glycoprotein. Life Sci 2002;71:1149–1160.
50. Malhotra S, Fitzsimmons ME, Bailey DG, Watkins PB. Use of Seville orange juice to "knock out" intestinal CYP3A4. Paper presented at the annual meeting of the American Association of Pharmaceutical Scientists (abstract 2091), New Orleans, LA, November 1999.
51. Hou YC, Hsiu SL, Tsao CW, Wang YH, Chao PD. Acute intoxication of cyclosporine caused by coadministration of decoctions of the fruits of Citrus aurantium and the Pericaps of Citrus grandis. Planta Med 2000;66(7):653–655.
52. Edwards DJ, Fitzsimmons ME, Schuetz EG, et al. 6′7′-Dihydroxybergamottin in grapefruit juice and Seville orange juice: effects on cyclosporine disposition, enterocyte CYP3A4, and P-glycoprotein. Clin Pharmacol Ther 1999;65:237–244.
53. Gurley BJ, Gardner SF, Hubbard MA, et al. In vivo assessment of botanical supplementation on human cytochrome P450 phenotypes: Citrus aurantium, Echinacea purpurea, milk thistle, and saw palmetto. Clin Pharmacol Ther 2004;76(5):428–440.
54. Keogh AM, Baron DW. Sympathomimetic abuse and coronary artery spasm. Br Med J 1985;291:940.
55. Suzuki O, Matsumoto T, Oya M, Katsumata Y. Oxidation of synephrine by type A and type B monoamine oxidase. Experientia 1979;35:1283–1284.

Chapter 16

Vitex agnus-castus

Margaret B. Artz

SUMMARY

Vitex agnus-castus is an herb that has been used for hundreds of years in Europe for female reproductive system disorders, is well-tolerated, and has established efficacy in helping with some symptoms associated with premenstrual syndrome. The major active constituents of *V. agnus-castus* are iridoid glycosides, flavonoids, alkaloids, and essential oils. Its dominant pharmacological effect on the body is inhibition of prolactin secretion. *V. agnus-castus* is available in a variety of dosage forms and its use is gaining popularity in the United States. Although it has a low adverse-effect profile, women should avoid ingesting the herb while trying to become pregnant, during pregnancy, or while nursing.

Key Words: Chasteberry; flavonoids; flavones; essential fatty acids; female reproductive system disorders; estrogenic herbs.

1. HISTORY

Vitex agnus-castus is a botanical plant that has the following National Oceanographic Data Center Taxonomic Code: Kingdom, Plantae; Phylum, Tracheobionta; Class, Magnoliopsida; Order, Lamiales; Family, Verbenaceae; Genus, *Vitex* L.; Species, *Vitex agnus-castus* L. The genus name *Vitex* is a Latin derivation for plaiting or weaving. The species name *agnus-castus* combines two Latin word origins: *"agnus,"* which means lamb, and *"castitas,"* which means chastity *(1)*.

V. agnus-castus is a large deciduous shrub, native to Mediterranean countries and central Asia, and is also used in America as an ornamental plant *(2)*.

From Forensic Science and Medicine:
Herbal Products: Toxicology and Clinical Pharmacology, Second Edition
Edited by: T. S. Tracy and R. L. Kingston © Humana Press Inc., Totowa, NJ

V. agnus-castus has long, finger-shaped leaves and displays fragrant blue-violet flowers in midsummer. Its fruit is a very dark-purple berry that is yellowish inside, resembles a peppercorn, and has an aromatic odor. Upon ripening, the berry is picked and allowed to dry *(1,3,4)*. The twigs of this shrub are very flexible and were used for furniture in ancient times *(2)*.

References to *V. agnus-castus* go back more than 2000 years, describing it as a healing herb *(2)*. Ancient Egyptians, Greeks, and Romans used it for a variety of health problems. In 400 BCE, Hippocrates recommended chaste tree for injuries and inflammation *(1,2)*. Four centuries later, Greek botanist Dioscorides recommended *V. agnus-castus* specifically for inflammation of the womb and lactation *(1,2)*. Use of *V. agnus-castus* continued into the Middle Ages, where folklore persists that medieval monks chewed *V. agnus-castus* tree parts to maintain their celibacy, used the dried berries in their food, or placed the berries in the pockets of their robes in order to reduce sexual desire; thus, the synonym of Monk's pepper *(2,4,5)*. Use of *V. agnus-castus* has persisted to modern times. Though its use was initially concentrated in the Mediterranean area, its popularity has increased in England and America since the mid-1900s *(2)*.

Traditional medicinal uses of *V. agnus-castus* lie predominantly around the oral ingestion of the shrub's fruit *(4–6)*; however, other plant parts such as leaves and flowers have been used in some preparations *(7,8)*. The dry or liquid extract of, or oils from, the berry have been used for a variety of symptoms, most commonly related to the female reproductive system *(5,9,10)*. Other uses include the treatment of hangovers, flatulence, fevers, benign prostatic hyperplasia, nervousness, dementia, rheumatic conditions, colds, dyspepsia, spleen disorders, constipation, and promoting urination *(5,10)*. Traditional topical medicinal uses of *V. agnus-castus* include acne, body inflammation, and insect bites and stings *(5)*. Use of *V. agnus-castus* is not commonly employed in traditional Chinese medicine or traditional Indian medicine (Ayurveda); however, other *Vitex* species (*negundo, trifoliata*) are used in these therapies.

2. CURRENT PROMOTED USES

Current promoted uses of *V. agnus-castus* relate to treatment of disorders of the female reproductive system such as short menstrual cycles, premenstrual syndrome (PMS), and breast swelling and pain (mastodynia/mastalgia). The Commission E has approved the use of *V. agnus-castus* for irregularities of the menstrual cycle, premenstrual complaints, and mastalgia *(10,11)*. Recent randomized, placebo-controlled studies have been conducted and found *V. agnus-castus* to be effective and well-tolerated for the relief of PMS symptoms, especially the physical symptoms of breast tenderness/fullness, edema, and headache *(12–14)*. *V. agnus-castus* is not considered effective for PMS-related symptoms of abdominal bloating, craving sweets, sweating, palpitations, or dizziness *(5)*.

   V. agnus-castus is not used in foods *(9)* and is not recommended for use in children, adolescents, pregnant women *(10,11,15)*, or women who are breastfeeding *(5,10,11,15,16)*. *V. agnus-castus* should be avoided in patients receiving exogenous sex hormones, including oral contraceptives *(16)*, as *V. agnus-castus* may counteract the effectiveness of birth control pills by its effect on prolactin.

3. SOURCES AND CHEMICAL COMPOSITION

   V. agnus-castus (L.), Agnolyt®, arbre chaste, chaste berry, chasteberry, chaste tree, chaste tree fruit, chastetree, chastetree berry, Cloister Pepper, *Fructus Agni Casti*, Fruit de Gattilier, Gattilier, Hemp Tree, Keuschlamm, Mönchspfeffer, Monk's Pepper, *V. agnus castus*, *V. agnus castus fructus*, Vitex.

   The major constituents of *V. agnus-castus* include the following. Flavonoids: flavonol (kaempferol, quercetagetin) derivatives, the major constituent being casticin. Additional flavonoids found include penduletin, orientin, chrysophanol D, and apigenin. Water-soluble flavones: vitexin and isovitexin. Alkaloids: viticin. Diterpenes: rotundifuran (labdane-type); vitexilactone; 6-β,7-β-diacetoxy-13-hydroxy-labda-8,14-diene; 8,13-dihydroxy-14-labden; X-hydroxy-y-keto-15,16-epoxy-13(16),14-labdadien;X-acetonxy- 13-hydroxy-labda-y,14-dien; cleroda-x,14-dien-13-ol; cleroda-x,y,14-trien-13-ol. Iridoid glycosides: In the leaf: 0.3% aucubin, 0.6% agnuside (the *p*-hydroxybenzoyl derivative of aucubin), and 0.07% unidentified glycosides. In the flowering stem (6'-O-foliamenthoylmussaenosidic acid [agnucastoside A], 6'O (6,7-dihydrofoliamenthoyl)mussaenosidic acid [agnucastoside B], and 7-O-trans-*p*-coumaroyl-6'-O-trans-caffeoyl-8-epiloganic acid [agnucastoside C], aucubin, agnuside, mussaenosidic acid, 6'-O-*p*-hydroxybenzoylmussaenosidic acid, and phenylbutanone glucoside [myzodendrone]. Essential oil of leaves and flowers: monoterpenes (major chemicals found: limonene, cineole, sabinene, and α-terpineol, linalool, citronellol, camphene, myrcene) and sesquiterpenes (major chemicals found: β-caryophyllene, β-gurjunene, cuparene, and globulol). Depending on the maturity of the fruits used and the distillation processes, the components of the essential oil can vary greatly. Other constituents: fatty acids (including stearic, oleic, linoleic, and palmitic acids), amino acids (glycine, alanine, valine, leucine), castine (a bitter principle), vitamin C and carotene, and trace amounts of hormones from leaves and flowers (progesterone and 17 α-hydroxyprogesterone). Main components of the volatile oil 0.5%: mixtures of monoterpenes and sesquiterpenes, cineol, and pinene *(4,8–11,17–23)*.

4. PRODUCTS AVAILABLE

V. agnus-castus is available as bulk berries, bulk powder, crushed fresh or dried berry, tea (loose or in tea bags), extract, tonic, elixir, or tincture. Chasteberry products may consist of the herb alone or in combination with other herbs and vitamins. Topically, it is used primarily as the essential oil, mixed in combination with other products in cream form.

A proprietary preparation (Agnolyt) containing an alcoholic extract of *V. agnus-castus* (0.2% w/w) has been available in Germany since the 1950s *(9)*. Other products using *V. agnus-castus* synonyms in their nomenclature are available by a myriad of manufacturers. Some single-entity brand names include Agnofem®, Agno-Sabona®, Agnucaston®, Agnufemil®, Agnuside®, Agnumens®, Agnurell®, Antimast N®, Antimast T®, Gynocastus®, Mastodynon®, and Vitex Extract®. Many combination products also contain *V. agnus-castus* and include, but are not limited to, the following: Herbal Premens®, Herbal Support For Women Over 45®, Dong Quai Complex, Emoton®, Feminine Herbal Complex (FM)®, Menosan®, Mulimen®, Phytoestrin®, Virilis-Gastreu S R41®, Femisana®, Lifesystem Herbal Formula 4 Women's Formula (FM) ®, PMT Complex (FM) ®, Presselin Dysmen Olin 3 N (FM) ®, Women's Formula Herbal Formula 3 (FM), and others *(24)*.

The amount of *V. agnus-castus* contained in oral tablets or capsules varies depending on whether the product contains crushed fruit or extract of the berry. For example, tablet or capsule formulations have included the following: Chaste Berry 450 mg/Chase Berry Extract 50 mg, Chastetree fruit 500 mg, Chaste Berry Dried Extract 1.6–3.0 mg corresponding to 20 mg *Vitex*, Chaste Tree Berry Extract (0.5% agnuside) 225 mg. Commercial extract forms of chasteberry are usually standardized to contain 6% agnuside constituent *(5,10)*. Chasteberry liquid extract may or may not contain alcohol.

4.1. Dosage

Generally, the Expanded Commission E Monographs reports the following total daily dosages:

- 30 to 40 mg of dry or fluid extracts of crushed fruit
- 0.03 to 0.04 mL of fluid extract 1:1 (g/mL), 50–70% alcohol (v/v)

- 0.15 to 0.2 mL of tincture 1:5 (g/mL), 50–70% alcohol (v/v)
- 2.6 to 4.2 mg of dry native extract (9.5–11.5:1 (w/w)

Other references have reported total daily dosages ranging from 20 to 1800 mg/day of crude *V. agnus-castus* extracts *(5,10)*. *V. agnus-castus* is considered safe when used orally and appropriately *(5)*.

5. PHARMACOLOGICAL/TOXICOLOGICAL EFFECTS

5.1. Prolactin Secretion

Evidence of varying levels of discrimination exists that demonstrate *V. agnus-castus* inhibits the secretion of prolactin by the pituitary gland. In a randomized, placebo-controlled, double-blind study, Milewicz et al. examined whether *V. agnus-castus* affected elevated pituitary prolactin reserve. Participants were 52 women with luteal phase defects caused by latent hyperprolactinemia. Intervention was *V. agnus-castus* 20 mg daily or placebo, for 3 months. Only 37 women (20 = placebo, 17 = *V. agnus-castus*) completed the study. Outcome measures were pre- and posthormonal analysis (blood draws taken on days 5–8 and day 20 of menstrual cycle) 1 month prior to treatment and after 3 months of treatment and latent hyperprolactinemia analysis (monitoring prolactin release 15 and 30 minutes after intravenous injection of 200 µg thyrotropin-releasing hormone (TRH). Results from this study showed that compared to preintervention, the *V. agnus-castus* group had statistically significant reduced prolactin release after 3 months, whereas the control group did not. The study's information came from an English abstract of a German publication. Information about inclusion/exclusion criteria, study specifics, study funding, or author disclosures were not available *(25)*.

In an open and intraindividual comparison study, Merz et al. conducted a clinical study of tolerance and prolactin secretion of *V. agnus-castus* using 20 healthy male subjects between the ages of 18 and 40 years. Placebo and three doses of *V. agnus-castus* (total daily dosages of 120, 240, and 480 mg were divided into 8-hour administration times) were given in an increasing sequence. Prolactin concentration profile after TRH stimulation was assessed by determining the maximum concentrations (C_{max}) and the area under the curve over a period of one hour ($AUC_{0–1h}$). These procedures were identical in all four study phases. Results for the $AUC_{0–24h}$ showed that as daily doses increased, prolactin levels decreased ([$AUC_{0–24h}$. {µIU · hour}/µΛ ± standard deviation]; placebo: 6182 ± 1827; 120 mg: 6874 ± 1790; 240 mg: 5750 ± 1594; 480 mg: 5998 ± 1664), with statistically significant findings for the 120-mg dosage only *(26)*.

Wuttke reported in a 1996 abstract the results of experiments demonstrating that 3 months of *V. agnus-castus* therapy (double-blind clinical study vs placebo) significantly reduced basal prolactin levels in patients. However, details of the experiments and study subjects were not outlined or referenced *(17)*.

5.2. Follicle-Stimulating Hormone, Luteinizing Hormone

There are a limited number of human studies regarding how *V. agnus-castus* directly effects luteinizing hormone (LH) or follicle-stimulating hormone (FSH) *(26,27)*. In a 1994 case report by Cahill et al., a 32-year-old woman undergoing unstimulated in vitro fertilization (IVF) treatment took *V. agnus-castus* for one cycle without consulting her physician. During this cycle, she had symptoms of mild ovarian hyperstimulation in the luteal phase. Her FSH and LH levels prior to day 13, the predicted day of LH surge in the IVF cycle, were reviewed and found to be much higher than normal. Reviewing five other cycles of this patient and finding normal pituitary gonadotrophin file and normal follicular ovarian responses, the authors suggest that *V. agnus-castus* was the causative agent *(27)*.

In the 1996 study by Merz et al. that primarily examined prolactin secretion in male subjects, initial hormone levels of FSH and LH were measured on days 1 and 13 (beginning and near-end of placebo phase) and from blood samples taken during the prolactin secretion profiling. The authors state that *V. agnus-castus* had no effect on FSH or LH levels, but no other details were provided *(26,28,29)*.

5.3. Progesterone/Testosterone Synthesis

In a randomized, placebo-controlled, double-blind study, Milewicz et al. examined the effect of *V. agnus-castus* on prolactin reserve and luteal phase progesterone synthesis in 52 women. The intervention was *V. agnus-astus* 20 mg daily or placebo, for 3 months. Results from the 37 women (20 = placebo, 17 = *agnus-castus*) who completed the study showed that compared to preintervention, the *V. agnus-castus group* had statistically significant increases in luteal phase progesterone synthesis. The study information came from a German publication with an English abstract. Information about inclusion/exclusion criteria, study specifics, study funding, or author disclosures were not available *(25)*.

In the 1996 study by Merz et al. that primarily examined prolactin secretion in male subjects, initial hormone level of testosterone was measured on days 1 and 13 (beginning and near-end of placebo phase), and from blood samples taken during the prolactin secretion profiling. The authors state that *V. agnus-castus* had no effect on testosterone levels, but no other details were provided *(26)*.

5.4. Infertility

Gerhard et al. studied the influence of a commercially available preparation of *V. agnus-castus* on infertility. Using a randomized, placebo-controlled, double-blind design, 96 women with fertility disorders (31 with luteal insufficiency; 38 with secondary amenorrhea; 27 with idiopathic infertility) received either *V. agnus-castus* or placebo twice a day for 3 months. The dose of V. *agnus-castus* was 30 drops of Mastodynon® twice a day (*agnus-castus* or casticin-standardization not mentioned). The outcome measures were: (1) pregnancy or spontaneous menstruation for women with secondary amenorrhea, and (2) pregnancy or improved luteal hormone levels in women with luteal insufficiency or idiopathic infertility. A total of 66 women were suitable for evaluation. No differences were noted between the placebo and V. *agnus-castus* groups with respect to effect. Information presented here came from the English abstract of the study, which was published in German *(30)*.

5.5. PMS and Menopausal Symptoms

In a 1997 multicenter, randomized, double-blind, controlled trial, Lauritzen et al. examined the efficacy and tolerability of a commercially available capsule formulation of *V. agnus-castus* (Agnolyt) compared with pyridoxine in women with PMS. Inclusion criteria were females aged 18 to 45 years, PMS symptoms in luteal phase of menstrual cycle, PMS symptoms with each cycle, PMS symptoms affecting quality of life, and no drug therapy for PMS in 3 months preceding the study. Of 175 participants, 85 were in the *V. agnus-castus* group (took one capsule twice a day, with one capsule containing 3.5 to 4.2 mg of *V. agnus-castus*, the second capsule containing placebo), and 90 were in the pyridoxine group (days 1–15, took one capsule twice a day, each capsule containing placebo; days 16–35, took one capsule twice a day, each capsule containing 100 mg of pyridoxine). Women in both treatment groups had equal reductions in PMS scores (*V. agnus-castus*: 15.2 to 5.1; pyridoxine: 11.9 to 5.1; $p = 0.37$), suggesting no differences in effect *(12)*.

In 2000, Loch et al. conducted an open label, uncontrolled study examining the efficacy and safety of a new oral *V. agnus-castus* treatment (Femicur®) for PMS complaints. Suffering from PMS was the only inclusion criterion and pregnancy was the only exclusion criterion. A questionnaire on mental and somatic PMS symptoms was completed by 857 gynecologists after interviewing 1634 females at the start of Femicur therapy (20 mg daily), and after a period of three menstrual cycles under therapy. Physicians reported that 42% of women reported that they had no more PMS symptoms, 51% showed a decrease in symptoms ($p < 0.001$), and 1% had an increase in number of symptoms. After 3 months of treatment, both psychic and somatic complaints were dramatically lowered. Although 30% of the women still complained about mastodynia after *V. agnus-castus* treatment, most reported complaints of lower intensity. Physicians described the patients' tolerance of this *V. agnus-castus* product as good or very good in 94% of women. Although one of the authors works for the pharmaceutical company that makes Femicur, the article did not contain funding disclosure statements *(6)*.

In 2000, Berger et al., using a prospective, multicenter trial design, examined the efficacy of an oral, casticin-standardized *V. agnus-castus* therapy on 43 women diagnosed with PMS. Treatment phases consisted of baseline (two cycles, pretreatment), treatment (three cycles), posttreatment (three cycles, no treatment). The dose was 20 mg, but no placebo control was included. At the end of the trial, Moos' menstrual distress questionnaire (MMDQ) scores were reduced by 43% compared with start (statistically significant, $p < 0.001$), but that improvement decreased gradually in the posttreatment phase. At the end of the posttreatment phase, patients had improved compared to the start of therapy ($p < 0.001$) and for up to three cycles thereafter. A group of 20 women had baseline MMDQ scores that were reduced by at least 50% at the end of treatment phase. Visual analogue scale (VAS) and global efficacy scales showed similar findings ($p < 0.001$ and global efficacy rated excellent by 38 women). Areas of improvement included symptoms related to pain, behavior, negativity, and fluid retention. The most frequent adverse events were acne, headaches, and menstrual spotting *(31)*.

In 2001, using a prospective, randomized, double-blind, placebo-controlled, parallel-group comparison design, Schellenberg studied the efficacy and tolerability of *V. agnus-castus* extract on PMS *(13)*. Participants were female outpatients, 18 years of age or older, of six general medicine clinics and had a PMS diagnosis according to the *Diagnostic and Statistical Manual of Mental Disorders, third edition, revised*. Dose of *V. agnus-castus* was 20 mg daily for 3 months. Results showed that the group receiving *V. agnus-castus* had significant improvements ($p < 0.001$) in all symptoms except bloating compared to the placebo group. Sensitivity analyses removing women taking contraceptives did not alter results. Tolerability was good with acne, itching, and mid-cycle bleeding as the adverse events noted *(13)*.

An uncontrolled study in 2002 by Lucks examined the effects of 3-month dermal application of *V. agnus-castus* essential oil (oil distilled at some point in the shrubs' development of the fruit but while some leaves were still on the plant) on menopausal and perimenopausal symptoms. A 1.5% solution of the essential oil was incorporated in a bland cream or lotion and applied once a day, 5 to 7 days/week, for 3 months. Descriptive outcome measures were self-report via a survey of symptomatic relief (major, moderate, mild, none, worse) and side effects. A total of 33% of women reported major improvement in symptoms, with the most often area of improvement being hot flashes/night sweats. Both improvement and worsening occurred in the areas of emotions and menstruation flow. Subjects who were also on progesterone supplementation reported breakthrough bleeding *(7)*.

5.6. Mastodynia

In a 1987, Kubista et al. reported results from a placebo-controlled study comparing the effects of lynestrenol, *V. agnus castus* (Mastodynon), and placebo therapy in women with severe mastopathy with cyclic mastalgia. More women in the lynestrenol and *V. agnus-castus* groups than placebo group reported good relief of PMS symptoms (82, 54, 37%, respectively). Information presented here came from the English abstract of the study, which was published in French *(32)*.

In 1998, Halaska and colleagues examined the tolerability and efficacy of *V. agnus-castus* extract on mastodynia/mastalgia (breast pain, breast tenderness). The study was a double-blind, placebo controlled, parallel-group (50 women each) design. Length of treatment (*V. agnus-castus* [60 drops daily dose] or placebo) was 3 months. Efficacy was determined using a VAS. Results of study showed that the intensity of mastodynia diminished more quickly in the *V. agnus-castus* group with low incidence of side effects (statistical significance not stated). Information presented here came from the English abstract of the study, which was published in Czech *(33)*.

5.7. Luteal Phase Length

In the 1993 randomized, placebo-controlled, double-blind study where Milewicz et al. examined the effect of *V. agnus-castus* on pituitary prolactin reserve, they also examined luteal phase length. Of the 37 women (20 = placebo, 17 = *V. agnus-castus*) who completed the study (1 month prior to, and 3 months treatment), the shortened luteal phases of the *agnus-castus* group became normal. The study information came from a German publication with an English abstract. Information about inclusion/exclusion criteria and study specifics were not available *(25)*.

5.8. Premenstrual Dysphoric Disorder

Premenstrual dysphoric disorder (PMDD) is characterized by markedly depressed mood, anxiety, affective lability, and decreased interest in daily activities during the last week of luteal phase in menstrual cycles of the last year. In 2002, Atmaca et al. conducted an 8-week, randomized, single-blind, rater-blinded, prospective- and parallel-group, flexible-dosing trial to compare the efficacy of fluoxetine with *V. agnus-castus* for the treatment of PMDD in 42 females. Both fluoxetine and *V. agnus-castus* had dose ranges from 20 to 40 mg. Outcome measures included the Penn daily symptom report, Hamilton depression rating scale, clinical global impression (CGI)-severity of illness scale, and CGI-improvement scale. Both drugs were well tolerated. No statistically significant differences between groups were found. The authors concluded that fluoxetine was more effective (a decrease of more than 50% in rating symptoms) for psychological symptoms (depression, irritability, insomnia, nervousness), whereas *V. agnus-castus* helped with physical symptoms (irritability, breast tenderness, swelling, cramps). Lack of placebo-control and short duration of treatment were significant limitations *(14)*.

5.9. Toxicological Effects

No systematic toxicological studies have been conducted, according to the Expanded Commission E Mongraphs *(10)*.

6. PHARMACOKINETICS

Pharmacokinetic and toxicokinetic information for *V. agnus-castus* is not available.

7. ADVERSE EFFECTS AND TOXICITY

Throughout years of use, *V. agnus-castus* has shown only mild adverse effects. Pruritus, rash (unspecified), urticaria, increased menstrual blood flow, persistent headaches, and gastrointestinal discomfort have been reported *(18,29)*. Few adverse events related to chasteberry have been reported to the Food and Drug Administration *(24)*.

7.1. Case Reports of Toxicity Caused By Commercially Available Products

An extensive search of all standard references, as well as reports of studies in humans, shows that there have been no case reports of toxic exposure to *V. agnus-castus* use. However, one case of nocturnal seizures, possibly attributed to *V. agnus-castus,* has been reported. The patient was taking concomitantly black cohosh root (*Cimicifuga racemosa*), *V. agnus-castus*, and evening primrose as well *(34)*. Thus, attribution of effect to a specific agent was not possible.

In animals, an adverse influence on nursing (lactation) performance has been observed; *V. agnus-castus* could potentially interfere with proper lactation *(15,16)*.

8. INTERACTIONS

According to the German Commission E Monographs, drug interactions with *V. agnus-castus* are unknown *(10,11)*. However, with animal experiments showing evidence of a "dopaminergic effect," it is generally recommended that the effect of *V. agnus-castus* can be diminished in cases when there is concurrent ingestion of dopamine-receptor antagonists (e.g., haloperidol). Similarly, because *V. agnus-castus* inhibits the secretion of prolactin via a dopamine-agonist action, drug interactions may occur with the D_2 family of dopamine-receptor agonists (bromocriptine, pergolide, pramipexole, ropinirole, cabergoline) *(5,16)*.

Some liquid formulations contain large percentages of alcohol ($\geq 50\%$ vol); health risks from ethanol may exist, and in certain populations, use of the dried extract formulations instead would be advisable.

9. REPRODUCTION

Although there are no known case reports of toxicity in human reproduction, because of possible endocrine effects, *V. agnus-castus* could disrupt fetal development or proper gestation. A case of ovarian hyperstimulation syndrome and multiple follicular development resulting in no pregnancy occurred in a woman who took *V. agnus-castus* prior to one of her IVF protocol cycles. Tests showed that her serum gonadotropin and hormone evels were out of the desired range *(27)*.

10. REGULATORY STATUS

V. agnus-castus is available for use without a prescription in all Member States of the European Union and the United States. In the United States, *V. agnus-castus* is categorized as a dietary supplement. Throughout the world, *V. agnus-castus* is available through pharmacies, health-food shops, mail order companies, supermarkets, and department stores.

REFERENCES

1. Brown DJ. Herbal research review: vitex agnus castus. Clinical monograph. Q Rev Nat Med 1994;2:111–121.
2. Hobbs C. The chaste tree: *Vitex agnus castus*. Pharm Hist 1991;33:19–24.
3. Foster S. *A Field Guide to Western Medicinal Plants and Herbs*. Boston: Houghton Mifflin Co., 2002.
4. Du Mee C. Medicinal plant review: vitex agnus castus. Aust J Med Herbalism 1993;5:63–65.
5. Jellin JM, Gregory, P, eds. Natural medicines comprehensive database. Stockton: Therapeutic Research Faculty, 2003. www.naturaldatabase.com (Last updated, December 13, 2003). Date accessed on June 16, 2006.
6. Loch EG, Selle H, Boblitz N. Treatment of premenstrual syndrome with a phytopharmaceutical formulation containing Vitex agnus castus. J Women's Health Gend Based Med 2000;9:315–320.
7. Lucks BC. Vitex agnus castus essential oil and menopausal balance: a research update. Complement Ther Nurs Midwifery 2003;9:157–160.
8. Males Z, Blazevic N, Antoli A. The essential oil composition of *Vitex agnus-castus* f. *rosea* leaves and flowers. Planta Med 1998;64(3):286–287.
9. Barnes J, Anderson LA, Phillipson JD, eds. *Herbal Medicines, second edition*. London: Pharmaceutical Press, 2002.
10. Blumenthal M, Goldberg A, Brinckmann, eds. Chaste Tree Fruit. In: *Herbal medicine, expanded Commission E monographs*. Austin: American Botanical Council; Boston: Integrative Medicine Communications, 2000.
11. Blumenthal M, Busse WR, Goldberg A, et al., eds. *The Complete German Commission E Monographs. Therapeutic Guide to Herbal Medicines*. Austin: American Botanical Council; Boston: Integrative Medicine Communications, 1998.
12. Lauritzen C, Reuter HD, Repges R, Böhnert KJ, Schmidt U. Treatment of premenstrual tension syndrome with *Vitex agnus castus*. Controlled, double-blind study versus pyridoxine. Phytomedicine 1997;4:183–189.
13. Schellenberg R. Treatment for the premenstrual syndrome with agnus castus fruit extract: prospective, randomised, placebo controlled study. BMJ 2001;322:134–137.
14. Atmaca M, Kumru S, Tezcan E. Fluoxetine versus *V. agnus-castus* extract in the treatment of premenstrual dysphoric disorder. Hum Psychopharmacol Clin Exp 2003;18:191–195.
15. Robbers JE, Tyler VE, eds. *Tyler's Herbs of Choice: The Therapeutic Use of Phytomedicinals*. Binghamton: Haworth Herbal Press, Inc., 1999.

16. Drug monograph: chastebeerry, chaste tree fruit, villexagnus-castus. AccessMedicine, copyright © 2006. The McGraw-Hill companies. Last revised: July 31, 2001.
17. Wuttke W. Dopaminergic action of extracts of Agnus Castus. Forsch Komplementarmed 1996;3:309–330.
18. Houghton PJ. Agnus castus. Pharm J 1994; 253:720–721.
19. Jarry H, Spengler B, Porzel A, Schmidt J, Wuttke W, Christoffel V. Evidence for estrogen receptor beta-selective activity of Vitex agnus-castus and isolated flavones. Planta Med 2003;69:945–947.
20. Sørensen JM, Katsiotis ST. Parameters influencing the yield and composition of the essential oil from Cretan Vitex agnus-castus fruits. Planta Med 2000;66:245–250.
21. Kuruuzum-Uz A, Stroch K, Demirezer LO, Zeeck A. Glucosides from Vitex agnus-castus. Phytochemistry 2003;63:959–964.
22. Liu J, Burdette JE, Sun Y, et al. Isolation of linoleic acid as an estrogenic compound from the fruits of Vitex agnus-castus L. (chaste-berry). Phytomedicine 2004;11:18–23.
23. Saden-Krehula M, Kutrak D, Blazevic N. Δ^4-3-ketosteroids in flowers and leaves of Vitex agnus-castus. Acta Pharm Jugosl 1991;41:237–241.
24. MARTINDALE PRODUCT INDEX® (intranet database). Version 5.1. Greenwood Village, CO, Thomson Micromedex.
25. Milewicz A, Gejdel E, Swøren H, et al. Vitex agnus castus-Extrakt zur Behandlung von Regeltempoanomalien infolge latenter Hyperprolaktinämie [Vitex agnus castus extract in the treatment of luteal phase defects due to latent hyperprolactinaemia. Results of a randomized placebo-controlled double blind study.] Drug Res 1993;43:752–756.
26. Merz PG, Gorkow C, Schrödter A, et al. The effects of a special Agnus castus extract (BP1095E1) on prolactin secretion in healthy male subjects. Exp Clin Endocrinol Diabetes 1996;104:447–453.
27. Cahill DJ, Fox R, Wardle P Harlow CR. Multiple follicular development associated with herbal medicine. Hum Reprod 1994;9:1469–1470.
28. Girman A, Lee R, Kligler B. An integrative medicine approach to premenstrual syndrome. Clin J Women's Health 2002; 2:116–127. Reprinted in Am J Obstet Gynecol 2003;188:S56–S65.
29. Israel D, Youngkin EQ. Herbal therapies for perimenopausal and menopausal complaints. Pharmacotherapy 1997;17:970–984.
30. Gerhard I, Patek A, Monga B, Blank A, Gorkow C. Mastodynon® bei weiblicher Sterilität. Randomisierte, plazebokontrollierte, klinische Doppelblindstudie. [Mastodynon® for female infertility. Randomized, placebo-controlled, clinical double-blind study. Forsch Komplementärmed 1998;5:272–278.
31. Berger D, Schaffner W, Schrader E, Meier B, Brattström A. Efficacy of Vitex agnus castus L. extract Ze 440 in patients with premenstrual syndrome (PMS). Arch Gynecol Obstet 2000;264:150–153.
32. Kubista E, Müller G, Spona J. Traitement de la mastopathie avec mastodynie cycliqu. Resultats cliniques et profiles hormonaux. [Treatment of mastopathy with mastalgia. Clinical results and hormonal profiles]. Rev Fr Gynécol Obstét 1987;82:221–227.

33. Halaska M, Raus K, Beles P, Martan A, Paithner KG. Lecba cyklicke mastodynie pomoci roztoku s extraktem z Vitex agnus castus: vysledky dvojite slepe studie ve srovnani s placebem [Treatment of cyclical mastodynia using an extract of Vitex agnus castus: results of a double-blind comparison with a placebo]. Ceska Gynekologie 1998;63:388–392.
34. Shuster J. ISMP Adverse drug reactions. Heparin and thrombocytopenia. Black cohosh root? Chasteberry tree? Seizures! Hospital Pharmacy 1996;31:1553–1554.

Chapter 17

Bilberry

Timothy S. Tracy

SUMMARY

Although bilberry has been used for a variety of conditions, it has only been shown to be moderately effective in the treatment of retinopathy. No other clinical studies have demonstrated bilberry's effectiveness for any other conditions. Fortunately, adverse effects from ingesting bilberry are minimal.

Key Words: *Vaccimium myritillus*; glycosamino glycans; cancer prevention; vascular permeability; visual acuity.

1. HISTORY

Vaccinium myrtillus L. (Ericaceae) is a shrub found in the mountains of Europe and North America *(1)*. It is related to the North America's blueberry and huckleberry *(2)*. The shrub produces a blue-black or purple berry with purple meat from July through September, depending on the elevation *(3)*. This berry is the part of the plant of interest. In addition to its use as a food, it was documented as being used to treat kidney stones, biliary problems, scurvy, coughs, and tuberculosis in the 1500s *(3)*. It has also been used to make a traditional tea to treat diabetes, and purportedly has a hypoglycemic effect *(1)*. Little is known about bilberry's active constituents and their pharmacology *(1)*, although it has been studied since at least 1964 for ophthalmological and vascular disorders *(4)*. Most of these studies were performed in Europe, and many are published in non-English or obscure journals. Stories of British Royal Air Force pilots eating bilberry jam during World War II to improve their night vision may have prompted some of these studies *(2)*.

From Forensic Science and Medicine:
Herbal Products: Toxicology and Clinical Pharmacology, Second Edition
Edited by: T. S. Tracy and R. L. Kingston © Humana Press Inc., Totowa, NJ

2. CURRENT PROMOTED USES

Orally, bilberry is used for improving visual acuity including night vision, degenerative retinal conditions, varicose veins, atherosclerosis, venous insufficiency, chronic fatigue syndrome (CFS), and hemorrhoids. It is also used orally for angina, diabetes, arthritis, gout, dermatitis, and prevention and treatment of gastrointestinal (GI), kidney, and urinary tract symptoms and diseases. Topically, it is used for mild inflammation of the mouth and throat mucous membranes.

3. SOURCES AND CHEMICAL COMPOSITION

Tegens® (Inverni della Beffa, Milan) is a standardized Italian bilberry product that contains Myrtocyan® (*V. myrtillus* L., fresh fruits, 25% anthocyanidins and 35% anthocyanosides) *(5,6)*.

4. PRODUCTS AVAILABLE

In the United States, bilberry is usually sold in capsule form as an antioxidant and to promote eye health. It is sometimes combined with other vitamins or herbs purported to be beneficial to the eye, such as lutein or eyebright.

5. PHARMACOLOGICAL/TOXICOLOGICAL EFFECTS

Bilberry's ability to stimulate synthesis of connective tissue glycosaminoglycans may be the mechanism underlying its beneficial effects in several pathologies. Its gastroprotective, vasoprotective, and healing properties may all be tied to this action *(5)*.

5.1. Antiulcer Activity

A bilberry extract containing 25% anthocyanidins (Myrtocyan) demonstrated antiulcer activity in several rat models *(5)*. Efficacy was measured using an "index of ulceration" and means were compared using the Mann-Whitney U test. Minimum doses producing statistically significant benefit were 100 mg/kg for ulcers induced by pyloric ligation, 25 mg/kg for reserpine ulcer, and 100 mg/kg for phenylbutazone ulcers ($p < 0.01$ compared to control). For acetic acid ulcers, efficacy was determined by measuring the ulcer surface area. The minimum effective dose was 50 mg/kg ($p < 0.01$ compared to control, Dunnett *t*-test). For restraint ulcer, efficacy was determined by comparing the number of ulcers in the control vs bilberry groups. The minimum effective dose was 100 mg/kg ($p < 0.01$ compared with control, Mann-Whitney U-test). Effects on volume or pH of gastric secretion were ruled out as a mechanism of action. Histological examination of the gastric mucosa showed increased mucus production in treated rats. Bilberry's beneficial effects were attributed to an increase in production of mucopolysaccharides.

Based on the promising results of this study, the antiulcer activity of the individual anthocyanidins was studied. One anthocyanidin, 3,5,7-trihydroxy-2-(3,4-dihydroxyphenyl)-1-benzopyrylium chloride (IdB 1027, Inverni della Beffa S.p.A., Milan) showed particular promise, and so it was produced synthetically so that its effects on several animal models of acute and chronic stomach ulcers could be studied further *(4)*. IdB 1027 administrered orally or intraperitoneally was able to inhibit acute gastric ulceration induced by pyloric ligation, stress (cold plus restraint), phenylbutazone, indomethacin, reserpine, ethanol, and histamine, as well as duodenal ulceration induced by cysteamine and chronic gastric ulcers induced by acetic acid. Results of additional experiments suggest the mechanism of action involves stimulation of protective gastric mucosal secretions. A drawback of this study is that the severity of ulceration caused by phenylbutazone, indomethacin, ethanol, and histamine was assessed using nonvalidated ordinal scales.

5.2. Vascular Permeability

In a rat model of hypertension, animals pretreated with *V. myrtillus* dry extract (Merck-Sharp and Dohme, Chibret, France) "rich in anthocyanin glucosides" at a dose of 50 mg/100 g body weight for 12 days prior to aortic ligation and for 14 days thereafter showed less permeability of the aorta, blood-brain barrier, and blood vessels of the skin to a tracer dye (1% tryptan blue solution), compared to untreated animals on day 7 after ligature *(7)*. The effect was most pronounced in the brain and least pronounced in the blood vessels of the skin. The investigators proposed, based on previous experiments, that anthocyanins in bilberry extract interact with collagen to make it more resistant to the effects of collagenase, thus preserving the integrity of the basal lamina and its control of vascular permeability.

In a subsequent experiment *(5)*, intraperitoneal injection of Myrtocyan ameliorated histamine-induced capillary permeability measured by permeability to Evans blue. The minimum effective dose was 50 mg/kg ($p < 0.01$ compared to control, Dunnett *t*-test). At a dose of 200 mg/kg Myrtocyan was effective in improving capillary resistance to vacuum-induced petechiae in rats fed a flavonoid deficient diet ($p < 0.01$ compared to baseline, Dunnett *t*-test).

5.3. Antiangiogenic Effects

Billberry extract was able to inhibit vascular endothelial growth factor expression by human keratinocytes in vitro *(8)*. This suggests that bilberry or its constituents may have a role in cancer prevention or treatment.

5.4. Antilipemic and Hypoglycemic Effects

Because bilberry has traditionally been used to treat diabetes, which is associated with alteration of lipid metabolism, the effects of a bilberry leaf extract on plasma glucose and triglycerides were studied in various rat models *(1)*. The study preparation was made by percolation of bilberry leaf powder with ethanol 40% (Indena, Milano, Italy). All statistical comparisons were done using unpaired *t*-tests.

In Sprague-Dawley rats made diabetic with streptozocin, plasma glucose levels were 26% lower ($p < 0.05$) 4 days after streptozocin administration, and 26.6% lower ($p < 0.01$, unpaired *t*-test) 3 weeks after streptozocin administration in rats treated with bilberry extract at a dose of 3 g/kg twice daily for five doses compared with untreated diabetic rats. Triglycerides were also 38.8% ($p < 0.05$) lower in the treated rats. The extract did not affect glucose levels in control animals, and did not affect weight *(1)*.

In rats with diet-induced hyperlipidemia, triglycerides were lower in rats treated with the extract at doses of 1.2 g/kg ($p < 0.05$) and 3 g/kg ($p < 0.01$) than in untreated rats. Weight was not affected. In genetically hyperlipidemic Yoshida rats, and in rats with alcohol-induced hypertriglyceridemia, triglycerides were 31.8% ($p < 0.05$) and 61.5% ($p < 0.01$) lower, respectively, in bilberry-treated rats than in untreated rats *(1)*.

The antilipemic effect of bilberry may be caused by improved breakdown of triglyceride-rich lipoproteins, as evidenced by findings of a third part of this study *(1)*. Rats were administrered Triton WR-1339 to induce hypertriglyceridemia. This agent has an acute effect of blocking lipoprotein clearance, and a secondary effect of stimulation of liver lipoprotein synthesis. Bilberrry was able to attenuate only the acute effect of Triton on triglycerides, suggesting that bilberry improves lipoprotein clearance, but does not affect lipoprotein production.

Although bilberry's effect on triglycerides is similar to that of the fibric acid derivatives (e.g., gemfibrozil, fenofibrate) used therapeutically to treat hypertriglyceridemia, bilberry did not affect thrombus size or composition, suggesting that it does not possess antithrombotic activity, as has been demonstrated with some fibric acid derivatives *(1)*.

Oxidized low-density liporpotein (LDL) is known for its ability to stimulate inflammatory processes involved in the formation of atherosclerotic plaques. For this reason, there has been interest in the use of antioxidants such as bilberry to protect against LDL oxidation.

In an in vitro study *(9)*, the ability of a bilberry extract (Leurquin-Mediolanum Laboratories, Neuilly sur Marne, France) containing 74.2 ± 4.9 mg/g polyphenols with $17.3 \pm 3.3\%$ catechin (mean ± standard deviation [SD] of 10 preparations) to attenuate copper-mediated LDL oxidation was studied. LDL was taken from six volunteers with normal lipid levels, and markers of oxidation were measured in the presence of bilberry extract at concentrations of 0, 5, 10, 15, 20, and 30 µg/mL. All comparisons were made using the Mann-Whitney test. Compared with control, bilberry was able to prolong the time to conjugated diene formation at concentrations of 20 and 30 µg/mL ($p < 0.01$), decrease production of lipoperoxides and malondialdehyde at concentrations of 10 µg/mL or higher for at least 1 hour and 0.5 hour ($p < 0.05$), respectively, and attenuate change in the net negative charge of LDL at concentrations of at least 15 µg/mL ($p < 0.01$).

5.5. Ocular Effects

Bilberry was reported to have beneficial effects on retinal vascular permeability and tendency to hemorrhage in 31 patients with retinopathy in a German study *(10)*. Benefits were particularly pronounced in patients with diabetic retinopathy, according to the abstract, which was the only part of the study published in English. A study published in Italian *(11)* showed ophthalmoscopic improvement in 11 and angiographic improvement in 12 of 14 patients with retinopathy caused by diabetes and/or hypertension, according to a review article *(3)*.

Bilberry jam purportedly improved night vision in Royal Air Force pilots within 24 hours of eating bilberry jam, and at least five European studies showing the beneficial effect of bilberry on night vision were published prior to 1970 *(2,3)*. A 1997 Israeli study published as an abstract *(12)* found negative results, as did a more recent study performed in 15 Navy Seals. In this trial, Muth and colleagues *(2)* studied the effect of bilberry extract (25% anthocyanocides) 160 mg taken three times daily for 3 weeks on night visual acuity and night contrast sensitivity in subjects with visual acuity correctable to at least 20/20. An independent laboratory verified the composition of the extract used. Eight subjects were given placebo and seven were given the extract in double-blind fashion. After a 30-day washout, the subjects were crossed over the alternate treatment arm. Nighttime visual acuity and contrast sensitivity were measured under lighting conditions simulating full moon-light (i.e., a luminescence of 0.005 candelas/m^2). To measure visual acuity, subjects were presented with Landolt C targets (computer-generated black Cs on a white background) with the opening of the C facing one of eight directions. Each subject was given five tries to correctly identify the direction of the C. If the subject was correct three out of five times, the subject was presented with another five targets of smaller size. Three incorrect responses ended the test. Contrast sensitivity was measured in the same fashion, except that instead of decreasing the size of the targets, the contrast between the target and the background was decreased. Baseline visual acuity and contrast sensitivity were measured three times during the first week of the study; 24–36 hours after beginning treatment, 4–6 days after beginning treatment; once between days 12–14; and once between day 19–21. Testing was performed once each week during the 4-week washout. After the second treatment phase, measurements were again taken weekly for 4 weeks. Repeated measures analysis of variance was used to compare the median of the three pretreatment visual acuity and contrast sensitivity measurements to the mean of those obtained during each of the two treatment periods, and to the last measurement taken in each of the two treatment periods. Subjects were also placed into one of four categories depending on whether they showed improvement with both treatments, neither treatment, placebo only, or bilberry only. These results were compared using McNemar's test. No difference between bilberry and placebo was detected. The investigators describe a previous French study (13) in which improvement in five of 14 subjects with poor pretreatment night vision was noted. A larger sample size or use of subjects with poor night vision at baseline may have yielded more promising results.

In a subsequent review (3), the results of two other early studies published in French (14,15) are described. These studies showed that bilberry improved night visual acuity, adaptation to darkness, and recovery of visual acuity after glare. Other articles published during the late 1960s in Italian and German showed beneficial effects of bilberry on retinitis pigmentosa (16) and quinine-induced hemeralopia (17), according to this review. The review also mentions a study published in an Italian journal (18) that purportedly showed that a single dose of bilberry anthocyanosides 200 mg improved electroretinographic findings in eight patients with glaucoma, purportedly by stabilizing the collagen of the trabecular network, thus improving aqueous humor outflow.

A review by Head (19) describes a study published in Italian (20) in which bilberry extract (25% anthocyanosides) 180 mg and d,l-tocopheryl acetate 100 mg twice daily for 12 weeks stabilized cataract growth in 96% of 25 treated patients vs 76% of 25 controls ($n = 50$).

6. PHARMACOKINETICS

The anthocyanins present in bilberry are thought to cross the blood-brain barrier *(21)*. To date, no human studies have been published regarding the pharmacokinetics of the anthocyanins present in bilberry. However, studies have been conducted using other sources of anthocyanins, such as blueberries, elderberries, and blackcurrant juice. Mazza and colleagues *(22)* studied the absorption of anthocyanins from a freeze-dried blueberry preparation in five human subjects. Following administration of 100 g of blueberry supplement containing 1.2 g of anthocyanins, serum concentrations of 11 anthocyanins were measured at 1, 2, 3, and 4 hours postdose. Serum concentrations of each of the anthocyanins ranged from 0.23 to 3.68 ng/mL, suggesting very low absorption of anthocyanidins from this preparation. However, urinary excretion was not measured, precluding an accurate assessment of absorption. The concentration of total anthocyanins ranged from 6.6 ng/mL at 1 hour to 9.6, 12.1, and 13.1 ng/mL at 2, 3, and 4 hours, respectively.

In a more complete pharmacokinetic study, Cao et al. *(23)* studied the plasma and urine pharmacokinetics of 720 mg of anthocyanins from an elderberry extract in four elderly women. Blood and urine samples were collected over 24 hours. These investigators found primarily two anthocyanins (cyanidin 3-sambubioside and cyanidin 3-glucoside) present in both blood and urine. The time to maximum concentration (t_{max}) of total anthocyanidins was approx 1 hour and the C_{max} was 97 nmol/L. The average half-life for total anthocyanidins was approx 130 minutes, with the mean half-life of cyanidin 3-sambubioside being 170 minutes and that of cyanidin 3-glucoside being 100 minutes. In concurrence with the results of Mazza et al. *(22)*, less than 0.001% of the dose of anthocyanins was recovered in the urine within 24 hours, again indicative of poor absorption of these compounds. However, no attempts were made to analyze for potential glucuronide or sulfate conjugate metabolites of these compounds, and thus a more complete absorption picture is not available.

Another study in six healthy volunteers studied the pharmacokinetics and bioavailability of anthocyanidin-3-glycosides from either blackcurrant juice or elderberries *(24)*. As with other studies listed previously, these investigators found very little (~0.04% from blackcurrant juice to 0.4% from elderberry extract) of the dose recovered in the urine, suggesting low bioavailability. Though the half-life of the anthocyanidins was similar regardless of preparation, the area under the curve and amount of anthocyanidins excreted were approx 10-fold higher in the subjects administered the elderberry extract as compared with those receiving blackcurrant juice. The half-life (~1.7 hours) was comparable to that in the aforementioned study by Cao et al. *(23)*, but the recovery of anthocyanins in the previous study by Cao et al. administering elderberry extract was substantially lower (~400-fold lower) than that observed in this study. The reason for this discrepancy in amount recovered is unclear from the information given in the articles, but may relate to how the elderberry extract is formulated.

7. ADVERSE EFFECTS AND TOXICITY

The lethal dose of 50% of the population (LD_{50}) of Myrtocyan in rodents is over 2000 mg/kg, and in dogs the only adverse effect from a dose of 3000 mg/kg was dark urine and feces. There is no evidence of mutagenicity, or of teratogenicity or impaired fertility in rats. According to unpublished data of 2295 patients taking Tegens, most of whom took 160 mg twice daily for 1–2 months, 94 subjects complained of adverse effects involving the GI, dermatological, and nervous system *(25)*.

There are currently no reported adverse effects from the consumption of bilberry or related compounds. When the fruit is consumed in amounts normally contained in foods, bilberry falls under the "Generally Recognized as Safe" category according to the US Food and Drug Administration *(26)*. However, death has been reported with the chronic consumption or high doses of the leaf (1.5 g/[kg · day]) *(27)*.

8. INTERACTIONS

Bilberry extract 200 mg/(kg · day) administered intraperitoneally to euthyroid rats increased radiolabeled triiodothyronine (T3) transport into the brain, compared to vehicle only *(21)*. Postulated mechanisms include central or peripheral inhibition of L-thyroxine's (T4) deiodination to T3; inhibition of T3 protein binding; or enhanced T3 binding to carrier proteins in the brain capillary wall *(21)*. Whether bilberry could interact with thyroid replacement therapy remains to be seen.

Studying a diabetic rat model, Cignarella and colleagues *(1)* found that administration of an extract from the leaves of blueberry plants caused a 26% reduction in plasma glucose. This was also accompanied by a 39% decrease in plasma triglycerides. Whether this reduction in glucose levels might result in a clinically significant interaction in patients taking antidiabetic agents is unknown, because the glucose lower effects have not been studied in humans. Likewise, whether the triglyceride-lowering effects of hypolipidemic agents used in humans would be enhanced by coadministration of anthocyanidins is unknown.

9. REPRODUCTION

There is insufficient information to determine the safety of bilberry consumption during pregnancy or lactation.

10. REGULATORY STATUS

Bilberry is classified as a dietary supplement.

REFERENCES

1. Cignarella A, Nastasi M, Cavalli E, Puglisi L. Novel lipid-lowering properties of Vaccinium myrtillus L. leaves, a traditional antidiabetic treatment, in several models of rat dyslipidaemia: a comparison with ciprofibrate. Thromb Res 1996;84:311–322.
2. Muth ER, Laurent JM, Jasper P. The effect of bilberry nutritional supplementation on night visual acuity and contrast sensitivity. Altern Med Rev 2000;5:164–173.
3. Monograph. Vaccinium myrtillus (bilberry). Altern Med Rev 2001;6:500–504.
4. Magistretti MJ, Conti M, Cristoni A. Antiulcer activity of an anthocyanidin from Vaccinium myrtillus. Arzneimittelforschung 1988;38:686–690.
5. Cristoni A, Magistretti MJ. Antiulcer and healing activity of Vaccinium myrtillus anthocyanosides. Farmaco [Prat] 1987;42:29–43.
6. Bonati A. How and why should we standardize phytopharmaceutical drugs for clinical validation? J Ethnopharmacol 1991;32:195–197.
7. Detre Z, Jellinek H, Miskulin M, Robert AM. Studies on vascular permeability in hypertension: action of anthocyanosides. Clin Physiol Biochem 1986;4:143–149.
8. Roy S, Khanna S, Alessio HM, et al. Anti-angiogenic property of edible berries. Free Radic Res 2002;36:1023–1031.
9. Laplaud PM, Lelubre A, Chapman MJ. Antioxidant action of Vaccinium myrtillus extract on human low density lipoproteins in vitro: initial observations. Fundam Clin Pharmacol 1997;11:35–40.
10. Scharrer A, Ober M. [Anthocyanosides in the treatment of retinopathies (author's translation)]. Klin Monatsbl Augenheilkd 1981;178:386–389.
11. Perossini M, Guidi G, Chiellini S, Siravo D. Diabetic and hypertensive retinopathy therapy with Vaccinium myrtillus anthocyanosides (Tegens) double blind placebo-controlled clinical trial. Ann Ottalmol Clin Ocul 1987;113:1173–1190.
12. Zadok D, Levy Y, Glovinsky Y. The effect of anthocyanosides in a multiple oral dose on night vision. Eye 1999;13(Pt 6):734–736.
13. Belleoud L, Leluan D, Boyer Y. Study on the effects of anthocyanin glycosides on the nocturnal vision of air traffic controllers. Rev Med Aeronaut Spatiale 1966;18:3–7.

14. Jayle GE, Aubert L. Action des glucosides d'anthocyanes sur la vision scotopique et mesopique du subjet normal. Therapie 1964;19:171–185.

15. Terrasse J, Moinade S. Premiers reultats obtenus avec un nouveau facteur vitaminique P "les anthocanosides" extraits du Vaccinium myrtillus. Press Med 1964;72:397–400.

16. Gloria E, Peria A. Effect of anthocyanosides on the absolute visual threshold. Ann Ottalmol Clin Ocul 1966;92:595–607.

17. Junemann G. [On the effect of anthocyanosides on hemeralopia following quinine poisoning]. Klin Monatsbl Augenheilkd 1967;151:891–896.

18. Caselli L.Clinical and electroretinographic study on activity of anthocyanosides. Arch Med Intern 1985;37:29–35.

19. Head KA. Natural therapies for ocular disorders, part two: cataracts and glaucoma. Altern Med Rev 2001;6:141–166.

20. Bravetti G. Preventive medical treatment of senile cataract with vitamin E and anthocyanosides: clinical evaluation. Ann Opthalmol Clin Ocul 1989;115:109–116.

21. Saija A, Princi P, D'Amico N, De Pasquale R, Costa G. Effect of Vaccinium myrtillus anthocyanins on triiodothyronine transport into brain in the rat. Pharmacol Rev 1990;22:59–60.

22. Mazza G, Kay CD, Cottrell T, Holub BJ. Absorption of anthocyanins from blueberries and serum antioxidant status in human subjects. J Agric Food Chem 2002;50:7731–7737.

23. Cao G, Muccitelli HU, Sanchez-Moreno C, Prior RL Anthocyanins are absorbed in glycated forms in elderly women: a pharmacokinetic study. Am J Clin Nutr 2001;73:920–926.

24. Bitsch I, Janssen M, Netzel M, Strass G, Frank T. Bioavailability of anthocyanidin-3-glycosides following consumption of elderberry extract and blackcurrant juice. Int J Clin Pharmacol Ther 2004;42:293–300.

25. Morazzoni P, Bombardelli E. Vaccinium myrtillus L. Fitoterapia 1996;68:3–29.

26. FDA. Bilberry. In: Center for Food Safety and Applied Nutrition OoPAEAfad, 2004. Available at http://www.cfsan.fda.gov/~comm/ds-econ4.html. Last accessed: Sept. 2006.

27. Blumenthal M, ed. *The Complete German Commission E Monographs: Therapeutic Guide to Herbal Medicines* (translated by S. Klein). Austin: American Botanical Council Boston, 1998.

Index

Printed in the United States of America.